Transdisciplinarity for Transformation

Barbara J. Regeer · Pim Klaassen ·
Jacqueline E. W. Broerse
Editors

Transdisciplinarity for Transformation

Responding to Societal Challenges through Multi-actor, Reflexive Practices

palgrave
macmillan

Editors
Barbara J. Regeer
Athena Institute, Faculty of Science
Vrije Universiteit Amsterdam
Amsterdam, The Netherlands

Pim Klaassen
Athena Institute, Faculty of Science
Vrije Universiteit Amsterdam
Amsterdam, The Netherlands

Jacqueline E. W. Broerse
Athena Institute, Faculty of Science
Vrije Universiteit Amsterdam
Amsterdam, The Netherlands

ISBN 978-3-031-60973-2 ISBN 978-3-031-60974-9 (eBook)
https://doi.org/10.1007/978-3-031-60974-9

This Palgrave Macmillan imprint is published by the registered company Springer Nature Switzerland AG
The registered company address is: Gewerbestrasse 11, 6330 Cham, Switzerland

Preface

In good Humboldtian fashion, our job as academics consists of both doing research and teaching. As for the research in which we engage: this tends to be dedicated not simply to developing new knowledge, but (also) to solving complex and persistent societal challenges and supporting 'just' social transformation through multi-actor collaborations. This type of research has come to be known as transdisciplinary research—or, as we prefer to call it: *transdisciplinarity for transformation*. As for our teaching: in many of the courses we have taught over the years, we saw that our master's and PhD students grappled with the intricacies of comprehending and engaging in such transdisciplinary work dedicated to social transformation. This book emerged first and foremost in response to the needs these students expressed. And indeed, an important aim of this book is to offer focused support to early-career researchers, as they navigate the complex winding road of becoming transdisciplinary scholars. An additional benefit of putting this book together, of course, is that it allows us to share with wider audiences what transdisciplinarity for transformation can be.

We do not presume to know all there is to it, but we do think we know many of the pertinent challenges, needs and conditions that anyone exploring the territory will be confronted with at some point or another. To include a wide variety of experiences, we teamed up with colleagues from all around the globe, with whom we share the conviction that, as much as anything else, transdisciplinarity for transformation also requires transforming not only how we do research, but ourselves too. We have tried to make much of this explicit in this book, and we hope it will be of help to all who are willing to go on this learning journey.

As for the journey so far, we would like to thank all our fellow travellers. To start with, of course, all the students and PhD candidates who over the years have taught us so much about what it means to engage in transdisciplinarity for transformation—what it demands of individuals who come from so many different places geographically, disciplinarily, professionally, culturally or personally, and everything it takes to navigate between life-worlds and systems, disciplines and professions, distinct roles they themselves feel less or more comfortable with and other roles they might be expected to fulfil in some transdisciplinary constellation or other, by some actor, for whatever reason. Without their teachings, we would not have known where to start, nor where to go.

And then there are all the authors and co-authors who have contributed to the many chapters this book draws together, many of whom also actively participated in two online workshops we held to find common threads, structures to think with, productive tensions and the collaborative spirit we needed to successfully finish this journey. Within this group there are some who deserve special mention, starting with Kris and Callum. When we just started this project, they were like the tour guides who made sure everyone was there at the right place and at the right time, with all materials needed in place. Both conceptually and practically they played a pivotal role in getting things going and we cannot thank them enough for this. Teun, Evelien and Anne, each in their own unique way, all helped us think through, consider and reconsider what this journey was all about in the first place, what it needed to be completed successfully and what (intellectual) luggage to better leave behind. We are certain that by sharing their critical and their inspiring insights they have made editing this book a better and a more interesting experience and the book itself a better and more interesting book. Whatever faults remain are all ours, as the saying goes.

Our colleagues Justine, Lotte and Evelyne also deserve a special thanks, for the many ways in which they have materially, mentally and practically supported us over the years, not in the least by organizing and overseeing events we think are crucial for building and nurturing the type of transdisciplinary community we aspire to be at the Athena Institute—and that we hope this book will inspire to build elsewhere too.

We would also like to honour the invaluable support of our co-teachers in the master's course "Advanced Methodology: Interactive Learning and Action" and the PhD course "Inter- and Transdisciplinary Research" run in Amsterdam and Chennai, where the learning questions around which the book is structured saw the day of light: Aishwarya, Callum, Claudia, Deepika, Durwin, Elena, Emmy, Jessica, Joske, Mirjam, Mrinalini, Nienke, Ramesh, Teun, Tjerk Jan, and Vandana.

We would furthermore like to thank all participants of the interactive panel sessions "Adopted, aspired, ascribed, resisted: dealing with roles in transdisciplinary processes" and "Transdisciplinary research design—the need for reflective standards" at the 2021 International Transdisciplinarity Conference, that formed the basis for Chapters 3, 14 and beyond.

Finally, we would like to thank Deborah, our text editor, for the precision, the helpfulness and the incredible pace with which she has supported us and all the authors contributing to this volume when we had the end of our journey in sight, but had not quite reached it yet. And of course, Ashwini and colleagues at Palgrave Macmillan, who always had an answer when we had a question and always reminded us to keep going and to keep to the deadlines we had to meet to reach the end.

Amsterdam, The Netherlands
February 2024

Barbara J. Regeer
Pim Klaassen
Jacqueline E. W. Broerse

Contents

Notes on Contributors

Lena Bendlin main interest as a political scientist is how to enable multi-level systems to foster sustainable transformations. While her journey started with a focus on European integration, she became increasingly interested in local and municipal action, particularly on climate change. Her research into soft modes of European governance and urban transformations revealed to what extent effective change depends on favourable administrative environments. She thus joined a change management unit at the Berlin city administration where she now advises on improved processes and administrative reform, always in view of unleashing the city's full potential for co-creating a sustainable urban future.

Alanya C. L. den Boer holds an MSc in Applied Communication Science, with specialization in Health and Society, Wageningen University & Research (WUR). During her studies she conducted research at the University of California, San Francisco on the topic of environmental sustainability and health. She obtained her PhD at the Athena Institute, Vrije Universiteit Amsterdam (2023), on the potential of transformative research in accelerating sustainable food system transformation. She currently works at the department of value creation at WUR. She coordinates and contributes to the further professionalization of the Wageningen Science Shop and works on the topic of Citizen Science.

Jacqueline E. W. Broerse is professor of Innovation and Communication in health and life sciences, Director Athena Institute, Vrije Universiteit

Amsterdam. She has a disciplinary background in Biomedical Sciences and Science, technology and society studies (STS). Her research aims to understand how research and innovation can better contribute to solving societal challenges in an equitable and responsible way. She particularly focuses on methodology development for multi-stakeholder engagement in research and innovation (e.g. through Living Labs), and the governance of food or health system transformation.

Hilde J. H. Brouwers is a postdoctoral researcher focusing on sustainability transition and transformation research with particular interests in how humans shape the places they live in and work in, in the face of change, as well as the role of policy and research in such processes. Her research focuses empirically on the Dutch countryside, including topics such as wildfire prevention, sustainable agriculture, nature policy and the provision of public services in rural areas.

Thomas Bruhn is a physicist leading the transdisciplinary research group "transformative spaces and mindsets" at RIFS, the Research Institute for Sustainability—Helmholtz Centre Potsdam. His work focuses on designing and researching transdisciplinary communication processes that allow for integrating diverse perspectives and worldviews for sustainability-related transformation. He is a trained facilitator in various group formats and expert on co-creativity from the perspective of psychodynamic leadership. Before joining RIFS, he did research on semiconductor nanomaterials in Berlin, Rome and Marseille, and he is currently serving as a member of the board of directors to the German Association to the Club of Rome.

Joske Bunders is professor em. of Biology and Society, received her PhD at the Vrije Universiteit Amsterdam for research on Participative Strategies for Science-Based Innovations, focusing on biotechnology and small-scale farmers in low- and middle-income countries. She was the founder of the Department of Biology and Society of the VU university, which was later renamed as the Athena Institute. She designs, conducts and guides research on methodology development to stimulate and facilitate interactive and transdisciplinary processes of problem-solving. Presently, she is also the director of the "Co-Creation Institute for Problems Solving in Society" from MIT World Peace University in Pune (India).

Rico Defila is a lawyer and a senior researcher at the University of Basel in the Department of Social Sciences Social Transitions Research

Group (STR) and deputy leader of the Research Group Inter-/Transdisciplinarity. His research covers the theory and methodology of inter- and transdisciplinarity, sustainable consumption and quality of life. He has been providing training courses as well as coaching and support for a wide number of diverse inter- and transdisciplinary and transformative projects.rico.defila@unibas.ch

Azza Dirar holds a PhD from the School of International Development at the University of East Anglia for her research on the land politics of dam displacement in Northern Sudan. She also holds an MSc in Sustainable Agriculture and Food. Currently, she is employed by Norfolk and Waveney Mind as project coordinator on the 'Sustain project', providing support for eco-anxiety and climate grief including a co-designed and co-facilitated six-week Active Hope and Mindfulness Course. She has facilitated numerous workshops based on the work of Joanna Macy, including in the Blue Zone of COP 26 in Glasgow.

Antonietta Di Giulio has a PhD in philosophy and is a senior researcher at the University of Basel in the Department of Social Sciences Social Transitions Research Group (STR) and leads the Research Group Inter-/Transdisciplinarity. Her research and teaching cover the theory and methodology of inter- and transdisciplinarity, sustainable consumption and quality of life. She has been providing training courses as well as coaching and support for a wide number of diverse inter- and transdisciplinary and transformative projects.antonietta.digiulio@unibas.ch

Linde R. Egberts works as Chief Science Officer for the Cultural Heritage Agency of the Netherlands. Previously she worked as Associate Professor in Heritage Studies at Vrije Universiteit Amsterdam. She is trained as a human geographer and cultural historian at Utrecht University and Vrije Universiteit Amsterdam, where she defended her PhD thesis Chosen Legacies in 2015. Her research is focused on the relationship between heritage and landscape transformations in the context of societal challenges, including climate change and democratization. Connecting heritage studies with other ways of doing research forms a red thread through her writings, teaching and strategic work.

Man Fang is an affiliate scholar at RIFS Potsdam. Trained as psychologist and psychotherapist both in China and Germany, she has been developing methodologies and approaches inspired by psychodynamic group therapy,

Satir-model-based transformative family therapy and body and movement therapy, for the application in trans-cultural and trans-disciplinary communication groups. As a facilitator, she is devoting herself to developing capacities for intercultural discourse and promoting the integration of multicultural identities. She holds a PhD in public administration and has worked as policy advisor in the field of risk governance and emergency management for the Beijing municipal government.

Karen Finn is a Dprof student in Systemic Practice at University of Bedfordshire—Known to friends as Finn, proud member of the Taos Institute and about to gather material for post-qualitative research regarding the emergence of a relational approach to Earth within systemic practice. Participation in the dialogical space provided by IASS at COP26 was valuable in witnessing the importance of a collective approach to difficult yet important stories regarding human relationship with Earth.

Aafke Fraaije is a postdoctoral researcher in Art for Climate Solidarity at the Technical University Delft. In her research, she employs transdisciplinary methodologies to explore how the visual and performing arts may foster inclusive deliberation on science and technology. Her PhD work at the Athena Institute, Vrije Universiteit Amsterdam, focused on engaging diverse citizens in smart city developments through improvisation theatre. Her current research is about promoting empathy and solidarity in the climate change debate through the visual and performing arts.

Carolin Fraude an educationalist, works on the RIFS project "Transformative Spaces and Mindsets" since 2018 and initiated the transdisciplinary research project "Co-Creative Reflection and Dialogue Space" at the UN Climate Change negotiations. She designs, facilitates and researches dialogue formats based on a relational understanding of transformation. Since 1998, she is teaching meditation and other reflective approaches as transformative practices to change consciousness, attitudes and behaviours. As part of her NGO work with the Brahma Kumaris, she is actively involved in the processes of the UN climate negotiations since 2015 and the Commission on the Status of Women since 2013.

Halleh Ghorashi is full professor of Diversity and Integration in the Department of Sociology at the VU Amsterdam. She combines a multidisciplinary background in social sciences (anthropology, sociology, gender-studies, (political) philosophy and organization studies) with a critical approach in diversity and refugee/migration studies and narrative co-creative methodology. In 2017, she received a prestigious VICI grant on *Engaged Scholarship and Narratives of Change* from NWO. In 2018, she is appointed as Crown Member of the SER (Dutch Social Economic Council) and in 2020 as member of KNAW (The Royal Netherlands Academy of Arts and Sciences).

Mads Dahl Gjefsen is a senior researcher at Studio Apertura, NTNU Social Research in Norway. His research explores large-scale technological, transition and transformation efforts with an emphasis on their promissory narratives, and he has worked on multiple technology domains including food, energy, climate and digitalization. He holds a PhD in Science and Technology Studies.

Maureen Goodman is the Programme Director for the Brahma Kumaris UK and their representative at the UN, Vienna. She works with diverse communities building spiritual capacity in times of transition. She participates in COP Climate change and biodiversity, exploring spiritual and ethical perspectives and the connection between our inner state of being and the state of our Earth. She represents the Brahma Kumaris on the Board of World Religious Leaders of the Elijah Interfaith Institute and is on the Executive Circle of the Spirit of Humanity Forum which explores spirituality in leadership and governance.

Callum Gunn completed a PhD at the Athena Institute, VU University Amsterdam in 2023. His research drew on Science, Technology and Innovation Studies (STIS) and reflexive governance scholarship to experiment with patient engagement practices in the context of an EU pharmaceuticals innovation partnership. This work explored the epistemic and political challenges of promoting constructive forms of technology assessment in the case of institutional health technology evaluation. He moved to University Medical Center Utrecht to contribute to the interdisciplinary study of responsible innovation in biomedicine and health care, including societal and public engagement in regenerative medicine research and translation.

Jaron Harambam (PhD) is assistant professor of Media, Truth Politics and Digitalization at the Sociology Department of the University of Amsterdam. His research deals with public disputes over truth in a digitalized public sphere. More specifically, he studies conspiracy theories, news and platform politics, and AI (content moderation, search/recommender systems). Central to his research is the participation of multiple stakeholders to design our (future) digital worlds along democratic and public values. His monograph "Contemporary Conspiracy Culture: Truth and Knowledge in an Era of Epistemic Instability" (2020) is out at Routledge.

Richard Helliwell is a senior researcher at Ruralis based in Trondheim, Norway. Richard's research covers a variety of topics at the interface of environmental, agricultural, science and policy issues.

Sabine Hoffmann is Group Leader of Inter- and Transdisciplinary Research at the Swiss Federal Institute of Aquatic Science and Technology (Eawag) and lecturer at the Swiss Federal Institute of Technology in Zurich (ETH Zurich). Her research focuses on integration and integrative leadership in large inter- and transdisciplinary research programmes in both theory and practice. She currently co-leads the Global Alliance for Inter- and Transdisciplinarity (ITD Alliance) Working Group on "Integration Experts and Expertise".

Fabian Holle (they/he) is a PhD candidate in the Department of Sociology at VU Amsterdam and part of the research project: *Engaged Scholarship and Narratives of Change*. They graduated in theatre making at the University of the Arts Utrecht in 2004. They worked with directors and choreographers, co-founded theatre collective Ponies (2005–2015) and performed in drag. In 2020, they completed a Master's in Sociology (cum laude) at VU Amsterdam. Their Master's thesis was the recipient of the 2020 ARC-GS MA Thesis Prize (from Amsterdam Research Centre for Gender and Sexuality at University of Amsterdam).

Evelien de Hoop is a teacher and transdisciplinary researcher with roots in STS, geography, post- and decolonial studies and connected history. Empirical foci include, e.g. landscape conservation and transformation, multi-sited socio-ecological change connected by trade in major commodities such as soy and palm oil, food systems (including food self-provisioning), biofuels, smart urbanism, policy evaluation and more. She is involved in both small- and large-scale projects in which she collaborates with a wide variety of societal stakeholders, in which she aims to co-produce transformative knowledge for more just futures across the world.

Annemarie Horn is an assistant professor at the Liberal Arts and Sciences department at Utrecht University. Here she teaches courses on interdisciplinary methods and integration and conducts research on inter- and transdisciplinary research practices. Her research is action-oriented and focuses on integration processes and the roles of early career researchers in collaborative research projects. Her current research builds onto the work that she has done in the context of her PhD at the Athena Institute in Amsterdam when she studied inter- and transdisciplinary collaboration and knowledge integration among master students in a higher education context.

Pim Klaassen worked as assistant professor *Reflexive Practices for Just Sustainability Transitions* at the Athena Institute, Vrije Universiteit Amsterdam when preparing this book. His work focuses on the societal meaning of science and innovation, and on governance arrangements and transdisciplinary research methodologies conducive to responsible research and innovation for sustainability transitions. He holds a PhD from the University of Amsterdam and an MPhil from the University of Cambridge. Per January 2024, he works as director of *Stadslab*, a foundation supporting institutions in secondary education with citizenship education, training teenagers the attitudes, skills and knowledge supportive of (bottom-up) democratic participation.

Kristiaan P. W. Kok is an assistant professor of Politics and Governance of Sustainability Transitions at the transdisciplinary Athena Institute (VU Amsterdam). His work focuses on the political dynamics of sustainable transformation in fields like food and agriculture, pertaining to issues of power, justice and democracy. He is interested in understanding how transdisciplinary experimentation can best be supported to catalyse transition dynamics, with particular attention to the roles that innovation systems and transformative policies play therein. He holds a PhD on the politics of sustainability transitions, and master's degrees in Physics and Political Sciences.

Stefanie Kunkel is a researcher in the project "Digitalisation and Sustainability Transformations" at the Research Institute for Sustainability (RIFS), Helmholtz Centre Potsdam. Holding degrees in economics and sociology, she uses social science methods to explore the effects of digitalization in industry on environmental sustainability. She previously worked in various contexts in the field of international environmental policy and economics, including in the United Nations Environment Programme and in the European Parliament. She is particularly interested in understanding if and how transdisciplinary engagement of stakeholders from politics, business, civil society and academia can foster more sustainable outcomes of decision-making.

Daniel J. Lang is professor at the Karlsruhe Institute of Technology (KIT), DE, since November 2022, and head of the research group Designing Real-World Laboratory Research at the Institute for Technology Assessment and Systems Analysis (ITAS). Before joining KIT, he was professor for Transdisciplinary Sustainability Research at Leuphana University of Lüneburg, DE, where he was also dean of the School of Sustainability and president's special advisor for sustainability for several years. Founding member of the Society for Transdisciplinary and Participatory Research (GTPF).daniel.lang@kit.edu

Mark G. Lawrence is scientific director at RIFS, the Research Institute for Sustainability—Helmholtz Centre Potsdam, and an integrative scientist who addresses key challenges of the Anthropocene by bringing together and applying a wide range of academic expertise in his team, including his own background in atmospheric science, as well as going beyond this interdisciplinary foundation to also involve societal actors in a transdisciplinary research approach. His main focal topics include air pollution, climate change, climate geoengineering, environment-related governance of vulnerable regions like the oceans, the Arctic and the Himalayas, and the interfaces of spirituality and the arts with sustainability. mark.lawrence@rifs-potsdam.de

Anne Loeber is an associate professor in governance and sustainability at the Athena Institute, Vrije Universiteit Amsterdam (VUA). She holds a PhD in Political Science (University of Amsterdam, 2004), based on a thesis on interactive Technology Assessment and its potential contribution to making the "sustainable development" concept operational in public policy through knowledge co-production. Her research explores the relation between knowledge, power and agency in the governance of highly complex societal issues. Current projects include analyses of, and action research on, the governance of food system transformation and (advice on) the evaluation of a variety of experimental, transition-oriented policy programmes.

Frans van Lunteren is a historian of science at the Vrije Universiteit Amsterdam and Leiden University. His work focuses on discipline formation, nineteenth-century laboratories, early international collaboration in science, the relationship between science and religion, and the history of Dutch physics. He is now supervising a project on agricultural history in the Netherlands.

Alexandra Lux is an economist by training with expertise conceptualizing and conducting integrative transdisciplinary research processes. She has been a research scientist at ISOE—Institut for Social-Ecological Research, Frankfurt/Germany since 2000. Since April 2023, she has been a member of the Executive Board and is co-head of the Knowledge Processes and Transformations hub. Having received her doctorate in the field of infrastructure development for a thesis dealing with public water supply and demographic shrinkage, Alexandra was active in the fields of integrated water research and social-ecological biodiversity research, which are brought together in conceptual and theoretical considerations in her current work on transdisciplinary integration and knowledge transfer. alexandra.lux@isoe.de

Kathleen A. Mar joined the RIFS in 2012 and leads the group "Climate Action in National and International Processes (ClimAct)". ClimAct focuses on participation in and understanding of political forums that aim to drive climate action, with emphasis on the United Nations Framework Convention on Climate Change (UNFCCC) and the Climate and Clean Air Coalition (CCAC). She holds a PhD in atmospheric chemistry from the University of California, Berkeley and worked at the United States Environmental Protection Agency (US EPA) prior to joining the RIFS. Her perspective integrates scientific expertise with practical experience in the world of policy and administration.

Shantanu Mandal is a student at the Brahma Kumaris (BK) World Spiritual University since 2000, focusing on media, education, youth engagement and environmental initiatives. With a background in fashion design from NIFT and Mod'art International Paris, he shared insights by teaching design and sustainability in Indian institutes. Serving as the BK youth representative for UN bodies from 2015 to 2024, Shantanu, now is the Co-Chair for ReSpECC (Religion, Spirituality, Environmental Conservation, and Climate Justice) a Specialist Group in IUCN. Merging storytelling, design and compassionate activism, he creates avenues for individuals to reconnect with impactful experiences, supporting their spiritual and environmental journey.

Natalie Mangondo is a climate change economist dedicated to fostering multi-stakeholder partnerships at the nexus of the environment, economy and equity. With expertise in climate finance, Natalie has led initiatives supporting mobilization of finance to emerging markets and developing economies. Currently serving as Carbon Trading and Climate Finance Consultant at the Ministry of Environment, Climate and Wildlife in Zimbabwe, she has been involved in Carbon Trading negotiations and legislations, as well as regularising sustainable projects that contribute to national and global climate action. Passionate about youth inclusion, Natalie actively engages in various initiatives related to the global youth climate movement.

Marjoleine G. van der Meij is an assistant professor in Design for Participatory Transformation at the Athena Institute, Faculty of Science, VU University in Amsterdam. She holds Master's degrees in Industrial Design Engineering and Science Communication, and a PhD in designing playful methods for reflection in responsible research and innovation contexts. Her research centres around playfulness, creativity and design thinking in the facilitation and scaffolding of system transformation in (governance of) research and innovation, urban and landscape design, food systems and healthcare. She also trains and supports professionals in transition contexts in knowledge exchange and (frame) reflection processes.

Barbara van Mierlo is associate professor who works as a sociologist at the Knowledge, Technology and Innovation Group of Wageningen University. She studies and supports processes of transformative, systemic change towards sustainability and their intersection with everyday social practices. Being actively engaged in these processes, special interests include the significance and features of interactive learning and discursive strategies, the value of change-oriented evaluation, emergence of reflexivity, responsible innovation and transdisciplinary collaboration. She has co-developed the methodology of Reflexive Monitoring in Action (RMA), which has a wide international uptake in among others sustainable agriculture, natural resource management, renewable energy and health.

Christine Peringer is a facilitator, mediator and expert in multi-stakeholder planning and problem-solving. She works in sectors including the environment, sustainable development, justice and health. Christine holds a law degree and is the author of "How We Work for Peace" Peace Research Institute – Dundas (1997). She is a founding partner of the Better Climate Governance Project (www.betterclimategovernance.com) and a member of the Climate Change Project of Mediators Beyond Borders International. She has been an observer at the UNFCCC negotiations since 2019. She has served for twenty-five years through her consultancy, *Group Facilitation and Mediation Services*. For more information see www.peringer.ca/

Elena Ponzoni is assistant professor at the Department of Sociology at the VU Amsterdam, she is part of the Refugee Academy and of the project *Engaged Scholarship and Narratives of Change*. Since 2023 she is director of the Co-Creation for Inclusive Knowledges Lab at the Faculty of Social Sciences (VU Amsterdam). A red thread in her work is the focus on how academic research can contribute to make visible and change implicit mechanisms of exclusion and inequality, through participatory methods of knowledge co-creation centring the experiences of the communities at stake.

Sarju Sing Rai is assistant professor in "Competencies Development for Societal Transformation through Public Engagement in Transdisciplinary Research and Education" with a focus on the themes Global Health, and Wellbeing at the Athena Institute, Vrije Universiteit Amsterdam. Sarju is a Global Health Pharmacist from Nepal. His research focuses on (1) education (competencies development through transdisciplinary education, critical service learning, internationalization and global citizenship) and (2) transdisciplinary Global Health (health-related stigma; intersectionality, health equity; mental wellbeing; inclusive living, access and mobility).

Sheila Ramaswamy has a background in Medical and Psychiatric Social Work and Public Health, works in SAMVAD, a country-wide Initiative located at the National Institute of Mental Health and Neurosciences, India, engaged in transdisciplinary work to enhance child protection and mental health services. Involved extensively in training, capacity building and research initiatives on child mental health and protection, her current research at Vrije Universiteit Amsterdam, focuses on transdisciplinary approaches to resolving complex problems in the context of the child and law, engaging stakeholders including judicial officers, mental health and child protection workers and children in contact with the law.

Barbara J. Regeer is professor of transdisciplinary research and purpose-led transformation in environmental sustainability and health at the Athena Institute, and holds a University Research Chair at, Vrije Universiteit Amsterdam. Her research interests are in emerging innovative strategies for (sustainable) development, with a focus on reflective practitioners, adaptive organizations and reflexive governance in sustainable food systems, integrated rural development, mental health care, child and youth care. She has co-developed the Reflexive Monitoring in Action approach. She is director of the Graduate School for Transdisciplinary PhD Education at Athena Institute and co-director of the India Centre for Transdisciplinary Research.

Morten Sager is senior lecturer in theory of science and research. He is the founder and director of a transdisciplinary master's programme in evidence-basing at the Department of Philosophy, Linguistics and Theory of Science, University of Gothenburg. Together with a group of several doctoral students and senior researchers, he combines the production of evidence in the form of systematic reviews with improvement work and epistemological analyses of evidence-based practices. Much of his research is done in close collaboration with societal actors in public health, health care, social work and education.

Niko Schäpke is an assistant professor at the Chair of Environmental Governance at the University of Freiburg and a senior research associate at the Research Institute for Sustainability, Helmholtz Centre Potsdam. He is an interdisciplinary social scientist situated in the field of solutions-oriented sustainability science. His work focuses on principles, settings and methods of transdisciplinary and transformative research. His specific interest are transdisciplinary labs (e.g. Real-world Labs), human agency and relational approaches. Here, he has co-authored more than 20 journal articles and diverse other publications, and is engaged in teaching and supervision.

Heike Schroeder is professor of Environmental Governance at the School of Global Development, University of East Anglia. Her work focuses on global environmental politics, indigenous peoples/knowledge and sustainable development, forest governance, the international climate negotiations, trust and sustainable food governance. She was PI of the

GCRR/ESRC-funded Indigenous-International Interactions for Sustainable Development (INDIS) project. She has previously held positions at the University of California, Santa Barbara and Oxford and been a senior research fellow at RIFS.

Shekhar Seshadri is a child psychiatrist with over 35 years of experience in the field of child mental health. As former senior professor in the Dept. of Child and Adolescent Psychiatry, at the National Institute of Mental Health and Neurosciences, India, his work has extended beyond the clinical population to children and service providers in communities across India, and in the South Asia region. In addition to preventive-promotive and curative child mental health interventions, he has undertaken various legal and policy-related initiatives. He is currently advisor to SAMVAD, a National Initiative & Integrated Resource for Child Protection and Mental Health.

Joseph A. Siegel teaches Environmental Dispute Resolution at Hofstra Law School and climate change and resilience seminars at Haub School of Law at Pace University. He is an environmental conflict resolution consultant, facilitator and mediator. In addition, he manages the Environmental Collaboration and Conflict Resolution program in the U.S. Environmental Protection Agency's New York regional office, serving as an impartial third party on climate change and a wide variety of other environmental matters, and co-chairs the office's Climate Change Workgroup. He is also a co-chair of Mediators Beyond Borders International's Climate Change Project, leading the group's Education Subcommittee. jsiegel@law.pace.edu

Laureline Simon is the founder and managing director of One Resilient Earth, with over 15 years of experience in international cooperation projects to address climate change. One Resilient Earth pioneers a regenerative and transdisciplinary approach to climate resilience, weaving together arts, science, ancient wisdom and new technologies, through diverse learning experiences. Previously, Laureline supported international climate negotiations on adaptation and loss and damage, at the secretariat of the United Nations Framework Convention on Climate Change; led an international research programme at the French Development Agency; and supported post-disaster reconstruction projects with the Indian NGO SEWA.laureline@oneresilientearth.org

Geert Somsen is a historian, interested in science and international relations. He has recently finished a European project on the phenomenon of "The Scientific Conference", resulting in a special issue of the *British Journal for the History of Science* (see here, OA). He is currently turning his attention to relations between scientific internationalism and imperialism.

Wolfgang Stark is professor em. of organizational and community psychology. He is founder and former director of the Organizational Development Laboratory and Center for Societal Learning and Social Responsibility, both based at the University of Duisburg-Essen (Germany). Since 2010, he is director and CEO of Steinbeis Center für Innovation and Sustainable Leadership. Since 2015 he serves as a guest researcher at the Strascheg Center for Entrepreneurship in Munich (Germany). His work has received various awards, e.g. the "Jimmy and Rosallyn Carter-Campus Community Partnership Award 2007".

Juliette Tronchon —After graduating in political science in 2019, Juliette Tronchon joined ProVeg International, now serving as Head of UN Affairs after several years with the organization. Collaborating closely with stakeholders, she advocates for a food systems approach that includes the consumption of healthy, sustainable diets. Leading the Food4Climate Pavilion at UNFCCC COPs, she emphasizes the urgency of transitioning to plant-rich diets and implementing resilient agricultural practices while addressing food loss and waste. Her goal is to deliver a clear message to negotiators and policymakers. Juliette also co-leads the ProVeg Youth Board, actively empowering the voices of young individuals globally.

Calder Tsuyuki-Tomlinson is a researcher, practitioner and artist exploring holistic interventions that aim to address planetary crises and global systems of inequity. They have previously worked at the intersection of global climate and biodiversity governance with the secretariats of the United Nations climate and biodiversity agencies, as well as the UAE COP28 Presidency and Climate Champions Team. They are currently a non-resident fellow at the United Nations University Center for Policy Research exploring governance innovations that promote regenerative economic systems, following previous work with the Future of Climate Cooperation initiative and the University of Oxford's Blavatnik School of Government.

Malika Virah-Sawmy is a facilitator, systems scientist and activist. She focuses on social transformation and provides leadership on systems change with institutions such as the UN, NGOs, with companies and with social movements. Her work is rooted in theory and practice in systems change, collective leadership, futures thinking, innovation, learning, embodiment and healing. She has worked for the last 20 years as a sustainability practitioner and systems scientist across the world from Asia, Australia, Africa to South America. Most of her action-based work has fostered transformation towards inclusive business models, sustainable production, consumption and trade, nature solutions, and community and indigenous rights.

Christine Wamsler is professor at Lund University Centre for Sustainability Studies (LUCSUS) and Director of the Contemplative Sustainable Futures Program. She is an internationally-renowned expert in sustainable development, climate action and associated (material and cognitive) transformation processes, with 25 years of experience, both in theory and practice. She has led many international projects and published more than 200 papers and book chapters on these issues. She draws inspiration from nature and her two children, who help her daily take better care of spiders, flowers and the world at large.

Esther de Weger (PhD) is a postdoctoral researcher at the Vrije Universiteit Amsterdam focusing on citizen participation in times of crisis with the aim of aligning science and policies more around citizens' interests and needs. Her work focuses on citizen participation and cross-sectoral collaboration for transformation in complex problems, specifically regarding health(care) and climate-related issues—aiming to improve the collaboration between the different actors.

Willemine Willems is an assistant professor Science Communication for Wicked Problems at the Athena Institute, Vrije Universiteit Amsterdam. She has a background in political philosophy, science and technology studies, and ethics. In her current research she focuses on science communication, more specifically on engaging publics in science in times of crisis, and the role of science in democracy.

Julia Wittmayer works as senior researcher and advisor at the Dutch Research Institute for Transitions and as Assistant Professor at the Erasmus School of Social and Behavioural sciences, Erasmus University Rotterdam. With a background in anthropology, she is interested in roles of and social relations and interactions between actors in sustainability transitions (governance). She has carried out a variety of projects related to action research and citizen science in the context of sustainability transitions and is editor of Action Research in Policy Analysis (2018, Routledge).

Hussein Zeidan is a PhD researcher and lecturer at Athena Institute, Vrije Universiteit Amsterdam. A Mechanical and Renewable Energy Engineer by training, he engaged with Science and Technology Studies (STS) to refine his technical expertise for societal transformation through a transdisciplinary approach. He has also worked with non-governmental organizations in the humanitarian and development sectors, enacting a participatory approach to the design and implementation of interventions. Currently, he is involved in research centred on cultivating transdisciplinary competencies in higher education students through co-creation.

Teun Zuiderent-Jerak is professor of transdisciplinary science & technology studies at the Athena Institute, Vrije Universiteit Amsterdam. His research focuses broadly on transdisciplinary social studies of health care and medicine and includes the study of evidence appraisal in relation to different knowledges, knowledge standardization and healthcare improvement, and technology development for inclusive health and employment. All his scholarship involves transdisciplinary research practices that explicitly aim to intervene in the practices studied. He is co-editor of "Making & Doing: Activating STS through Knowledge Expression and Travel" (2021, The MIT Press).

Marjolein B. M. Zweekhorst is professor of Innovation and Education in the health and life sciences at the Athena Institute, Vrije Universiteit Amsterdam. She wrote her PhD thesis on the institutionalization of an interactive approach to technological innovation (2004). Her current research is focused on (1) methodology development for inter- and transdisciplinary research, and (2) development of educational approaches and innovations aiming to prepare students for inter- and transdisciplinary research.

List of Figures

List of Tables

CHAPTER 1

What Is That Thing Called 'Transdisciplinarity for Transformation'?

Barbara J. Regeer, Pim Klaassen, and Jacqueline E. W. Broerse

Unprecedented challenges in the spheres of health, global justice, environmental degradation and climate change characterize our current era. This situation constitutes a call to all of humanity to respond. Arguably it comes with an even greater call to those in privileged positions—including, for instance, academics. In other words, today's complex problems call for societal transformation, and this in turn calls for not only new knowledge, but also new ways of producing knowledge and new ways of dealing with different knowledges. Ways that transgress age-old boundaries associated with epistemic and social hegemonic systems, like

B. J. Regeer (✉) · J. E. W. Broerse
Athena Institute, Vrije Universiteit Amsterdam, Amsterdam, The Netherlands
e-mail: b.j.regeer@vu.nl

J. E. W. Broerse
e-mail: j.e.w.broerse@vu.nl

P. Klaassen
Stichting Stadslab, Amsterdam, The Netherlands
e-mail: pim@hetstadslab.nl

B. J. Regeer et al. (eds.), *Transdisciplinarity for Transformation*,
https://doi.org/10.1007/978-3-031-60974-9_1

the boundaries between what is human and what is not, and between what counts as scientific and what does not. We are convinced that working towards the practical, material and just resolution of urgent, complex real-world problems requires *co-creating* better knowledges and better stories (Altınay & Pető, 2022), and that this should be done together with those affected by or interested in these problems—irrespective of whether they are researchers rooted in any one discipline or interdiscipline, or whether they belong to any other type of stakeholder group, featuring whatever form of relevant (experiential) expertise. We thus need ***transdisciplinarity for transformation.***

This book serves as a guiding beacon for early-career academics navigating the complexities of transdisciplinarity for transformation, offering diverse examples of what transdisciplinarity for transformation can be. Irrespective of whether you are interested in, for instance, environmental sustainability, health system transformation or queer or child rights, this volume will illuminate the power and challenges of transdisciplinarity in catalysing meaningful change and shaping a more resilient and equitable future in a still inhabitable world.

In this first chapter, we will first sketch in further detail what we take transdisciplinarity for transformation to mean, along the way also touching upon what is worthwhile and what is difficult about it. Subsequently, we turn to the centrality of *purpose* in transdisciplinarity. After that, we will provide a rough guide to the book's overall structure, and share a number of learning questions that we have collected over the years and that we have found to be pertinent to whoever tries to practice transdisciplinarity for transformation. We conclude the chapter with learning questions because in our view this aptly expresses the spirit which we aspire to convey with this book—one of celebrating openness to the new and unexpected, of daring to acknowledge our ineptitudes and blind spots, enthusiasm for experimentation and sharing, while recognizing that there are many, and that there is much that we can already learn from and build on.

1.1 Transdisciplinarity: A Response to Persistent Problems

To understand what transdisciplinarity for transformation means and why it would be worthwhile, let us start with a question: '***But what is the problem you are responding to?***'. In our capacity as academic researchers and teachers, we consistently pose this question to our graduate (master's and PhD) students, our academic peers and ourselves. It serves as a reminder that our research should be firmly rooted in tangible *real-world problems*. Problems that extend beyond the scope of academic research and reflection, and demand *action*—problems urging (societal) transformation.

Of course, 'problem' is too generic a concept, given that much of our research and the chapters in this edited volume tend to revolve around problems of a specific type: urgent, complex and persistent. Such problems might involve anything from child abuse (Ramaswamy et al., 2024, Chapter 10, this volume) to the environmental risk of wildfires (Brouwers et al., 2024, Chapter 9, this volume) and from the exclusive nature of conventional deliberations on mitigating climate change (Bruhn et al., 2024, Chapter 7, this volume) to the socio-political exclusion of refugees, queers and those at the intersection of these groups (Holle et al., 2024, Chapter 11, this volume). Problems, moreover, that are almost universally rooted in structural injustices associated with some form of infringement of safe and just (Earth) systems (Rockström et al., 2023), and that—paradoxically—tend to be reproduced partly through attempts to resolve (aspects of) them (Schuitmaker, 2012). Problems, lastly, not seldomly associated with enactments of racism, classism, capitalism, sexism, extractivism, colonialism, ableism and comparable notions, which denote some form of (explicit or implicit) belief in superiority and inferiority. With *transdisciplinarity for transformation,* we intend to express that our orientation towards these kinds of problems is coupled with the conviction that these take different forms—that what they look like, what they feel like, how they affect different actors, or actants, depends on who one is, where one is, how one identifies, and how this intersects or interacts with which values and needs and which knowledge one brings—among other things.

This book continues a line of work going back at least to 2009, when we defined transdisciplinary research as 'an umbrella term for all kinds of efforts towards reflexive co-evolution of science, technology and society' (Regeer & Bunders, 2009, p. 42). Echoing the spirit of the Zurich conference in 2000, where 800 people came together from a broad range of academic strands and social practices, we wanted to acknowledge the diversity of efforts that could be recognized under the 'umbrella' of transdisciplinary research, ranging from those initiated in academia, such as constructive or interactive technology assessment (Grin et al., 1997; Rip et al., 1995), Interactive Learning and Action (Bunders, 1994), and patient participation (Abma & Broerse, 2010) to efforts initiated by emerging 'intermediaries', with no primary academic embedding, that likewise attempt to construct interactive interfaces between science and society to address major societal challenges. These efforts, in different ways, build on decades of work across academic strands that culminated in wide calls to revise the contract between science and society by developing new interfaces that encompass 'attempts at reflexive co-evolution' (e.g. Rip, 2005). We feel it is important to highlight a few of these earlier calls as a means to open up the black box of 'transdisciplinary research as a given', as today, it is sometimes considered a monolithic concept that is self-explanatory and beyond critical consideration.

1.1.1 Science–Society Relations

One way of understanding what transdisciplinarity does or is, is in terms of its conception of relations between science and society. Arguably, it brings science into society and society into science. As such, it can be seen as one in a series of calls for changing the contract between science and society. Other such appeals also tend to be firmly rooted in the idea that today's problems are highly persistent and require other modes of problem-solving than those 'normally' employed by policymakers and researchers alike. Many have followed the line of thinking of policy scientists Rittel and Webber (1973) in their seminal work on 'wicked problems', or have defined highly persistent societal problems as 'intractable' (Rein & Schön, 1996), 'unstructured' (Hoppe & Hischemoller, 1998) or as 'grand challenges' (Ulnicane, 2016), to refer to contemporary complex and persistent issues that defy any straightforward planning response, based on the so-called best available knowledge. These types of issues require different responses, which will be as unstructured as the problems themselves and that will also reshape relationships between citizens and government, between policy scientists and policy practitioners, and between science and society more broadly. These responses recognize that scientists do not have the monopoly on knowledge and knowledge production, and that, on the contrary, the knowledges of policymakers, practitioners and citizens should be included in attempts to resolve the major problems facing contemporary societies.

This resonates with another set of calls, this time emphasizing that our knowledge systems need rethinking. Some who argue for this do so because of the way our knowledge systems reproduce epistemic injustices (Herzog & Lepenies, 2022), others because they are based on a misconstrued self-understanding, according to which science and society are perceived to be far more distinct and neatly separated than in fact they are (Latour, 2012). Originating in different scholarly fields, others emphasize that local knowledge (Fischer, 2000), practice-based knowledge (Dampier, 2009), professional knowledge (Schön & Rein, 1994), citizen knowledge (Ostrom et al., 1978), patient knowledge (Epstein, 1996), or farmers' knowledge (Bunders, 1994) have been excluded for too long, and that these exclusionary practices deeply pervade our current knowledge systems.

From within a wide array of academic fields or traditions, with roots in social and political sciences or the humanities, ideas such as these have been developed into full-fledged research programmes and methodological approaches. Think, for instance, of diverse research fields as Science and Technology Studies (STS), feminist science studies and care ethics, or decolonial and post-colonial studies. Although they all have their own distinctive emphases and research focus, we are more interested in the comparable sensitivities they share than in the obvious differences between them. Thus, where much of STS has been devoted to establishing the intricate interwovenness of ways of scientific knowing and social structures (Latour, 1987; Latour & Woolgar, 1979; Shapin & Schaffer, 1985), including the analysis of relations of power, knowledge and politics (Foucault, 2020; Mirowski, 2002), feminist science scholars have looked particularly at the ways in which gender biases play out in science and how the situatedness of knowledge practices matters, contributing substantially to reconceptualizing relations between identity and gender in the context of biology and technology and valuing how care offers a multi-faceted lens on relationships (Oyěwùmí, 1997; Puig de la Bellacasa, 2017; Van der Velden & and Sjåfjell, 2022). Scholars of decolonialism and post-colonial STS have analysed colonial legacies in scientific discourse and practice and argue for ways of getting beyond this, for instance through acknowledging science's practices of exclusion and valuing centuries-old knowledges from Indigenous and traditional practices (Harding, 2011; Pollock & Subramaniam, 2016). From within all these fields, then, arguments are emerging in favour of challenging hegemonic norms and crossing epistemic and social boundaries in order to foster co-creation between actors from different intellectual, practical, professional and societal spheres.

1.2 Transdisciplinary: What Is It (Not)?

One of the most common themes to emerge in discussions with anyone newly introduced to the topic of transdisciplinary research concerns its demarcation: What is it, and what is it not? How, for instance, does transdisciplinary research relate to approaches and schools of thinking and doing, like participatory action research, engaged scholarship or appreciative inquiry? Or, to research approaches such as sustainability science, mode-2 science or post-normal science? There is a growing group of people who consider themselves as part of the transdisciplinary community—they label their work as transdisciplinary in academic papers, they organize or attend conferences under the banners of transdisciplinarity, and they use the concept in their daily practice.[1] Thus, it is through these associations and uses that the concept of 'transdisciplinarity' acquires meaning, not through its definition. At the same time, there is a group of people (within and outside academia)—notably a much larger group—that engages in co-creative practices around complex societal issues, with roots that long predate the current upsurge of transdisciplinary academic literature, and also far beyond the hegemony of the Western academic community, but that do not self-identify as 'transdisciplinary' researchers or practitioners. And yet again, it is the practices, rather than the naming of them, that are of relevance here, particularly because they seem to share a philosophy of embracing epistemological plurality, valuing the importance of contextualisation, facilitating multi-stakeholder collaboration across boundaries (whether they are disciplinary, sectoral, or multi-level), and experimental governance within an overarching orientation towards a more sustainable, just and equitable society—notwithstanding the pluriform normativity that is inherently part of each transdisciplinary endeavour.

With this book, we aim to resist the natural inclination to try to reach conceptual closure on what falls inside or outside the confines of 'transdisciplinary research', and instead stay close to the spirit of cutting across (academic) divides, celebrating diverse knowledges, understandings and normative plurality that is at the heart of transdisciplinary ways of working. That is also why, featured in the three parts of this book,

[1] See, for example, Vermeulen and Witjes (2020) for a graph showing the rapid increase of academic publications containing the terms 'transdisciplinary research' or 'transdisciplinar*' after 2000 (p. 9, Figure 1.1).

you will find a diverse range of understandings and practices of transdisciplinarity. More important to us than underscoring the knowledge production that takes place in transdisciplinary practices, is the connection between knowledge and action in transdisciplinarity. Moreover, in this book terms like *transdisciplinarity*, *transdisciplinary research* or *transdisciplinary approaches* are used fairly colloquially and are treated as interchangeable. Indeed, we prefer to mobilize this conceptual leniency to underscore that not all transdisciplinary work is motivated principally by research, and has been done satisfactorily only when predetermined research aims have been achieved. Researchers neither have to be at the start of transdisciplinary projects nor at the core of (often transient) transdisciplinary collectives, as transdisciplinary work can, at least theoretically, also be undertaken from within different realms, such as policy, societal or entrepreneurial circles (Cummings et al., 2013).

In this section, we will reflect on different shades of transdisciplinarity in the segment of scholarly literature that employs the 'transdisciplinary' terminology before considering some of the larger volume of relevant scholarly literature that does not.

1.2.1 Different Shades of Transdisciplinarity

As Somsen and van Lunteren's chapter in this volume eloquently elaborates, in scholarly literature on transdisciplinary research, an 'Other' is routinely staged to articulate transdisciplinarity's self-identification. This Other tends to be the amalgam of approaches embraced by researchers and practitioners who feel at home in fields of mono-, multi- or interdisciplinary knowledge production. In recent years, rather than a juxtaposition, we see monodisciplinary forms of research (in all their plurality) as part of a transdisciplinary continuum. For instance, Jahn and colleagues (2022) on the basis of 59 sustainability-oriented projects identify five (transdisciplinary) research modes: (1) purely academic research; (2) practice consultation; (3) selective practitioner involvement; (4) ideal–typical transdisciplinary research; and (5) practice-oriented research, where the first mode (purely academic research) covers projects with strictly academic research (i.e. the research question was oriented towards academic problems), while aiming to realize substantial societal impact (i.e. the production of societally applicable knowledge or results is described as a goal of the project), with no non-academic actors involved in research design and execution. Similarly, Chambers and colleagues (2021), on the basis of an analysis of 32 co-production initiatives to address complex sustainability challenges, identify six modes of co-production: (1) researching solutions; (2) empowering voices; (3) brokering power; (4) reframing power; (5) navigating differences; and (6) reframing agency, in which the first mode (researching solutions) is focused primarily on generating evidence, or producing practical scientific knowledge, with the goal of informing and influencing policies and interventions, with relatively low inclusion of societal actors. Thus, recent classifications have embraced rather than juxtaposed non-participatory, non-co-creative, forms of research.

Rather than as one among multiple modes of transdisciplinarity or co-production, we can also envisage non-participatory, or non-co-creative forms of, knowledge production as an essential part of any transdisciplinary endeavour. There will be many instances where specific questions arise in transdisciplinary collaborations that demand further knowledge generation and that do not necessarily require participation or co-creation. We can thus envisage a nested structure in which transdisciplinarity encloses non-participatory, mono-, multi- or interdisciplinary research (see Fig. 1.1a). Surrounding non-participatory modes of research by transdisciplinarity can be interpreted in multiple ways. First, transdisciplinary collaboration for research agenda-setting or research design may result in a research agenda or research design including a mixed palette of sub-projects of which some could be primarily focused on non-participatory knowledge generation (e.g. in large research consortia in response to funding calls that demand that research explicitly works on societal impact through multi-stakeholder engagement, such as EU Horizon Europe, see Fig. 1.1b). In these large projects, often multiple iterations and integrative processes are included (outer circle), and still distinct mono-, multi- or interdisciplinary sub-projects can be identified (inner circle). Jahn and colleagues (2022) describe projects that come close to ideal–typical transdisciplinary research (Lang et al., 2012) as projects in which active interaction with practitioners takes place, primarily in early project stages by (co-)defining the research problem, and in the later stages of the assessment, dissemination and implementation of research results. One can only assume that between the early and later stages, some 'conventional' research activities also take place.

Second, we want to highlight that, especially in the case of smaller, master's or PhD research projects, transdisciplinarity might take a different shape. Transdisciplinary research, which is generally motivated

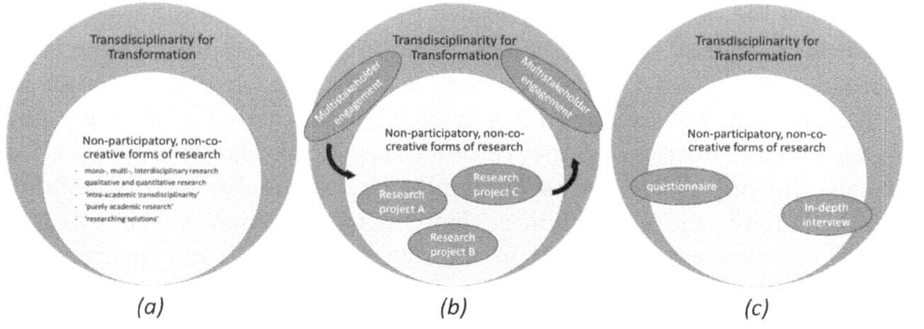

Fig. 1.1 **a** Generic nested relationship between transdisciplinarity and non-transdisciplinary (non-participatory, non-co-creative) forms of research. **b** Non-transdisciplinary sub-projects are nested in a larger transdisciplinary programme, with multiple sub-projects. **c** Non-transdisciplinary methods are nested in a single study, transdisciplinary project

by its capacity to handle complex, real-world issues, by integrating various knowledge types, stimulating reflexive learning processes and producing durable solutions to urgent societal problems, can also be more demanding in terms of resources and time, and require skills and knowledge that are not always amply available to PhD students (Rogga & Zscheischler, 2021). Van Breda and colleagues (2016, p. 152), based on the experience of three PhD students in South Africa, concluded that 'individual transdisciplinary research effort cannot necessarily tackle the aforementioned societal challenges in the same way as large transdisciplinary research teams'. Within smaller, sometimes even individual, transdisciplinary projects (see Fig. 1.1c), one can and will most likely employ conventional research methods (e.g. in-depth interviews, focus group discussions (FGDs), questionnaires). Being embedded in a context with a transdisciplinary intent, these regular research methods can be applied with more sensitivity to questions of power (e.g. sensitivity to the perpetuation of extractivism; considering 'what's in it for them?'; shifting relations from researcher-respondent to partners), as well as sensitivity to questions of agency (e.g. by employing techniques that evoke mutual sensemaking; by asking not only about problem perceptions, but also about strategies for responding to these problems, thus making use of the participants' innovative capacities and tapping into a sense of hope, in the midst of despair that might also be there[2]). These differences can be subtle; they may relate to which location to choose for an interview, whether to sit across from each other or next to each other (or not sit at all, but walk or do something together while talking), whether the researcher takes notes in private, or notes are taken collaboratively, such as through drawing, large-size note taking (e.g. on flipchart). It is through such seemingly negligible choices that a researcher makes that substantial differences can be realized in the degree to which one manages to stimulate reflection on action and tap into people's empathy and genuine engagement. Through this, in turn, such design choices can make a big difference in whether change is realized at the individual and collective levels—varying from something small like a mother deciding to buy an air-fryer as result of a well-facilitated FGD on healthier snacks (e.g. Iqbal et al., 2023) to multiple stakeholders coming together in their problem definition of a sustainability issue of some sort (see, e.g., Brouwers et al., this volume).

Figure 1.1 means to convey that transdisciplinarity does not preclude other types of non-transdisciplinary knowledge production—on the contrary. Mono-, multi-, or interdisciplinary forms of research, with which transdisciplinary research is often contrasted (see Chapter 2, this volume, for a problematization of this narrative), explicitly can play an important part in transdisciplinary research, as also indicated by the fact that one of the roles that is often distinguished in the context of transdisciplinary research is that of 'traditional researcher' (Bulten et al., 2021), or 'pure scientist' (Vinke-de Kruijf et al., 2022 term borrowed from Pielke, 2007, also see Gunn et al., Chapter 15, this volume, for a nuanced narrative on roles, including the 'scientific role'). In Fig. 1.1, besides the non-participatory research activities that have been labelled 'purely academic research' by Jahn and colleagues (2022) and 'researching solutions' by Chambers and colleagues (2021), we have also included 'intra-academic transdisciplinarity', which Vermeulen and Witjes (2020) have used to refer to strands of transdisciplinarity that pursue a unifying theory, or forms

[2] See Zachariah and colleagues (2023) for a practice of listening *for* rather than listening *to* perspectives or views. Listening *for* entails listening for concerns and hopes, leading to feelings of being validated and being heard, and listening for strengths and capacities, leading to recognizing the capacities to care, form a community, change, hope and embody leadership.

of complex modelling that enable better understanding to inform forecasting and scenario building. Stakeholder involvement is limited, and like the other mentioned examples in the inner circle, intra-academic transdisciplinarity builds on the idea that a lack of knowledge is the principal barrier to change (Chambers et al., 2021). Many variants and in-betweens, besides 1.1b and 1.1c, can naturally be imagined. The specifics of Fig. 1.1c (i.e. the slight overlap of inner circle activities with the outer circle) can of course also be applied to Fig. 1.1b, so that sub-projects are designed and conducted with more attention and sensitivity to context, direction, power and agency. The specifics of Fig. 1.1b can also be applied in a smaller, even individual, project, for instance by involving relevant stakeholders in the design phase of the study and by co-creating recommendations, with stakeholders, based on preliminary research findings. In all cases, the 'purely academic research' that is situated in the inner circle will take on a different form than it would have done had the outer circle not been there; the inner circle activities are performed with a transdisciplinary 'intent'.

Box 1.1: Burn survivor participation in research agenda-setting for burn research (based on Broerse, 2013; Broerse et al., 2010)
In 2006, burn survivors, researchers, and healthcare providers collectively formulated a research agenda for the Dutch Burns Foundation, following the transdisciplinary 'Dialogue Model' for research agenda-setting.

Phase 1: Exploration
Exploratory interviews ($n = 10$) were held with burns survivors, care and research coordinators and meetings were held with relevant organizations in the field of burn research, including patient organizations. Literature research and a desk study were undertaken.

Phase 2: In-depth study
In this phase, burn survivors and professionals were consulted separately and group-specific lists of research priorities were established. Burn survivors were consulted through FGDs ($n = 37$) and two additional interviews to add children's perspectives through proxy respondents. Professionals (prevention, health care, research) were consulted through three thematic FGDs ($n = 21$).

A notable outcome was that during interviews (phase 1) and FGDs with burn survivors, the issue of itching frequently arose. People are driven mad by it: 'as if millions of tiny ants are crawling under your skin'. In discussions with researchers and healthcare providers, however, itching was not mentioned. Nor was any research being conducted on this topic in the Netherlands. Healthcare providers saw it as collateral damage and patients rarely brought up itching with their doctor: 'It's something for home', and there's nothing that can be done about it anyway'. Patients thus identified a 'blind spot' in burn research.

Phase 3: Prioritization
The insights from phase 2 were translated into 60 topics for research and clustered in 10 research themes, which was, after pilot testing, sent out as a questionnaire to burn survivors (n = 224). The questionnaire revealed that there was broad consensus on the topic of itching. Professionals were asked to prioritize research themes and topics through various written and oral rounds (n = 12).

Phase 4. Integration
In a dialogue meeting, 14 burn survivors and 15 professionals met to further discuss and prioritize the research topics through dialogue and voting. During this multi-actor dialogue session, various researchers and healthcare providers were persuaded by patients that itching deserves much more attention. Scar itching emerged as the second priority on the joint research agenda.

Box 1.1 presents an example that illustrates how conventional social science methods can become embedded in a transdisciplinary framework. There is nothing transdisciplinary about conducting interviews, holding FGDs, or sending out a questionnaire. This could even be considered a form of data extractivism; burn survivors share their experiences and views on a very painful aspect, and period of their lives, researchers conduct their analysis, through their own lenses, which may only to some extent reflect the original narratives, and then publish about it. What is different here is that, while often patient participation in agenda-setting concerns only consulting patients about their problems and needs through a questionnaire, interview or focus group, or including patients as members in a research programming committee, here a multi-actor approach was used throughout the project, from research design to interpretation of findings. It thus combines consultation, using regular research methods (inner circle), with collaboration (outer circle), which is considered a fruitful way of approaching patient participation (Abma & Broerse, 2010). It does not leave integration of knowledge inputs from patients up to certified experts, nor does it completely transfer power to patients. As Abma and Broerse have argued, it is difficult to see how the perspectives of patients will be accepted and used by researchers if control is simply shifted from researchers to patients. Separate trajectories of consultation were conducted, because power asymmetries usually prevent meaningful interaction right from the start (Broerse et al., 2010). The consultation with burn survivors was designed and implemented in collaboration with the Association of Burn Survivors. Efforts were made to empower patients by supporting them in articulating previously unvoiced experiences and views and by supporting them in a preparatory meeting prior to the dialogue meeting with professionals. Similarly, in the various research encounters, professionals were sensitized to the relevance and importance of experiential knowledge. The transdisciplinary project thus combined 'engaging powerful actors to reduce their own and peers power over marginalized actors (influencing powerful actors)' with 'engaging actors to iteratively shift power relations with powerful actors (empowering marginalized actors)' (Chambers et al., 2021, p. 986). The conversations that took place during the dialogue meeting between patients, researchers and healthcare providers shifted priorities (mutual learning[3] took place) and scar itching emerged as the second priority on the joint research agenda.

Thus, what makes us consider this project 'transdisciplinary' is not so much the research methods that were employed, but the collaboration and mutual learning that was fostered through a carefully prepared and facilitated multi-actor session. A second relevant 'transdisciplinary' aspect of this case only transpired afterwards. While this case can be seen as a project, with a clear beginning and an end, transdisciplinarity for transformation is not a project; it has no clear beginning and end. In this particular case what happened is that scar itching was included in the research programme of the Burns Foundation and a 'call for proposals' was issued in 2007. However, the foundation received no proposals

[3] The Dialogue Model that was employed in this case (Abma & Broerse, 2010) mentions *mutual learning* as an essential characteristic. It is defined as: 'a genuine dialogue implies that participants learn in the process and may change their opinion through listening to each other and learning about each other's experiences. Mutual learning is fostered through face-to-face meetings. In such meetings participants ask questions, probe, argue and deliberate about their experiences and opinions' (Broerse et al., 2010, p. 219).

on the topic—while researchers had been willing to learn, the identified 'blind spot' proved difficult to fill in. Several years later, in 2012, the Biotechnology and Genetics Forum took the initiative of organizing an expert meeting on research into 'scar itching' together with the Association for Burn Survivors and a pharmaceutical company. The central question was: What needs to be done to develop a remedy for scar itching? involved a so-called 4P partnership—public sector, private sector, patients and practitioners. Since the 'call for proposals' in 2007, scarcity of resources was certainly no longer the limiting factor—the changes needed to develop an effective innovation in the field of such a 'blind spot' require not only financial resources but also a shared vision, commitment and concerted action from a larger number of stakeholders. The meeting was an important step towards establishing a partnership to develop an effective remedy for itching in burn injuries.

Thus, rather than as a project, transdisciplinarity for transformation can be seen as a movement or development in which all kinds of projects, in an unplanned manner, may play a part. From the perspective of a specific project, one could say 'don't start from scratch—there is always a "before the beginning" and an "after the end"' (Regeer et al., 2011, p. 29). Ideas for transdisciplinary projects usually result from vague notions, previous experiments, and especially from actively building on the ideas of others and recognizing and acknowledging the complementary opportunities a multi-stakeholder environment offers. Ideas that may develop out of one project may settle in the heads of individuals who take them into new networks and practices, waiting for an opportunity to continue on the path of transformation (Regeer et al., 2011).

1.2.2 *Transdisciplinarity, by Any Other Name*

Many fields of research and practice can be discerned that do not self-identify as transdisciplinary but that share in the generic inclusive and transformative intent described above. There are, of course, too many to include here and we will leave many untouched, including Ground Up Inquiry (Verran et al., 2022), appreciative inquiry (Whitney & Cooperrider, 1998), Theory U (Scharmer, 2009), reflective structured dialogue (DeTemple & Sarrouf, 2017), knowledge management for development (KM4Dev, Boyes et al., 2023), reflexive interactive design (Elzen & Bos, 2019), pragmatic complexity (Ansell & Geyer, 2017), responsive (evaluation) methodologies (Guba & Lincoln, 1989), Integration and Implementation Sciences (I2S) (Bammer, 2013) and futures studies (Sardar, 2010). As students often ask about the differences between transdisciplinary research and, depending on their educational training, either participatory action research or engaged scholarship, we will discuss these traditions here, but only briefly. However much space we might dedicate in this chapter, it would still not do justice to their rich and important legacies. We will also very briefly touch upon other pleas for different modes of knowledge production.

Participatory Action Research (PAR) has multiple roots—like transdisciplinary research, it is better described as a family of approaches than as one tangible and clearly demarcated approach. We do want to recognize two of its roots, however, including one strand that is from Latin America and builds on the works of the Brazilian educator and philosopher Paolo Freire and the Colombian critical sociologist Orlando Fals Borda; and a second strand that builds on the works of German/North American social and organizational psychologist Kurt Lewin. Lewin is known to have introduced the term *action research* in 1946: a form of problem-solving through consecutive cycles of planning, action and reflection, that was initially mostly used in democratic settings to democratize workplaces (Breda, 2014). Lewin proposed that communities usually excluded from the research process should join researchers to study 'real-life' situations and collaboratively produce knowledge to effect social change (Torre, 2014). Freire's ground-breaking 'pedagogy of the oppressed' (1968) reflects a revolutionary pedagogic method aimed at empowering marginalized, impoverished and oppressed populations through developing a critical consciousness about their circumstances, awareness of the need for social change and a recognition of their own wisdom and knowledge (see also Freire, 1973). PAR is then not so much a methodology as a movement. A movement that induces a praxis of the oppressed to liberate themselves and their oppressors (Torres, 2021). A movement that includes a new way of thinking and doing science (Breda, 2014). Or, as Fals Borda put it 'PAR could be considered not only as a methodology for research to be taken into account by the institutions, but also as a *philosophy of life*. Those who practised it were feeling-thinking people ready to struggle for changes and understand them better' (Fals Borda, 2013, p. 162 [our emphasis]). More recently, PAR, or rather *critical* PAR, has been 'rediscovered' in connection to feminist, queer, critical race and Indigenous theories (Torre et al., 2012).

As a family of 'approaches' PAR encompasses community action research, appreciative inquiry and co-operative inquiry (Reason & Bradbury, 2001), and has extended to applications in natural resource management, agriculture and food security, among others in the form of *participatory rural appraisal* (Chambers, 1992). More recently, participatory *health* research approaches have gained popularity as encouraging pathways to cross epistemic and social boundaries and foster co-creation to drive innovation and improve healthcare practice and policy. Participatory health research, again, encompasses a range of approaches (Roura, 2021), including Community-Based Health Research (Blumenthal & DiClemente, 2004), Community-Based Participatory Research (Cacari-Stone et al., 2014) and the Dialogue Model for patient participation (Abma & Broerse, 2010). Besides PAR, these participatory health research approaches build on collaborative inquiry (Bray, 2000), patient and public involvement (Brett et al., 2014) and public, citizen or community engagement (Irwin, 2006) and share the 'celebrat[ion of] participation and democracy in the research process' (Bray, 2000, p. 3).

We do not present (the family of approaches of) PAR here as different from transdisciplinary research, but as one of the ways of thinking and doing science from which transdisciplinary research takes inspiration and builds upon (see, for instance, Neuhauser, 2018), or, as stressed by others, could more strongly build upon. Jones and Loeber (under review) observe that in recent forms of funded transdisciplinary research (such as 'Living Labs'), *'despite dedicated formulations in subsequent funding programmes to researchers to include the voices of civil society actors, or even make them full partners, in practice is still a long way to go into making knowledge and intervention with social movements in shared ownership in line with guiding concepts of liberatory PAR'*. They invite a reconsideration of early accounts of PAR, in Colombia and the broader global South, to serve *'as an inspiration in making knowledge a transdisciplinary resource for liberatory action in the context of EU-funded R&I projects'*. This indeed is a relevant point. A difference between transdisciplinary research and PAR may be that, in its application, the former more explicitly includes researchers from different academic fields and actors in positions of power, while sharing the commitment 'to enable local people to share, enhance and analyse their knowledge of life and conditions, to plan and to act' (Chambers, 1992, p. 1); i.e. transdisciplinary research is often associated with multi-stakeholder, or multi-actor,[4] engagement. While PAR aims to create meaningful change, not only in social contexts, but also in the realms of policy, science and corporations ('liberating the oppressed as well as the oppressor'), its starting point is empowerment of the oppressed. It could be questioned if indeed a multi-actor approach, like transdisciplinarity, does not always risk perpetuating the inequalities it was committed to fighting; for a start, it requires critical self-reflection of those external to the marginalized or oppressed, 'the academicians' (Fals Borda, 2013). With this, let's turn to engaged scholarship.

In Chapter 11 of this volume, Holle and colleagues coin the term *transformative engaged scholarship* to refer to forms of scholarship that enable more inclusive practices, both in society and in academic institutions. It builds on critically engaged scholarship, which moves away from earlier readings of engaged scholarship that put primacy on providing social analyses that would be of direct relevance to urgent public issues without questioning the authority of academia (Franklin, 2022; Smets et al., 2020). Critically engaged scholarship, by contrast, does question the authority of academia by 'knowing that engagement also means becoming complicit in the processes [sociologists] try to criticize and change' (Smets et al., 2020, p. 287). Transformative engaged scholarship, then, is not only about creating spaces for mutual learning with different social actors, particularly those from marginalized communities, but, importantly, it challenges normalized power dynamics in knowledge production and implies that scholars acknowledge power relationships,

[4] We prefer *multi-actor* over *multi-stakeholder* for two reasons. First, because all stakeholders are actors, but not all actors are stakeholders. Sometimes the knowledge of some actor or group of actors is pertinent to reach an in-depth understanding of the relevant situation, even though those actors do not have an obvious stake in the matter at hand. An example is ecotoxicological knowledge in the context of understanding environmental hazard associated with pollution by a chemical plant. Ecotoxicologists, unlike local residents, farmers, environmental protection activists, policy-makers working at various levels or institutes of government and the chemistry company at hand, do not have stake per se in any findings concerning the situation at hand and do not (obviously, straightforwardly) have an interest in one or another proposed course of action in relation to the analysis made. Yet, ecotoxicologists' knowledge is highly valuable in such a situation and rightly deserves to be included in transdisciplinary work aiming at transformation. Second, we wish to underscore that parties engaged in transdisciplinary processes, more than only having a stake that engaged parties can defend or promote, have the ability to *act*, i.e., to do something and, through that, to make a difference.

including their own position and their emotional, practical and structural capacities and boundaries (Holle et al., this volume, Chapter 11; see also Strumińska-Kutra, 2016). Zuiderent-Jerak (2015) offers a similar critique of engaged scholarship in sociology. He argues that it is often presented as producing scholarship that is both rigorous and relevant, with a weak connection between the two. This version of engaged scholarship risks overlooking what scholars can add *sociologically* to practices, by simply combining activism with scientific authority. This leads Zuiderent-Jerak to introduce *situated intervention* as a form of scholarship that locates normativity 'in the many attachments that actors in the field, including scholars, sort out in practice' (ibid. p. 23). Moreover, through intervening, this scholarly approach aims to produce sociological insights (rather than change practice on the basis of sociological knowledge). Situated interventions could thus be argued to share with PAR a focus on learning through change, by all those involved, and it may differ from PAR because of its explicit positioning as a 'scholarly' approach, aiming to produce sociological insights.

We have not yet mentioned the many other pleas for different modes of knowledge production better able to respond to persistent, societal problems, upon which transdisciplinary researchers often build. Descriptive accounts of the intertwining of the scientific and the social in the field of STS have also led to prescriptive appeals for doing science and technology differently. Funtowicz and Ravetz (1994), for instance, express the need for '*post-normal science*' in cases characterized by high uncertainties and conflicting values and introduced the *'extended participation model'* (Liberatore & Funtowicz, 2003) as a potential way to do this. Analysis of the 'triple helix' dynamics between universities, government and business (Etzkowitz & Leydesdorff, 2000; Leydesdorff & Etzkowitz, 1996) have inspired '*quadruple helix models*' that include civil society actors in social transformation (e.g. Bunders et al., 1999; Nguyen & Marques, 2022). More generally, there are broad calls for revising the contract between science and society, notably the call for '*mode-2 knowledge production*' (Gibbons et al., 1994; Nowotny et al., 2001); a mode of knowledge production where people have a place, not only as end users, but also at the core of knowledge production; where diverse values are not added to science but integrated into its practice. Other calls for revising the contract between science and society argue for the social contextualization of knowledge production (Rip, 2011) by developing new interfaces that encompass attempts at '*reflexive co-evolution*' (e.g. Rip, 2005).

These and comparable ways of rethinking science and research and their relationship with 'real-world' phenomena and communities not only find their way into arguments in favour of transdisciplinarity as an approach to research as such, but are arguably also foundational for the governance approach to research and innovation called Responsible Research and Innovation (RRI) (see, e.g., Klaassen et al., 2018; Owen et al., 2012). Also inspired by the work of feminist thought generally and care ethics specifically, RRI recognizes the need to listen to and incorporate the needs of all relevant parties in developing knowledge and engaging in innovation, for instance because '[t]echnologies [as products of research and innovation] should be treated [...] as elements of practices of care that both serve intended ends and that mediate our changing conceptions of these ends' (Macnaghten et al., 2014, p. 196), where, for multiple reasons, deciding on ends is not to be left only to scientists and innovators.

More recently, *knowledge co-production*, or more generically *co-production,* is increasingly used. Turnhout and colleagues (2020), in their discussion on participation, power and transformation use the term co-production as a shorthand for participatory modes of knowledge production. Others speak about the *co-creation* of knowledge for sustainability (Mauser et al., 2013). Still others broaden the scope beyond knowledge production and speak about the co-production of knowledge, action and social change; for instance, Chambers and colleagues (2021) in the earlier mentioned article distinguishing six modes of co-production for the sustainable development of ecosystems. Similarly, Miller and Wyborn (2020) speak about co-production guiding the design and implementation of sustainability research and action (see also Bremer & Meisch, 2017 for a review of the use of co-production in climate change research, as well as Wyborn et al., 2019). It is interesting to see how scholars continue to grapple with the question of 'naming' attempts at reflexive co-evolution. For instance, in an article that advocates a different response to climate change from the research community (Fazey et al., 2018), the sketched response, consisting of ten essentials, is referred to as '*action-oriented and second order transformation and energy research*'. The term transdisciplinary research is almost absent from these writings, but in our view they are part of the same family of ways of thinking and approaching persistent problems.

Recent works on *collaborative governance* also use the terminology of co-production to point to governance arrangements that involve collaborations between a wide range of actors, including citizens, that produce new forms of knowledge, values and social relations and contribute to innovating public services and social innovation (Osborne & Strokosch, 2013; Sorrentino et al., 2018). This emerging field of interest manifests conceptual and analytical affinity with earlier works from another domain, in which policy scientists argued that in our present society, relationships between citizens and government as well as between policy scientists and practice need revising. Schön and Rein (1994) argued that traditional approaches to policy analysis are not appropriate for understanding persistent problems (or 'intractable controversies' as they call them), and, more importantly, they do not aid in their resolution. The dichotomy between reflection by academic scholars and the practice of policymaking should be resolved by collaboration between policy academics and policy practitioners. They emphasized the role of policymakers in knowledge co-creation: 'Policy researchers should focus on the substantive issues with which policymakers deal, the situations within which controversies about such issues arise, the kinds of inquiry carried out by those practitioners who participate in a controversy or try to help resolve it, and the evolution of the policy dialectic within which practitioners play their roles as policy inquirers' (Schön & Rein, 1994, p. 193). Schön and Rein introduced the idea of *collaborative frame reflection* as a concrete approach to co-production of knowledge in the context of policy analysis for intractable problems. Laws and Hajer (2006) have also moderated the claim that knowledge by itself can guide policymaking, and have argued for cooperation. The corresponding idea that the units within which policy has to be made coincide ever less with the constitutionally defined settings (Hajer & Versteeg, 2005) has given rise to a new role for citizens in our deliberative democracy and an emphasis on local knowledge and participative inquiry in the search for new forms of knowledge (e.g. Fischer, 2000, 2003). Collaborative governance today is often shaped in experimental settings such as governance experiments, urban laboratories and living labs (see Bulkeley & Castán Broto, 2013; Vandenbussche et al. 2024). Again, these scholarly works do not make explicit associations with transdisciplinary work—but we feel they can provide inspirations to those new to the field, especially because they may resonate with the diverse (epistemic) backgrounds and motivations of readers who wish to explore transdisciplinarity.

1.3 Transdisciplinarity for Transformation—A Multi-actor, Reflexive Practice Approach

It would be hard to find a source that talks about transdisciplinarity without talking about societal change or transformation. And if that is the case, then why title this book *Transdisciplinarity for Transformation*? How could transdisciplinarity *not* be for transformation?

The scholarly field of transdisciplinary research was given a great impulse at the 2000 Zurich conference titled: 'Transdisciplinarity: joint problem-solving among Science, Technology and Society' (Klein, 2001). The importance of 'how to' knowledge (Fazey et al., 2018), or transformation knowledge (Hadorn et al., 2008), in addition to widely available and ever-growing bodies of knowledge on the evidence for complex societal problems, has since increasingly been recognized. More than two decades after that landmark conference, we observe an increased engagement with transdisciplinary research, and multiple advances along epistemological, methodological and ethical lines. Since then, methodologies for organizing meaningful knowledge integration have been experimented with and iteratively improved (e.g. Horn et al., 2022; Tell et al., 2017), as is also the case for principles and heuristics to guide citizen participation in science and technology (e.g. Chilvers & Kearnes, 2020) and multi-stakeholder innovation processes (e.g. van de Poel et al., 2020); strategies to address power dynamics and to overcome systemic barriers (e.g. Kok et al., 2021); and approaches to sustain and upscale processes and outcomes (e.g. Aalbers & Sehested, 2018).

1.3.1 An End in Itself, Rather than a Means to an End?

However, the increasingly urgent need to create impact in view of ever more devastating health and sustainability challenges and forms of societal injustice indicates a gap in the current focus of co-creation literatures—as these tend primarily to be dedicated to processes of (knowledge) co-creation. With the evident need for in-depth insight into the dynamics of (and approaches to) multi-actor innovation processes comes the risk that understanding processes of collaboration, co-creation, social learning and reflexivity, including their political and power dimensions (Turnhout et al., 2020), becomes an end in itself, rather than a means to an end. Or, phrased differently, the (undeniably important) focus on process criteria for transdisciplinarity tends to blur the 'ends' of the endeavour, and hence takes the 'transformation' out of transdisciplinary research. It is telling that a quite recent in-depth study of 16 transdisciplinary research projects concerned itself with 'explor[ing] how to proactively generate potential for societal effectiveness in TDR via the adaptive shaping of TDR processes' (Lux et al., 2019, p. 184). While one might think of transdisciplinary research as being from the outset concerned with fostering certain societal effects, it appears that there is some light between shaping transdisciplinary research processes on the one hand and bringing about societal effects on the other. Therefore, there is a need to understand how to bring the intended societal effects centre stage in shaping transdisciplinarity. The intention of transdisciplinary research is to start with a complex societal issue and co-create an approach that fits the continuously re-negotiated purpose.[5] This is markedly different from any disciplinary, multidisciplinary or interdisciplinary way of working, in which either one or more disciplinary perspectives, or the perspective of a (funding) policy institution, or one particular actor's version of what the pertinent issue is, decides the course of action.

However, this is highly challenging on many accounts. For instance, the imagined effects of transdisciplinary endeavours will often only become visible after longer periods of time, beyond the conventional duration of a project. This makes it challenging to keep aspired societal effects at centre stage and to keep collaborators engaged throughout. Another reason this might be challenging has to do with the different rules and institutions within which transdisciplinary collaborators might work. That is to say, the approaches devised in co-creation do not necessarily align with what is expected from different collaborators in the institutional settings in which they operate on a day-to-day basis—their home base. While this is recognized in the literature on transdisciplinary research (e.g. by developing *context-sensitive* transformative approaches, Van Breda & Swilling 2019), we would like to reiterate that there is much emphasis on multi-actor engagement, co-creation or 'participatory contact zones' (Torre, 2005). And while the many benefits of growing insights into processes of co-creation are invaluable, we want to stress

[5] Note that we intentionally use 'purpose' here, because it better communicates the sense of an *internalized* mindset, or philosophy, as a point of departure, rather than *external* (future) goals, vision or societal effects as aspired end points.

the importance of also better understanding participants' multiple attachments (Jensen, 2007), especially those in their professional capacities, as people often experience only a narrow space for transformative action, given the (implicit) rules and institutions that guide the practices in their home base.

1.3.2 On the Power of Shared Practices

A major challenge for co-creation-oriented approaches has to do with the many ways we are bounded—by our professional roles as policymakers, researchers, entrepreneurs and the explicit and implicit rules and institutions that come with these; by the access we do or do not have to different knowledges; by the cultural repertoires that we embody; by our racial and gender identity; and so on (Knapp et al., 2019; Marguin et al., 2021). These multiple identities deeply shape the ways in which we perceive, theorize, think, reflect and act—in other words, our multiple identities carve out our room for manoeuvre, both scientifically and practically. This might, however, present multi-actor approaches with a huge challenge.

This topic has been studied elaborately, both by scholars of positionality (e.g. Baur, 2021) and by practice theorists. At the basis of this work is the notion that we acquire knowledge, and assign meaning to the world around us, through participation in shared social practices. Lave and Wenger (1991)[6] have referred to these shared practices as 'communities of practice', whose members share a repertoire of resources to give meaning and make sense, including routines, words, instruments, ways of acting, stories, symbols and gestures. Such communities of practice can be groups of professionals, such as claims assessors of insurance companies, as Wenger (1999) explored in depth. Comparably, the sociologist of science Knorr-Cetina refers to 'epistemic cultures', which she explains as 'those amalgams of arrangements and mechanisms [...] which, in a given field, make up how we know what we know' (Knorr-Cetina, 1999, p. 1). As both concepts reflect, members of a professional group or (scientific) field share a practice or culture; that is, the social and the scientific are mutually constitutive—or, as Jasanoff puts it with specific focus on the practice of science: '[scientific knowledge], both embeds and is embedded in social practices, identities, normal, conventions, discourse, instruments

[6] Building on the work of philosophers like John Dewey (1910) and Ludwig Wittgenstein (1953).

and institutions – in short the social' (Jasanoff, 2004, p. 3). This explains why it is so hard for scientists who are rooted in different epistemic cultures to co-develop a systemic, integrated approach to 'a problem'. First of all, because 'a problem' does not exist out there in a realm separate from perceptions, norms and identities. But also because the room for manoeuvre a researcher may experience is shaped by the norms, rules, conventions and routines that constitute their epistemic culture. Whitley (1982) conceptualized the possibility for 'outsiders' to exercise influence on knowledge production within a particular discipline.[7] Whitley found that this depends on the degree of mutual dependency between scientists and the degree of (un)certainty about the position, tasks and intentions of the discipline: the higher the dependency and task certainty, the harder it is to exercise influence from the outside. For instance, in the example of patient involvement in research agenda-setting for burns survivors, the task certainty of the (biomedical) researchers that could potentially apply for funding for research on scar itching was initially too high to include this new topic into their scope of relevance, or their 'reputational system', as Whitley calls this.

In the same vein, institutional theory establishes institutions as 'sets of public norms that condition the interaction between subjects' (Salet, 2018, p. 1), such as formal and informal rules, behavioural norms, practices and narratives (Lowndes & Roberts, 2013). Furthermore, the mutual dependency of actors (persons and organizations) contributes to the stability of a system (Arkesteijn et al., 2015), and thus prevents the inclusion of new actors and insights, or any sudden change of rules. Lastly, materialities (Grin, 2020; see also Huitzing et al., 2020), such as soil quality, or landscape, but also infrastructures and material components of technologies (Arkesteijn et al., 2015) can be seen as part of the mesh of institutions.

By presenting the concepts of communities of practice, (epistemic) cultures, appreciative systems, and institutions almost in a single brushstroke, we certainly discard all sorts of bigger and smaller conceptual and empirical differences. However, what is more important to us than the nuances of each of these and the differences between them, is that they help articulate what we mean by the 'boundedness of practice' (cf. Nicolini, 2009).

[7] See also Bunders (1987) for an empirical example of the same in a community of biologists.

Our 'membership' in a diversity of practices, and the multiple identities we thus hold, is manifold and transgresses professional and academic boundaries. As emphasized before, this includes any and all aspects that make up our positionality, our (multiple) self, including in terms of ethnicity, nationality, age, gender, disability, sexual orientation, and social and economic status (Staffa et al., 2022). Thus, we can, for instance, be a biologist-by-training, a policymaker, a mother and a volunteer at the same time. What we often see, however, is that institutional logics are so dominant that most of our identities are marginalized when we enter the professional realm; there, the formal and informal rules of our professional and academic practices start to dominate in all kinds of intricate and scarcely visible ways. Again, we can refer here to the burns case, where even those biomedical researchers who were part of the dialogue meeting in which research on scar itching was collectively prioritized did not apply for any funding for this research subsequently. Crossing the boundaries of their academic field (including the boundaries to what are considered relevant research topics) might come with great personal costs, in terms of reputation, career opportunities or even the loss of a sense of belonging.

This is why the above poses such a magnificent challenge to transdisciplinarity. For when we act, speak, perceive and choose in ways that strengthen rather than weaken our belonging to our professional communities—whether it is by spending our time on publishing in high-impact journals as an academic, preparing policy briefs to advocate for a certain concern from a single-issue non-government organization (NGO), or navigating the internal politics at a ministry—the chances are high we disavow many of our other identities. And often the more powerful actors within a transdisciplinary process, such as a government official or a scientist, tend to be the ones that are *more* bounded by their professional affiliation. A single multi-actor event that might be part of transdisciplinary process will not change the way these actors do science or make policy once they are back at their 'laboratory' or office (Lynch et al., 2017). Thus, while co-production of knowledge and action, and even destabilizing power relations, may take place in certain spaces at certain instances, structurally addressing injustices and the lack of institutional listening (Scudder et al., 2021) in prevailing knowledge and governance systems takes much more.

1.3.3 *Balancing Means and Ends Through Reflection*

It is recognized that it is extremely difficult to keep one's eye on the prize and remain focused on the purpose from which one is working. This requires a constant balancing act among the actors involved in emerging transdisciplinary spaces. Moreover, transdisciplinarity for transformation requires an interplay between emerging multi-actor practices, *reflective* practitioners, *adaptive* organizations and *reflexive* governance.[8] In the words of Rakesh Kapoor (2007, p. 475), '[a]ll social transformation can be seen as a dialectical play between three sets of oppositions: between (individual) biography and (social) history, between theory and praxis, and between the micro/local and the macro/global levels of organization'. Although enacting the dialectic at all three levels simultaneously is highly demanding, it is crucial for true social transformation. We know that if demarcated, co-creative efforts are not accompanied by systemic transformation, we are left with little more than individual pilot projects and programmes that come and go, without leaving behind any lasting impact (Felt, 2017). Such is the predicament in which we find ourselves.

1.4 Dynamic Learning Agenda, Learning Questions and How the Book Is Structured

This book is organized into three parts, the first focusing on the design and evaluation of transdisciplinary work, the second on whom to include and how, and finally one on roles and competencies requisite to engaging

[8] Reflexive governance may refer to an alternative orientation for governance, needed to direct society's course towards more sustainable outcomes. It is a response to the unintended side-effects of modernization that are inherently generated by the practices of modernization (hence, reflexive modernization), and it proposes strategies that incorporate rather than eliminate uncertainty, ignorance, heterogeneity, ambiguity and unintended effects (see Beck, 1992; Voss et al., 2006). Here we mean, with reflexive governance, not only the transient strategies required for transformation, but also the integration of experimental ways of day-to-day governance and accountability mechanisms at all levels, that may stimulate reflexive practice. We define reflective practice as the practice that incorporates reflection-in and reflection-on actions undertaken in that practice, so as to engage in continuous cycles of adaptation and learning that build on an understanding of the practical values and theories figuring in these (everyday) practices (Argyris & Schön 1974). Reflective practice is often explained as entailing both reflection and reflexivity—where reflexivity means something related but not identical to how it is used above in the context of reflexive governance (Hargreaves & Page, 2013).

in transdisciplinarity. For each part, different learning questions are relevant. In this final section, we will present these learning questions after first introducing the concept of a learning agenda.

There is an appeal to considering transdisciplinary research as an approach that can be taught, like any other research approach or methodology. At the same time, or perhaps first and foremost, it is a particular mode of thinking and doing; a philosophy or a mindset. Getting accustomed to a new mode of thinking and doing generally requires becoming part of a community, learning the unspoken rules, starting to acquire a new vocabulary, developing sensitivities for certain observations above others. In particular for early-career researchers this may be quite challenging; supporting continuous reflection is hence seen as pivotal in any transdisciplinary endeavour.

There are different approaches to organizing this type of reflective practice, and even tools to support it. One of them is the so-called Dynamic Learning Agenda (DLA) that was developed in the context of supporting transdisciplinary projects that aimed to contribute to system innovation in the context of the sustainable development of Dutch agriculture (Regeer et al., 2009). It has since been incorporated in a reflexive monitoring approach that accompanies transdisciplinary endeavours, Reflexive Monitoring in Action (Van Mierlo et al., 2010), and employed in empirical domains, ranging from perinatal care (Schuitmaker-Warnaar et al., 2021), the micro-politics of urban food governance (Luger et al., submitted), agricultural innovation (Kilelu et al., 2014) and sustainability transitions (Svare et al., 2023). We will delve into the DLA here, because we have used it to provide some structure to the learning process of master's and PhD students in our transdisciplinary education over the past decade. The learning questions that the students themselves have formulated have been the starting point for this volume, as we will see below.

The Dynamic Learning Agenda builds on the idea that conditions for a transdisciplinary strategy are never in place and that, as a consequence, the strategy itself should focus on creating or dealing with these conditions, as Broerse (1998) formulated on the basis of her transdisciplinary research in Zimbabwe. For instance, an unmet condition may be 'the project team includes all relevant expertise, experience, and other relevant "stakes" needed to tackle the sustainability problem [...]' (Lang et al., 2012, p. 30), or 'academic reward systems acknowledge and value knowledge co-creation with societal actors', or 'powerful stakeholders

[are] aware of how their privileges influence processes and outcomes' (Roura, 2021, p. 783). These unmet conditions inform the formulation of learning questions on the Dynamic Learning Agenda.

The formulation of learning questions follows two rules (Van Veen et al., 2014). First, it should convey a sense of agency. It is therefore typically formulated as 'How can I/we …?' rather than, for instance, 'Why do academic reward systems not ….?'. It thus helps to move beyond an initial aggravation or frustration that is part of any transformative process (e.g. 'they just don't understand!') and speaks to the importance of (self-) reflexivity in transdisciplinarity. Second, learning questions should convey a sense of 'toughness', thus bringing to light the difficult issues that are often 'swept under the rug' (Kleiner & Roth, 1996, p. 14). To do this it helps to add a dependent clause to the learning question, starting with 'while'. Taken together, learning questions are formulated as: 'How can I …. while …?'. So, a learning question may read: 'How can I, as an academic researcher, co-create tangible outputs that resonate with societal needs, while academic reward systems do not acknowledge or value these outputs?'. Or, 'How can I ensure that the often not heard voices are heard and taken seriously, while those in power are not aware of how their privileges influence processes and outcomes?'. Note that we have formulated these learning questions as a meta reflection on the transdisciplinary research process. In a project on sustainable agriculture, a learning question might read: 'How can we motivate farmers to produce biological tomatoes, while (they say) the market is demanding cheap and colourful tomatoes?'. And, in a project on patient empowerment in health care, a health professional may see their professional identity change and express their pain and grief about this as: 'How can I support the client to find their own way, while I deeply feel it is part of my professional identity to help them, to do things for them?'.

In a transdisciplinary project, a first learning agenda is formulated and monitored over time and changes are captured in the second, third, etc. learning agenda; hence the *Dynamic* Learning Agenda. Learning questions may disappear when they are resolved, they may need reformulating as insights into the issues evolve, or they may persist over time. Learning agendas may be formulated by a so-called reflexive monitor based on observations and fed back into the process (Regeer et al., 2009), they may result from conversations or reflection workshops within collaborative constellations (e.g. Van Veen et al., 2014), or they may be formulated and followed over a period of time by individual actors in the change efforts, e.g. midwives (Schuitmaker-Warnaar et al., 2021) or policymakers (Luger et al., subm.). The idea is that formulating a concern, or an unmet condition, as a learning question creates not only ownership, but also steering capacity (Van Veen et al., 2014). It gives clues for action, for casual inquiry, for trial and error. In this way, the unmet conditions that may hamper a development will no longer be seen as properties of an external system, but as points of leverage for the strategies that the team needs to develop (Regeer et al., 2011). It speaks to the need for 'how to' knowledge (Fazey et al., 2018) or transformation knowledge (Hadorn et al., 2008)—the kind of knowledge that is recursively developed through the Dynamic Learning Agenda.

The Dynamic Learning Agendas can be seen as a living archive of challenges and related strategies encountered in transformative change processes (Regeer et al., 2009). The idea is that questions that remain on the learning agenda only briefly, first-order learning questions, pertain to issues that lie within the capacities of practitioners to resolve, through single-loop learning (incremental improvement of existing routines) or through gaining experience and learning new skills. It is especially the tough issues reflected in persistent learning questions, those that stay on the learning agenda for a longer period of time, that are of interest. We have referred to these as second-order learning questions as they involve changes in underlying beliefs, norms and assumptions (Regeer et al., 2009), akin to the notion of double-loop learning as introduced by Argyris and Schön (1974, 1978). In a similar vein, humanist philosopher Kunneman (2006) noted in his account of the existential state of contemporary societies that although tough questions may be shoved away under the table, from this subordinate position, they will continue to give persistent signals. According to Kunneman, these signals can become visible when there is room for exploration and even acceptance of differences between people and positions (cited in Regeer et al., 2009).

Above, we spoke about the boundedness of practice (Nicolini, 2009), which is reflected in the learning questions by embracing the unmet conditions, or persistent signals, ('while ...?') that upon continuous neglect, will hamper development. At the same time, these learning questions reflect a message ('How can I/we ...') of hope, of agency, of willingness to 'staying with the trouble' (Haraway, 2016), of human creativity in persevering 'causal inquiry' (Schön, 1995), eloquently brought together by Nicolini in 'appreciating practice as *bounded creativity*' (2009, 1404, *our italics*).[9] Learning questions address trouble, tensions or discrepancies when and where they arise. Learning questions are specific and situational, and intended to contribute to the sensitization of all involved to issues that emerge as relevant and can be 'judged by the quality of the conversation they provoke' (Regeer et al., 2009, citing Kleiner & Roth, 1996, p. 20). Those conversations are situated neither in the 'scientific' nor the 'social'; rather they are situated in the here and now. Van Breda et al. (2016) use the notion of 'socio-epistemic relationships' that develop through transdisciplinary encounters. We can perceive the DLA as fostering these socio-epistemic relations through salient questions that have meaning in hybrid spaces, in 'transdisciplinary epistemic communities' (Regeer & Bunders, 2003; Van Breda et al., 2016).

By situating the conversations around learning questions firmly in the here and now and in locally specific transdisciplinary epistemic communities, they foster 'epistemic humility' in researchers, 'challenging intellectual rigidity, showing that to *hold on* to your position without understanding the benefits of humility [...] is a disadvantage' (Gardiner, 2020, p. 38). At the same time, academic researchers can perform alignment or translational work to align these socio-epistemic conversations with specific (monodisciplinary) epistemic communities or academic debates. Dynamic Learning Agendas can thus contribute to abductive theorizing (Dubois & Gadde, 2002; Stirling, 2015), for instance on conceptual spaces as distinct as organizational listening (Macnamara, 2018) and epistemic justice (see Ramaswamy et al., Chapter 10, this volume, Fricker, 2007) when the conversation revolves around a learning question such as 'How can I ensure that the often not heard voices are heard and taken seriously, while those in power are not aware of how their privileges influence processes and outcomes?'.

We have employed the Dynamic Learning Agenda as a way to structure the learning experiences of our master's and PhD students, and at the same time for them to become acquainted with a specific tool they might employ in their transdisciplinary research with participants to guide collaborative learning processes. We have distilled some patterns out of the hundreds of learning questions formulated by these students (see also Gunn et al., Chapter 15, this volume) and used them to structure this

[9] We would like to acknowledge Anne Loeber for making this connection between the DLA and Nicolini's work on practice as bounded creativity.

edited volume[10]: (1) Design and Evaluation; (2) Diversities and Inclusion; and (3) Roles and Competences. Of course, these are not mutually exclusive, but highly intertwined categories, which iteratively inform one another. We will introduce each of the three parts along the lines of the learning questions.

1.4.1 Part I: Design and Evaluation

Designing a transdisciplinary research project before starting is almost a contradiction in terms; transdisciplinary processes are, by definition, shaped in practice. At the same time, researchers need to spell out their problem understanding, suggested approach and expected outcomes in research proposals for supervisors and/or funders before the research commences. Evaluation is then the other side of the coin; on the basis of which evaluation frameworks will these proposals be assessed? And against which indicators can we judge whether transdisciplinary research projects have created the intended impacts? Learning questions that students have formulated regarding design and evaluation pertain to the specifics of individual research projects, the challenges of creating impact and the nature of emergent design in non-conducive contexts.

A first set of learning questions is about the specific framework conditions of graduate research projects. Students wonder:

> How can I create real impact in the area of my research, while I have limited time and resources available?

This is, of course, true for any graduate student, but in particular in the case of transdisciplinary research, engaging with relevant actors and communities requires time and attention and openness to unforeseen opportunities and developments. A number of studies conducted in the past 10 years into the particular challenges experienced by early-career researchers embarking on the path of transdisciplinary research confirm this. Enengel and colleagues (2012), for instance, report on the experience of four PhD students in Austria implementing transdisciplinary approaches within a traditional university setting. They highlight the tension between shifting responsibilities and control to non-academic

[10] The learning questions presented in this chapter are amalgams and slightly generalized versions of the specific learning questions formulated by students.

partners, and the dependencies this creates in the context of PhD projects 'that should result in externally reviewed doctoral theses after three years' (Enengel et al., 2012, p. 114).

Box 1.2: Example PhD research design

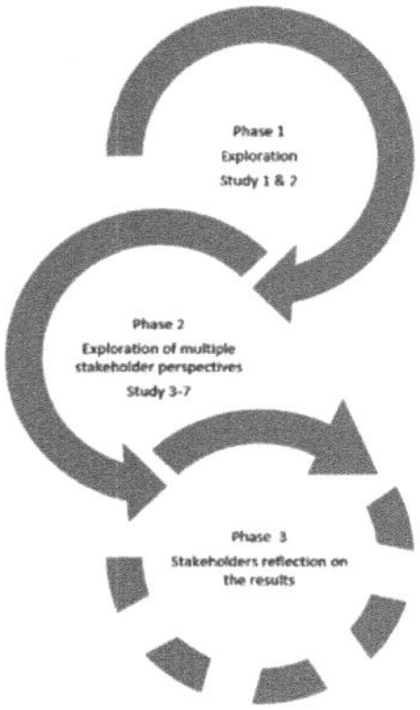

Phase 1: Exploration
In order to gain understanding of the problem field and understand potential barriers to and facilitators of the inclusion of people with mental disability, exploratory interviews were held and desk studies were conducted.

Phase 2: Exploration of Barriers and Facilitators Perceived by Different Stakeholder Groups
The second phase involved an exploration of stakeholder perspectives on the main study question. In his PhD, Ikenna Ebuenyi explored perspectives on barriers to and promoters of inclusion of persons with mental disability in employment in Kenya according to persons with mental disabilities, (potential) employers, mental healthcare providers and mental health/ DPOs.

Phase 3: Stakeholders' Reflection on the Results
The third phase involved exploring potential pathways to improved employability of persons with mental illness through reflection of the results from the previous phases with multiple stakeholders. This phase was commenced but not completed. The thesis presents the preliminary

analysis of a roundtable meeting where the findings of the first two phases were shared and discussed with stakeholders, followed by a discussion on pathways to improved employability for persons with mental disability in Kenya.

PhD thesis Ikenna Ebuenyi (2019).

While a PhD trajectory can be seen as a project, with a clear beginning and an end, as said before, transformation is not a project. But, there is always an 'after the end' (like there is always a 'before the beginning'). Within the restricted time frame of a master thesis or PhD study, the 'project' might be delineated to gaining insight into the perspectives of different stakeholders separately with only the beginning of a mutual learning process among them. See Box 1.2 for an example in which an integration phase (phase 3) was commenced but not completed. Phase 1 and phase 2 had already resulted in seven published research articles and the three-year term of contract had ended. However, we do see that the relationships that were formed during the research process continued; like the experiences of the South African PhD students mentioned above, 'these socio-epistemic relationships took on a "social existence" *beyond* the individual research project' (Van Breda et al., 2016, p. 161). Still, delineation in time, because of the researchers' pursuit of degrees and publications, comes with the risk of sustaining power inequalities between communities and researchers. An action researcher says: 'research topics that would have unreasonably extended the completion of my degree were taken off the table. However, what if those other options were more beneficial to the [community]?' (Dillon, 2014, p. 11, quoted in Strumińska-Kutra, 2016).

A second type of learning question is about the challenge of bringing about significant change through transdisciplinary research. This can be argued to be the overarching challenge of transdisciplinary research, encompassing all others. A learning question may read:

> How can I apply transdisciplinary research to bring substantial and meaningful change in relation to a specific complex problem, while there are many barriers and there is resistance in the system?

This is an important question because, as alluded to before, a lot of attention to better understanding processes of knowledge co-creation, power dynamics, mutual learning and reflexivity, in our engagement in transdisciplinary practices as scholars, carries the risk of foregrounding academic knowledge production, and backgrounding the end to which these transdisciplinary practices were set up in the first place. Recognizing the inseparability of understanding and change means not losing sight of the transformational intent of a transdisciplinary practice as it evolves, and can be argued to be the crux of a transdisciplinary governance strategy (De Wildt-Liesveld et al., 2015). Chapter 3 (Regeer et al., this volume) elaborates on the relationship between design, evaluation and reflexive governance in the context of transdisciplinary projects.

A final set of learning questions relates to possible tensions between the transdisciplinary research project and its institutional environment, and those of collaborating partners. Learning questions may then be:

> How do I allow for an emergent process that provides room for collaborative decision making and multiple iterations and adaptations, while I need to be specific before the start about activities and outcomes to increase chances to get funding, *or* while stakeholders involved may have to navigate rigid organizational procedures, *or* while people might be accustomed to typical project management thinking, or disinclined to take risks?

Being risk-averse might be an individual character trait (these are discussed under Part 3: Roles and Competencies), but might also be congruent with an organizational culture geared towards formalization and standardization, resonating with the new public management ideology (Gruening, 2001). In the research and innovation system, we also see a trend towards 'projectification' (e.g. Ika & Munro, 2022) that poses specific challenges to transdisciplinary research design (Gjefsen et al., this volume). Some researchers resolve this tension by making the highly intangible processes of deliberation, collaboration and mutual sensemaking, very tangible and concrete in their proposals through laying out clear processes of co-design through various multi-stakeholder workshops (see also Jones & Loeber, under review).

In sum, transdisciplinary research design is by definition emergent, as a 'purely deliberative strategy precludes learning once the strategy is formulated; emergent strategy fosters it' (Mintzberg, 1987, p. 66). As Chapter 3 will argue, transdisciplinary research cannot be done by simply designing the process based on one of the existing highly sophisticated frameworks. In earlier research we saw that the hard and pioneering work in the area of defining transformative approaches had resulted in a disposition of transdisciplinary, transformation or transition scholars to assess and critique transdisciplinary practice according to their theories of change. 'However, in interviews and informal conversations that we held with a range of programme managers and project leaders, they have expressed agitation and annoyance with being repeatedly confronted with the gaps between programme theory and their practice. They argue that they know about the theory but struggle with the implementation and have expressed the need for help and guidance' (Regeer et al., 2009, p. 522). We may need forms of *ex durante* reflexive governance of transdisciplinary practice, in which *ex ante* design frameworks and *ex post* evaluation frameworks become part of the conversation.

1.4.2 Part II: Diversities and Inclusion

The majority of learning questions participants formulate, pertain to the manifold diversities involved in the participatory process. A typical learning question may read as follows:

> How can I find a shared direction in my transdisciplinary research project, while perceptions of the problem may vary widely across the different stakeholder groups?

Students often stipulate differences in terms of interests, needs, power, language, knowledge, culture, history and values. Indeed, at the core of many transdisciplinary endeavours is a challenge of balancing directionality and stakeholders' perspectives, which often appear incommensurable (see also Kok et al., 2021). Taking complex and persistent problems as a starting point implies by definition that transdisciplinary research is concerned with issues that lack agreement on facts and values (Douglas & Wildavsky, 1982; Hisschemöller & Hoppe, 2018). Problem structuring (Dunn, 2015) has hence been advocated as an important place to start transdisciplinary research processes. Various methodologies have been introduced to support this process, including interpretative frame analysis (Schuitmaker, 2012; Van der Wilt & Reuzel, 2012), mapping of diverse argumentation trees (De Cock Buning, 2010), and frame reflection (Kupper & De Cock Buning, 2011; Van der Meij et al., 2018). Often, the process of collaborative problem structuring, involving multiple stakeholder groups, might take up the entire PhD trajectory, which can then be seen as the first cycle or stage of a larger transformative transdisciplinary process (see above and Box 1.2).

At the same time, stakes are high and decisions urgent where complex real-world problems are concerned (Funtowicz & Ravetz, 1994). This implies that processes of problem structuring and experimentation are closely intertwined and iteratively inform one another throughout. Furthermore, given the almost always plethora of diversities, problem formulations and associated aspired visions are to be considered as provisional rather than definitive and perhaps should be seen as temporary moments of closure of what tend to be potentially highly contested debates, which can flare up again with changing configurations of stakeholders involved and with evolving insights and events. It is argued that these provisional episodes of closure can speed the process of experimentation and consequently enhance a deeper understanding of the issues at stake—something that clearly resonates with Kurt Lewin's famous saying that 'If you want truly to understand something, try to change it'.[11]

Considering aspired visions as 'speculative commitments' (Jerak-Zuiderent, 2019, following Puig de la Bellacasa, 2017) implies a commitment to putting them to the test as the process evolves. Moreover, it is a response to the problematization of the idea of a *shared* vision or *consensus*, which is hard to imagine when recognising the width but also the sheer depth of the diversities; forms of knowing and being are deeply rooted in people's shared social and professional practices.[12] Rather than reaching consensus on problem definition or vision, scholars describe looking for congruency, alignment or convergence between different (value) frames of involved actors (Grin & Van de Graaf 1996; Hoes et al., 2008; Vandenbussche et al., 2024). Formulating learning questions on the plurality of understandings and perspectives of different stakeholders invites these kinds of considerations.

Students, secondly, formulate learning questions reflecting that the manifold diversities are not only challenging in terms of finding a shared direction, but also in terms of the expected interaction between different

[11] This quote is generally attributed to Lewin. Our source: Tolman, C.W. (1996). *Problems of theoretical psychology* (Vol. 6, p. 31). Captus Press.

[12] Reflected in concepts like 'form of life' (Wittgenstein, 1953), 'community of practice' (Wenger, 1999), 'epistemic culture' (Knorr-Cetina, 1999) and 'thought collective' (Fleck 1981 [1935]) and also implied in the concepts of 'paradigm' (Kuhn, 2012), and 'interpretative frame' (Schön & Rein, 1994).

people, who may not understand each other due to their different ways of understanding or knowing, the languages and jargon they do (not) master, the cultures they are enculturated in or the less or more substantial power differentials existing between them. Thus, many students formulate learning questions around power dynamics, which typically may read as follows:

> How can I effectively integrate diverse perspectives and knowledge of stakeholders relevant to my research, while knowledge hierarchies may favour (alleged) expert dominance in deliberative processes, reproducing hegemonic power structures?

Or, in a different variant, emphasis is placed on those excluded by existing (knowledge) hierarchies:

> How can we ensure equitable inclusion of (perspectives of) vulnerable or marginalized communities, while benefits of participation may not outweigh the burdens, *or* while there is unequal access to resources, self-confidence, or social capital, *or*, while trust is lacking?

Both variations of this learning question concerning power and hierarchies are pivotal to explore in the context of transdisciplinary research. For one because, while the notion of democratizing knowledge processes might have been at the historical roots of transdisciplinary research, we are only at the very start of the journey towards epistemic justice. Indeed, the difficulty of this journey becomes almost painfully manifest when we realize that speaking of powerful versus vulnerable or marginalized communities in itself constitutes a form of reproducing both stigmatization and power imbalances, as it implicitly reaffirms what counts as centre and what as margin and disallows some membership of the so-called centre. Consistent with this, students also recognize that the language of marginalized or vulnerable groups in itself almost automatically embodies a reproduction of existing hierarchies, as it tends to be powerful actors, who have no trouble in gaining access to places where decisions are made or research is funded, designed and implemented, nor in articulating their positions, who identify *others* as vulnerable or belonging to the margins, and thus as people who ought to be *given a voice*. Arguably, however, nobody needs to be *given a voice*—rather, it is hegemonic actors who need to learn how to listen to any and all pertinent voices, irrespective of whether or not they are raised by actors resembling them (Zachariah et al., 2023).

The more recent calls to take power differentials far more seriously than transdisciplinary scholarship has to date (e.g. Strumińska-Kutra & Scholl, 2022; Turnhout et al., 2020) are calls for recentring members of 'marginalized' communities. The learning questions formulated by students speak to the fact that participation so easily becomes instrumental or tokenistic (Ocloo & Matthews, 2016), habitually reinforcing hegemonic systems of knowledge in so many ways, and at so many levels (Roura, 2021). Recentring marginalized communities implies a profound rethinking of power relations, in order to bring about a 're-humanised world' (Ndlovu-Gatsheni, 2019, referring to Maldono-Torres' work on human rights).

Transdisciplinary practices attempting to foster epistemic justice and decolonising research practices are, for instance, found in the growing body of literature on Indigenous-led knowledge practices (e.g. Moewaka Barnes et al., 2021), particularly in environmental and sustainability research, which give examples of co-creating knowledge through power-sharing and creative action. In a different way, efforts to include and value patient knowledge in health policy, care and research practices (Pittens, 2013) have tried to navigate between, or iteratively accommodate both, academic (medical) discourses dominated by highly specific operationalizations of 'objectivity' that are informed first and foremost by empiricism and reductionism, and the lived experience and embodied and situated knowledges of both patients and care professionals (Lösch et al., 2023; Zuiderent-Jerak et al., 2012). The Dialogue Model (Abma & Broerse, 2010), for instance, is carefully designed to mitigate power differentials between patients, health professionals and researchers, by allowing patients to start articulating their lived experience and experiential knowledge with peers, which not only has substantive but also affective value, and enables the translation of an individual 'I'-voice into a shared 'we'-voice.

A third set of learning questions that pertain to the manifold diversities involved in transdisciplinary research reflects the experienced difference between a transdisciplinary way of working and thinking, and a 'non-transdisciplinary' way of working and thinking.[13] Students especially express this in relation to experiences with research internships in monodisciplinary academic or professional environments or in relation to their earlier academic training. A learning question may read:

> How can I meaningfully conduct transdisciplinary research, while working within a traditional university setting which persists to be discipline-centred and wherein recognition for one's work is still very much decided upon in terms of discipline-oriented performance indicators?

And this, of course, is true for all participants in a transdisciplinary research process: each of them may have to comply with existing guidelines, protocols or routines in their (professional) community. It is sometimes said that transdisciplinary research requires 'double work': besides instigating and supporting processes of co-creation, co-innovation, and transformation, there is also a continuous need for alignment work (De Wildt-Liesveld et al., 2015; Verwoerd et al., 2021) to not lose the connection with existing incentive structures, or rather to instigate processes of institutional reflexivity. If we do not attend to these institutional learning processes, transdisciplinary projects will just be projects, with no impact beyond their duration.

1.4.3 *Part III: Roles and Competences*

The third part of this volume is based on the premise that a reflexive approach, like transdisciplinary research, requires a great deal of self-reflection regarding diverse roles to navigate and competences to acquire. Here, we see learning questions that can be considered counterparts of some of the learning questions above, turning the gaze specifically to the 'self'. For instance, while a learning question regarding design (Part 1) might be about the tension between an emergent design process and existing, rigid organizational structures and procedures, the counterpart question that would fit in Part 3 would be:

[13] Like many binaries, the binary between transdisciplinary and non-transdisciplinary has been criticized—cf., 'We have never been modern' (Latour, 2012). Latour argues that the so-called practice of 'purification' gave rise to the rhetoric of 'modernity'. Similarly, acts of purification are at play that give rise to 'mode 1 research', or 'linear' policy processes, or 'non-transdisciplinary' ways of working and thinking.

> How can I embrace uncertainty and complexity while I feel a deep need for clarity, certainty and manageability, and feel more comfortable with mode 1 research?

These questions call for self-reflection and a deepening awareness of the importance of engaging with unarticulated, implicit and sometimes unknown aspects of oneself in being able to act meaningfully in a messy, unstructured process and engage with diverse other people, each with their own, often implicit, commitments and associations. Like in any qualitative research endeavour, the recognition that as a (transdisciplinary) researcher you are your own instrument (Dodgson, 2019) brings a responsibility to inquire into yourself, your normative commitments and the role of the manifold associations that shape the way you perceive.

Students, in the formulation of their learning questions, show a deep awareness of the need for continuous reflection on their positionality, particularly with regard to power dynamics and their position as a researcher, often from the global North. An illustrative example of how students have formulated such a learning question is:

> How can I engage in TDR processes with marginalized people and communities, and build a relationship of trust, while they may mistrust me because of my positionality as a privileged western researcher?

Like ourselves, many of our white global health students are wary of perpetuating the White Saviour Industrial Complex (WSIC) (Banerjee et al., 2023) and might even show a sense of paralysis. One way to escape this is by making the researcher's positionality, values and agenda, visible and open to negotiation.[14]

While students recognize their 'perceived authority' as a (sometimes 'white' or 'global north') researcher in relation, for instance, to marginalized people and communities, they also indicate doubts about how to handle power differentials between themselves, as young and inexperienced researchers, and those in positions of power. They ask, for instance:

[14] See also Strumińska-Kutra (2016) for a similar argument concerning paralysis resulting from constructivism. Struminska-Kutra argues this based on the idea that constructivism leads to relativism and hence paralysis. See Regeer and Bunders (2003) for a Wittgensteinian argument against equating constructivism with relativism.

> How can I engage with experienced researchers, professionals and those with lived experience, while I am a young and inexperienced researcher, *or* while I am not comfortable taking the lead, *or* while I tend to steer away from conflicts?

Chapter 14 (den Boer, 2024, this volume) presents a thick account of what it is like—as an early-career researcher—to conduct transdisciplinary research in the context of a City Lab, supporting food system transformation. She distinguishes a wide variety of roles she adopted over a period of three years and describes the synergies and tensions she encountered between various roles. The requirements of conducting both scientifically rigorous and societally relevant transdisciplinary research are generally experienced as highly stressful (Sellberg et al., 2021). PhD students fear falling behind in their careers compared to their monodisciplinary counterparts, as transdisciplinary research requires 'double work'; they need to invest in creating interactional spaces or materials that aid the process of co-creation as well as spend considerable time and energy in producing academic papers that may take longer to get published because of a lack of transdisciplinary reviewers. Doctoral students conducting transdisciplinary research projects (Enengel et al., 2012) reported that the time they spent on interacting with non-academic actors, and including local knowledge into the research process, yielded realizable solutions that accommodate conflicting interests and can hence more effectively contribute to addressing real-world problems. At the same time, however, these transdisciplinary interactions left them with less time for disciplinary exchange, and methodological and theoretical innovation (ibid.).

Furthermore, early-career researchers are still finding their grounds in their original scientific field and are uncertain what a move towards transdisciplinary research might mean in terms of their sense of 'feeling intellectually and socially "at home"' in an academic community, or 'epistemic living space' (Felt et al., 2013, p. 513). They may ask themselves:

> How do I develop myself to become a transdisciplinary researcher while I am still finding my position as a researcher within the structures of academia? *Or*, while I feel more comfortable in my original academic background?

From their experience in their graduate education and research internships, they have not encountered many academic communities in which this other mode of doing research is appreciated, rewarded or simply supported. So, what does that mean for their (academic) careers? Will they be lone advocates and pioneers in otherwise not very conducive academic environments? As Felt and colleagues (2013) describe, early-career researchers grapple to reconcile the demands of transdisciplinarity with other normative requirements in contemporary research. Developing attachment to transdisciplinarity at an epistemic level was experienced as especially difficult, and PhD students were inclined to re-attach to their 'home-disciplines' (Felt et al., 2013).

A third set of questions relate to roles and competences that pertain to the researcher's own normative orientation. Students question how to deal with the potential tensions between their own views and opinions and those of stakeholders. A learning question might read:

> How can I engage constructively with a diverse sample of stakeholders with different perspectives, while I am a highly politically engaged person and already have my own opinions, values and ideals on the research topic, and may strongly oppose or align with some of the stakeholders?

We see a heightened awareness of the potential contention between the ability for deep listening and one's own position, intensified by other students stating that they have a 'clear view and opinion' themselves, are 'a highly opinionated person' or that they 'find it difficult to accept that other points of view exits next to mine'. This is also related to having 'an anticipated outcome of my research' and hence relates to the earlier tension between directionality and stakeholder involvement. These learning questions solicit reflection on one's own normative commitments in transdisciplinary research and invite researchers to put these commitments to the test, to explore them, to be curious about how strong they are and why, and to observe how these commitments affect their research practice. It requires practising with turning strong commitments into speculative commitments; they are provisional, or tentative and can change as a result of the process. This relates to a wider search on the part of critically oriented action researchers on their own stance in transformative research: 'How and where should they locate themselves in relation to organizational change or even broader to the change of organizational constituencies' (Alvesson et al., 2009, cited in Strumińska-Kutra, 2016,

p. 864). Or 'how to be a genuine partner to a "community" and simultaneously [...] adopt a critical stance that presupposes the definition of their problem' (Strumińska-Kutra, 2016, p. 865).

In a conceptual article on power inequities across the social ecology of participatory health research, Roura (2021) identifies a number of interdependent areas at micro, meso and macro levels at which power inequities are at play. She formulates monitoring questions to guide the assessment of power dynamics in participatory health research that resonate well with some of the learning questions formulated by novice transdisciplinary researchers. At the micro-level, Roura's monitoring questions pertain, for instance, to self-reflexivity and cultural humility (where students have openly reflected on their positionality); at the meso-level, monitoring questions pertain, for instance, to reward systems and effective techniques for dialogue (where students have reflected on character traits, experience and also their rigid normative orientation, which might hamper effective dialogue), and monitoring questions at the macro level relate, for instance, to the distribution of power and resources (where students have shown an awareness of unfair resource distribution and the power of the academic system of which they are a part). Rather than monitoring power dynamics as externalized (e.g. 'are the most powerful stakeholders ready to give up power and the privileges that come with it?' (Roura, 2021, Table 1, p. 783), the learning questions bring power-related issues close to home and invite the kind of self-reflexivity that is essential both for novice and experienced participants of transdisciplinary research processes. The learning questions thus set an agenda for navigating roles, which essentially are about the pluriform relationships, with others and with self, that deliberately, but more often implicitly, take shape in a transdisciplinary process; and the continuous development of the capacity, of oneself and others, to navigate the many diversities while keeping the (continuously negotiated) direction of transformation centre stage.

References

Aalbers, C. B., & Sehested, K. (2018). Critical upscaling. How citizens' initiatives can contribute to a transition in governance and quality of urban greenspace. *Urban Forestry & Urban Greening, 29*, 261–275.

Abma, T. A., & Broerse, J. E. (2010). Patient participation as dialogue: Setting research agendas. *Health Expectations, 13*(2), 160–173.

Altınay, A. G., & Pető, A. (2022). Feminist+ solidarity as transformative politics. *European Journal of Women's Studies, 29*(4), 477–488.

Ansell, C., & Geyer, R. (2017). 'Pragmatic complexity' a new foundation for moving beyond 'evidence-based policy making'? *Policy Studies, 38*(2), 149–167.

Argyris, C., & Schon, D. A. (1974). *Theory in practice: Increasing professional effectiveness.* Jossey-Bass.

Argyris, C., & Schön, D. A. (1978). *Organizational learning: A theory of action perspective.* Addison Wesley.

Arkesteijn, M., van Mierlo, B., & Leeuwis, C. (2015). The need for reflexive evaluation approaches in development cooperation. *Evaluation, 21*(1), 99–115.

Bammer, G. (2013). *Disciplining interdisciplinarity: Integration and implementation sciences for researching complex real-world problems.* ANU Press.

Banerjee, A. T., Bandara, S., Senga, J., González-Domínguez, N., & Pai, M. (2023). Are we training our students to be white saviours in global health? *The Lancet, 402*(10401), 520–521.

Baur, N. (2021). Decolonizing social science methodology: Positionality in the German-language debate. *Historical Social Research, 46*(2), 205–43. https://doi.org/10.12759/hsr.46.2021.2.205-243

Beck, U. (1992). *Risk society. Towards a new modernity.* Sage.

Blumenthal, D. S., & DiClemente, R. J. (Eds.). (2004). *Community-based health research: Issues and methods.* Springer.

Boyes, B., Cummings, S., Habtemariam, F. T., & Kemboi, G. (2023). 'We have a dream': Proposing decolonization of knowledge as a sixth generation of knowledge management for sustainable development. *Knowledge Management for Development Journal, 17*(1/2), 17–41.

Bray, J. N. (2000). *Collaborative inquiry in practice: Action, reflection, and making meaning.* Sage.

Breda, K. L. (2014). Participatory action research. In M. de Chesnay (Ed.), *Nursing research using participatory action research* (pp. 1–11). Springer.

Bremer, S. & Meisch, S. (2017). Co-production in climate change research: Reviewing different perspectives. *Wiley Interdisciplinary Review (WIREs) Climate Change, 8*, e482. https://doi-org.vu-nl.idm.oclc.org/10.1002/wcc.482

Brett, J., Staniszewska, S., Mockford, C., Herron-Marx, S., Hughes, J., Tysall, C., & Suleman, R. (2014). Mapping the impact of patient and public involvement on health and social care research: A systematic review. *Health Expectations, 17*(5), 637–650. https://doi-org.vu-nl.idm.oclc.org/10.1111/j.1369-7625.2012.00795.x

Broerse, J. E. W. (1998). *Towards a new development strategy. How to include small-scale farmers in the biotechnological innovation process.* Delft, Eburon.

Broerse, J.E.W. (2013). *Naar een inclusief innovatieproces in de gezondheids- en levenswetenschappen. Inaugurele rede*. Vrije Universiteit Amsterdam.

Broerse, J. E. W., Zweekhorst, M. B. M., van Rensen, M. J. M., & de Haan, A. J. M. L. (2010). Involving burn survivors in agenda setting on burn research: An added value? *Burns, 36*(2), 217–231.

Bulkeley, H., & Castán Broto, V. (2013). Government by experiment? Global cities and the governing of climate change. *Transactions of the Institute of British Geographers, 38*(3), 361–375.

Bulten, E., Hessels, L. K., Hordijk, M., & Segrave, A. J. (2021). Conflicting roles of researchers in sustainability transitions: Balancing action and reflection. *Sustainability Science, 16*(4), 1269–1283. https://doi-org.vu-nl.idm.oclc.org/10.1007/s11625-021-00938-7

Bunders, J. F. G. (1987). The practical management of scientists' actions: The influence of patterns of knowledge development in biology on cooperations between university biologists and non-scientists. In S. Blume, J. F. G. Bunders, L. Leydesdorff, & R. Whitley (Eds.), *The social direction of the public sciences*. D. Reidel Publishing Company.

Bunders, J., Broerse, J., & Zweekhorst, M. (1999). The triple helix enriched with the user perspective: A view from Bangladesh. *The Journal of Technology Transfer, 24*(2), 235–246.

Bunders, J. F. (1994). Participative strategies for science-based innovations: the case of biotechnology for small-scale farmers in developing countries. In *Participative strategies for science-based innovations: The case of biotechnology for small-scale farmers in developing countries*. VU University Press.

Cacari-Stone, L., Wallerstein, N., Garcia, A. P., & Minkler, M. (2014). The promise of community-based participatory research for health equity: A conceptual model for bridging evidence with policy. *American Journal of Public Health, 104*(9), 1615–1623.

Chambers, R. (1992). *Rural appraisal: Rapid, relaxed and participatory* (IDS Discussion Paper 311). https://www.ids.ac.uk/download.php?file=files/Dp311.pdf

Chambers, J. M., Wyborn, C., Ryan, M. E., Reid, R. S., Riechers, M., Serban, A., Pickering, T., et al. (2021). Six modes of co-production for sustainability. *Nature Sustainability, 4*, 983–996.

Chilvers, J., & Kearnes, M. (2020). Remaking participation in science and democracy. *Science, Technology, & Human Values, 45*(3), 347–380. https://doi-org.vu-nl.idm.oclc.org/10.1177/0162243919850885

Cummings, S., Regeer, B. J., Ho, W. W., & Zweekhorst, M. B. (2013). Proposing a fifth generation of knowledge management for development: Investigating convergence between knowledge management for development and transdisciplinary research. *Knowledge Management for Development Journal, 9*(2), 10–36.

Dampier, S. (2009). Action research. *Nurse Researcher, 16*(2), 4–7.
De Cock Buning, T. (2010). Four steps to stimulate meaningful communication on sensitive issues in societal debate: The case of a research agenda for biotechnology and food in the Netherlands. *Knowledge Democracy: Consequences for Science, Politics, and Media*, 241–253.
DeTemple, J., & Sarrouf, J. (2017). Disruption, dialogue, and swerve: Reflective structured dialogue in religious studies classrooms. *Teaching Theology & Religion, 20*(3), 283–292.
Dewey, J. (1910). *How we think.* D.C. Heath & Co.
De Wildt-Liesveld, R., Bunders, J. F., & Regeer, B. J. (2015). Governance strategies to enhance the adaptive capacity of niche experiments. *Environmental Innovation and Societal Transitions, 16*, 154–172.
Dodgson, J. E. (2019). Reflexivity in qualitative research. *Journal of Human Lactation, 35*(2), 220–222.
Douglas, M., & Wildavsky, A. (1982). *Risk and culture: An Essay on the selection of technical and environmental dangers.* University of California Press.
Dubois, A., & Gadde, L. E. (2002). Systematic combining: An abductive approach to case research. *Journal of Business Research, 55*(7), 553–560.
Dunn, W. N. (2015). *Public policy analysis.* Routledge.
Ebuenyi, I. (2019) *Inclusive employment: Understanding the barriers to and facilitators of employment for persons with mental disability in East Africa* (PhD thesis), Vrije Universiteit Amsterdam.
Elzen, B., & Bos, B. (2019). The RIO approach: Design and anchoring of sustainable animal husbandry systems. *Technological Forecasting and Social Change, 145*, 141–152.
Enengel, B., Muhar, A., Penker, M., Freyer, B., Drlik, S., & Ritter, F. (2012). Co-production of knowledge in transdisciplinary doctoral theses on landscape development—An analysis of actor roles and knowledge types in different research phases. *Landscape and Urban Planning, 105*(1–2), 106–117.
Epstein, S. (1996). *Impure science: AIDS, activism, and the politics of knowledge* (Vol. 7). University of California Press.
Etzkowitz, H., & Leydesdorff, L. (2000). The dynamics of innovation: From National Systems and "Mode 2" to a Triple Helix of university–industry–government relations. *Research Policy, 29*(2), 109–123.
Fals Borda, O. (2013). Action research in the convergence of disciplines. *International Journal of Action Research, 9*(2), 155–167.
Fazey, I., Schäpke, N., Caniglia, G., Patterson, J., Hultman, J., Van Mierlo, B., Wyborn, C., et al. (2018). Ten essentials for action-oriented and second order energy transitions, transformations and climate change research. *Energy Research & Social Science, 40*, 54–70.

Felt, U., Igelsböck, J., Schikowitz, A., & Völker, T. (2013). Growing into what? The (un-)disciplined socialisation of early stage researchers in transdisciplinary research. *Higher Education, 65*(4), 511–524.

Felt, U. (2017). Under the shadow of time: Where indicators and academic values meet. *Engaging Science, Technology, and Society, 3*, 53–63.

Fischer, F. (2000). *Citizens, experts, and the environment: The politics of local knowledge*. Duke University Press.

Fischer, F. (2003). *Reframing public policy: Discursive politics and deliberative practices*. Oxford University Press.

Fleck, L. (1981). *Genesis and development of a scientific fact*. University of Chicago Press.

Foucault, M. (2020). Power/knowledge. In *The new social theory reader* (pp. 73–79). Routledge.

Franklin, A. (2022). *Co-creativity and engaged scholarship: transformative methods in social sustainability research* (p. 559). Springer Nature.

Freire, P. (1968/2001). *Pedagogy of the oppressed* (M. Bergman Ramos, Trans.). Continuum.

Freire, P. (1973). *Education for critical consciousness*. Seabury Press.

Fricker, M. (2007). *Epistemic injustice: Power and the ethics of knowing*. Oxford University Press.

Funtowicz, S. O., & Ravetz, J. R. (1994). Uncertainty, complexity and post-normal science. *Environmental Toxicology and Chemistry: An International Journal, 13*(12), 1881–1885.

Gardiner, P. (2020). Learning to think together: Creativity, interdisciplinary collaboration and epistemic control. *Thinking Skills and Creativity, 38*, 100749.

Gibbons, M., Limoges, C., Nowotny, H., Schwartzman, S., Scott, P., & Trow, M. (1994). *The new production of knowledge: The dynamics of science and research in contemporary societies*. Sage.

Grin, J. (2020). 'Doing' system innovations from within the heart of the regime. *Journal of Environmental Policy & Planning, 22*(5), 682–694.

Grin, J., & Van De Graaf, H. (1996). Technology assessment as learning. *Science Technology Human Values, 21*, 72–99.

Grin, J., Van De Graaf, H., & Hoppe, R. (1997). *Technology assessment through interaction. A guide*. Den Haag, Rathenau Instituut.

Gruening, G. (2001). Origin and theoretical basis of new public management. *International Public Management Journal, 4*(1), 1–25. https://doi-org.vu-nl.idm.oclc.org/10.1016/S1096-7494(01)00041-1

Guba, E. G., & Lincoln, Y. S. (1989). *Fourth generation evaluation*. Sage.

Hadorn, G. H., Biber-Klemm, S., Grossenbacher-Mansuy, W., Hoffmann-Riem, H., Joye, D., Pohl, C., Zemp, E., et al. (2008). The emergence of transdisciplinarity as a form of research. In *Handbook of transdisciplinary research* (pp. 19–39).

Hajer, M., & Versteeg, W. (2005). Performing governance through networks. *Europeans Political Science, 4*(3), 340–348.

Haraway, D. J. (2016). *Staying with the trouble: Making kin in the Chthulucene*. Duke University Press.

Harding, S. (Ed.). (2011). *The postcolonial science and technology studies reader*. Duke University Press.

Hargreaves, J., & Page, L. (2013). *Reflective practice*. Wiley.

Herzog, L., & Lepenies, R. (2022). Citizen science in deliberative systems: Participation, epistemic injustice, and civic empowerment. *Minerva, 60*(4), 489–508.

Hisschemöller, M., & Hoppe, R. (2018). Coping with intractable controversies: The case for problem structuring in policy design and analysis 1. In *Knowledge, power, and participation in environmental policy analysis* (pp. 47–72). Routledge.

Hoes, A. C., Regeer, B. J., & Bunders, J. F. (2008). TransFormers in knowledge production: Building science–practice collaborations. *Action Learning: Research and Practice, 5*(3), 207–220.

Hoppe, R., & Hisschemoller, M. (1998). Weerbarstige beleidscontroverses: een pleidooi voor probleemstructurering in beleidsontwerp en analyse. In In: R. Hoppe, A. Peterse (red.) *Bouwstenen voor Argumentatieve Beleidsanalyse* (pp. 51–74). Reed Business BV.

Horn, A., Urias, E., & Zweekhorst, M. B. M. (2022). Epistemic stability and epistemic adaptability: Interdisciplinary knowledge integration competencies for complex sustainability issues. *Sustainability Science, 17*(5), 1959–1976.

Huitzing, H., Loeber, A., & Regeer, B. (2020). *Lerende evaluatie IBP VP: Onderzoeksmethodiek voor het evalueren van transformerend leren en handelen. Evaluatiekader voor het Interbestuurlijk programma Vitaal Platteland*. Planbureau voor de Leefomgeving.

Ika, L. A., & Munro, L. T. (2022). Tackling grand challenges with projects: Five insights and a research agenda for project management theory and practice. *International Journal of Project Management, 40*(6), 601–607.

Iqbal, H., West, J., McEachan, R. R. C., & Haith-Cooper, M. (2023). Reflections from an insider researcher 'doing' feminist participatory action research to co-produce a research agenda with British Pakistani women; a seldom heard group. *Action Research, 21*(4), 456–471.

Irwin, A. (2006). The politics of talk: Coming to terms with the 'new' scientific governance. *Social Studies of Science, 36*(2), 299–320.

Jacobi, J., Llanque, A., Mukhovi, S. M., Birachi, E., von Groote, P., Eschen, R., Hilber-Schöb, I., Kiba, D. I., Fossard, E., & Robledo-Abad, C. (2022). Transdisciplinary co-creation increases the utilization of knowledge from sustainable development research. *Environmental Science & Policy, 129*, 107–115.

Jahn, S., Newig, J., Lang, D. J., Kahle, J., & Bergmann, M. (2022). Demarcating transdisciplinary research in sustainability science—Five clusters of research modes based on evidence from 59 research projects. *Sustainable Development, 30*(2), 343–357.

Jasanoff, S. (2004). Ordering knowledge, ordering society. In S. Jasanoff (Ed.), *States of knowledge: The co-production of science and social order*. Routledge.

Jensen, C. B. (2007). Sorting attachments: On intervention and usefulness in STS and health policy. *Science as Culture, 16*(3), 237–251.

Jerak-Zuiderent, S. (2019). How to care for our accounts? In *The Routledge companion to actor-network theory* (pp. 190–199). Routledge.

Jones, T., & Loeber, A. (under review) Towards a critical recovery of liberatory PAR for food system transformations: Struggles and strategies in collaborating with radical and progressive food movements in EU-funded R&I projects. *Journal of Responsible Technology*.

Kapoor, R. (2007). Transforming self and society: Plural paths to human emancipation. *Futures, 39*(5), 475–486.

Kilelu, C. W., Klerkx, L., & Leeuwis, C. (2014). How dynamics of learning are linked to innovation support services: Insights from a smallholder commercialization project in Kenya. *The Journal of Agricultural Education and Extension, 20*(2), 213–232.

Klaassen, P., Rijnen, M., Vermeulen, S., Kupper, F., & Broerse, J. (2018). Technocracy versus experimental learning in RRI: On making the most of RRI's interpretative flexibility. In R. Gianni, J. Pearson, & B. Reber (Eds.), *Responsible research and innovation: From concepts to practices* (pp. 77–98). Routledge.

Klein, J. T. (Ed.). (2001). *Transdisciplinarity: Joint problem solving among science, technology, and society: An effective way for managing complexity*. Springer Science & Business Media.

Kleiner, A., & Roth, G. (1996). *Field manual for the learning historian*. MIT, Center for Organizational Learning.

Knapp, C. N., Reid, R. S., Fernández-Giménez, M. E., Klein, J. A., & Galvin, K. A. (2019). Placing transdisciplinarity in context: A review of approaches to connect scholars, society and action. *Sustainability, 11*(18), 4899.

Knorr-Cetina, K. (1999). *Epistemic Cultures: How the sciences make knowledge*. Harvard University Press.

Kok, K. P., Gjefsen, M. D., Regeer, B. J., & Broerse, J. E. (2021). Unraveling the politics of 'doing inclusion' in transdisciplinarity for sustainable transformation. *Sustainability Science, 16*, 1811–1826.

Kuhn, T. S. (1962/2012). *The structure of scientific revolutions.* University of Chicago Press.

Kunneman, H. (2006). *Voorbij het dikke-ik. Bouwstenen voor een kritisch humanisme.* Huministics University Press.

Kupper, F., & De Cock Buning, T. (2011). Deliberating animal values: A pragmatic–pluralistic approach to animal ethics. *Journal of Agricultural and Environmental Ethics, 24*, 431–450.

Lang, D. J., Wiek, A., Bergmann, M., Stauffacher, M., Martens, P., Moll, P., Thomas, C. J., et al. (2012). Transdisciplinary research in sustainability science: practice, principles, and challenges. *Sustainability science, 7*, 25–43.

Latour, B. (1987). *Science in Action. How to follow scientists and engineers through society.* Open University Press.

Latour, B. (2012). *We have never been modern.* Harvard University Press.

Latour, B., & Woolgar, S. (1979). The construction of a fact. In B. Latour & S. Woolgar (Eds.), *Laboratory life: The social construction of scientific facts* (pp. 105–150). Sage.

Lave, J., & Wenger, E. (1991). *Situated learning: Legitimate peripheral participation.* Cambridge University Press.

Laws, D., & Hajer, M. (2006). Policy in Practice. In M. Moran, M. Rein & R. Goodin (Eds.), *The oxford handbook of public policy.* Oxford University Press.

Leydesdorff, L., & Etzkowitz, H. (1996). Emergence of a Triple Helix of university—industry—government relations. *Science and Public Policy, 23*(5), 279–286.

Liberatore, A., & Funtowicz, S. (2003). 'Democratising' expertise, 'expertising' democracy: What does this mean, and why bother? *Science and Public Policy, 30*(3), 146–150.

Lösch, L., Willems, W., Bongers, M., Timen, A., & Zuiderent-Jerak, T. (2023). Kaleidoscopic integration: Advancing the integration of incommensurable knowledge in healthcare guidelines. *Social Science & Medicine, 339*, 116360.

Lowndes, V., & Roberts, M. (2013). *Why institutions matter: The new institutionalism in political science.* Bloomsbury Publishing.

Luger, J. Van der Meij, M., Regeer, B. J., Loeber, A. M. C. (submitted). *The politics of food governance in European cities: the role of administrative staff and food justice implications.*

Lux, A., Schäfer, M., Bergmann, M., Jahn, T., Marg, O., Nagy, E., Theiler, L., et al. (2019). Societal effects of transdisciplinary sustainability research—How can they be strengthened during the research process? *Environmental Science & Policy, 101*, 183–191.

Lynch, D. H., Klaassen, P., & Broerse, J. E. (2017). Unraveling Dutch citizens' perceptions on the bio-based economy: The case of bioplastics, bio-jetfuels and small-scale bio-refineries. *Industrial Crops and Products, 106*, 130–137.

Macnaghten, P., Owen, R., Stilgoe, J., Wynne, A., Velho, L., et al. (2014). Responsible innovation across borders: Tensions, paradoxes and possibilities. *Journal of Responsible Innovation, 1*(2), 191–199. https://doi-org.vu-nl.idm.oclc.org/10.1080/23299460.2014.922249

Macnamara, J. (2018). Toward a theory and practice of organizational listening. *International Journal of Listening, 32*(1), 1–23.

Marguin, S., Haus, J., Heinrich, A. J., Kahl, A., Schendzielorz, C., & Singh, A. (2021). Positionality Reloaded. *Historical Social Research/historische Sozialforschung, 46*(2), 7–34.

Mauser, W., Klepper, G., Rice, M., Schmalzbauer, B. S., Hackmann, H., Leemans, R., & Moore, H. (2013). Transdisciplinary global change research: The co-creation of knowledge for sustainability. *Current Opinion in Environmental Sustainability, 5*(3–4), 420–431.

Miller, C. A., & Wyborn, C. (2020). Co-production in global sustainability: Histories and theories. *Environmental Science & Policy, 113*, 88–95.

Mintzberg, H. (1987, July–August). Crafting strategy. *Harvard Business Review*.

Mirowski, P. (2002). *Machine dreams: Economics becomes a cyborg science*. Cambridge University Press.

Moewaka Barnes, H., Harmsworth, G., Tipa, G., Henwood, W., & McCreanor, T. (2021). Indigenous-led environmental research in Aotearoa New Zealand: Beyond a transdisciplinary model for best practice, empowerment and action. *AlterNative: An International Journal of Indigenous Peoples, 17*(2), 306–316.

Ndlovu-Gatsheni, S. J. (2019). Discourses of decolonization/decoloniality. *Papers on Language and Literature, 55*(3), 201–226.

Neuhauser, L. (2018). Practical and scientific foundations of transdisciplinary research and action. In D. Fam, L. Neuhauser, & P. Gibbs (Eds.), *Transdisciplinary theory, practice and education: The art of collaborative research and collective learning* (pp. 25–38). Springer.

Nguyen, H. T., & Marques, P. (2022). The promise of living labs to the Quadruple Helix stakeholders: Exploring the sources of (dis) satisfaction. *European Planning Studies, 30*(6), 1124–1143.

Nicolini, D. (2009). Zooming in and out: Studying practices by switching theoretical lenses and trailing connections. *Organization Studies, 30*(12), 1391–1418.

Nowotny, H., Scott, P., & Gibbons, M. (2001). *Re-thinking science: Knowledge and the public in an age of uncertainty* (p. 12). Polity.

Ocloo, J., & Matthews, R. (2016). From tokenism to empowerment: Progressing patient and public involvement in healthcare improvement. *BMJ Quality & Safety, 25*(8), 626–632.

Osborne, S. P., & Strokosch, K. (2013). It takes Two to Tango? Understanding the C o-production of Public Services by Integrating the Services Management and Public Administration Perspectives. *British Journal of Management, 24*, S31–S47.

Ostrom, E., Parks, R. B., Whitaker, G. P., & Percy, S. L. (1978). The public service production process: A framework for analyzing police services. *Policy Studies Journal, 7*, 381.

Owen, R., Macnaghten, P., & Stilgoe, J. (2012). Responsible research and innovation: From science in society to science for society, with society. *Science and Public Policy, 39*(6), 751–760.

Oyěwùmí, O. (1997). *The invention of women: Making an African sense of western gender discourses*. U of Minnesota Press.

Pielke, R. A. (2007). *The honest broker: Making sense of science in policy and politics*. Cambridge University Press.

Pittens, C.A.C.M. (2013). *Knowledge co-production in research, policy and care practices* (PhD thesis). Uitgeverij BoxPress.

Pollock, A., & Subramaniam, B. (2016). Resisting power, retooling justice: Promises of feminist postcolonial technosciences. *Science, Technology, & Human Values, 41*(6), 951–966.

Puig de La Bellacasa, M. P. (2017). *Matters of care: Speculative ethics in more than human worlds* (Vol. 41). University of Minnesota Press.

Reason, P., & Bradbury, H. (Eds.). (2001). *Handbook of action research: Participative inquiry and practice*. Sage.

Regeer, B. J., & Bunders, J. F. (2003). The epistemology of transdisciplinary research: From knowledge integration to communities of practice. *Interdisciplinary Environmental Review, 5*(2), 98–118.

Regeer, B. J., & Bunders, J. F. (2009). *Knowledge co-creation: Interaction between science and society. A transdisciplinary approach to complex societal issues*. Advisory Council for Research on Spatial Planning, Nature and the Environment/ Consultative Committee of Sector Councils in the Netherlands [RMNO/ COS].

Regeer, B. J., Hoes, A. C., van Amstel-van Saane, M., Caron-Flinterman, F. F., & Bunders, J. F. (2009). Six guiding principles for evaluating mode-2 strategies for sustainable development. *American Journal of Evaluation, 30*(4), 515–537.

Regeer, B. J., Mager, S., & Van Orsouw, Y. (2011). *Licence to grow: Innovating sustainable development by connecting values*. VU University Press.

Rein, M., & Schön, D. (1996). Frame-critical policy analysis and frame-reflective policy practice. *Knowledge and Policy, 9*(1), 85–104.

Rip, A. (2005). Verschuivingen in het sociaal contract: wetenschappelijke en technologische ontwikkelingen in nieuwe maatschappelijke kaders. In *Innovatie en maatschappelijke ontwikkeling. Omgaan met een haat-liefde verhouding*. B. Elzen en W. De Ridder. Den Haag, SMO.

Rip, A. (2011). The future of research universities. *Prometheus, 29*, 443–453.

Rip, A., Misa, T. J., & Schot, J. (1995). *Managing technology in society*. Pinter.

Rittel, H., & Webber, M. (1973). Dilemmas in a general theory of planning. *Policy Sciences, 4*, 155–169.

Rockström, J., Gupta, J., Qin, D., Lade, S. J., Abrams, J. F., Andersen, L. S., Zhang, X., et al. (2023). Safe and just Earth system boundaries. *Nature*, 1–10.

Rogga, S., & Zscheischler, J. (2021). Opportunities, balancing acts, and challenges-doing PhDs in transdisciplinary research projects. *Environmental Science & Policy, 120*, 138–144.

Roura, M. (2021). The social ecology of power in participatory health research. *Qualitative Health Research, 31*(4), 778–788.

Salet, W. (2018). *Public norms and aspirations: The turn to institutions in action*. Routledge.

Sardar, Z. (2010). The namesake: Futures: Futures studies; futurology; futuristic; foresight—What's in a name? *Futures, 42*(3), 177–184.

Scharmer, C. O. (2009). *Theory U: Learning from the future as it emerges*. Berrett-Koehler Publishers.

Schön, D. A. (1995). Causality and causal inference in the study of organizations. In R. Goodman & W. Fisher (Eds.), *Rethinking knowledge: Reflections across the disciplines* (pp. 69–101). SUNY Press.

Schön, D., & Rein, M. (1994). *Frame reflection: Toward the resolution of intractable policy controversies*. Basic Books.

Schuitmaker, T. J. (2012). Identifying and unravelling persistent problems. *Technological Forecasting and Social Change, 79*(6), 1021–1031.

Schuitmaker-Warnaar, T. J., Gunn, C. J., Regeer, B. J., & Broerse, J. E. (2021). Institutionalizing reflexivity for sustainability: Two cases in health care. *Sustainability, 13*(21), 11712.

Scudder, M. F., Ercan, S. A., & McCallum, K. (2021). Institutional listening in deliberative democracy: Towards a deliberative logic of transmission. *Politics* [First Online]. https://doi-org.vu-nl.idm.oclc.org/10.1177/02633957211060691

Sellberg, M. M., Cockburn, J., Holden, P. B., & Lam, D. P. (2021). Towards a caring transdisciplinary research practice: Navigating science, society and self. *Ecosystems and People, 17*(1), 292–305.

Shapin, S., & Schaffer, S. (1985). *Leviathan and the air-pump*. Princeton University Press.

Smets, P., Reitsma, B., & Ghorashi, H. (2020). Community service learning and the issue of power: University students' engagement with disadvantaged neighbourhoods in Amsterdam. In M. Crul, L. Dick, H.Ghorashi, & A. Valenzuela, Jr., (Eds.), *Scholarly engagement and decolonisation: Views from*

South Africa, The Netherlands and the United States (pp. 277–299). Sun Media.

Sorrentino, M., Sicilia, M., & Howlett, M. (2018). Understanding co-production as a new public governance tool. *Policy and Society, 37*, 277–293.

Staffa, R. K., Riechers, M., & Martin-Lopez, B. (2022). A feminist ethos for caring knowledge production in transdisciplinary sustainability science. *Sustainability Science, 17*(1), 45–63.

Stirling, A. (2015). *Developing 'nexus capabilities': Towards transdisciplinary methodologies*. University of Sussex.

Strumińska-Kutra, M. (2016). Engaged scholarship: Steering between the risks of paternalism, opportunism, and paralysis. *Organization, 23*(6), 864–883.

Strumińska-Kutra, M., & Scholl, C. (2022). Taking power seriously: Towards a power-sensitive approach for transdisciplinary action research. *Futures, 135*, 102881.

Svare, H., Gjefsen, M. D., Den Boer, A. C., & Kok, K. P. (2023). Learning systems and learning paths in sustainability transitions. *Ecology and Society, 28*(1), article 22.

Tell, F., Berggren, C., Brusoni, S. & Van de Ven, A. (Eds.) (2017). *Managing knowledge integration across boundaries* (pp. 241–254). Oxford University Press.

Tolman, C.W. (1996). *Problems of theoretical psychology* (Vol. 6) Captus Press.

Torre, M. E. (2005). The alchemy of integrated spaces. In L. Weis & M. Fine (Eds.), *Beyond silenced voices: Class, race, and gender in United States schools* (pp. 251–266). SUNY Press.

Torre, M. E., Fine, M., Stoudt, B., & Fox, M. (2012). Critical participatory action research as public science. In P. Camic & H. Cooper (Eds.), *The handbook of qualitative research in psychology: Expanding perspectives in methodology and design* (2nd ed., pp. 171–184). American Psychological Association.

Torre, M. E. (2014). Participatory action research. In T. Teo (Ed.), *Encyclopedia of critical psychology*. Springer. https://doi-org.vu-nl.idm.oclc.org/10.1007/978-1-4614-5583-7_211

Torres, C. A. (2021). Introduction. In P. Freire, *Education for critical consciousness* (pp. vii–xl). Bloomsbury Publishing (M. Bergmann Ramos, Trans.).

Turnhout, E., Metze, T., Wyborn, C., Klenk, N., & Louder, E. (2020). The politics of co-production: Participation, power, and transformation. *Current Opinion in Environmental Sustainability, 42*, 15–21.

Ulnicane, I. (2016). Grand challenges' concept: A return of the 'big ideas' in science, technology and innovation policy? *International Journal of Foresight and Innovation Policy, 11*(1–3), 5–21.

Van Breda, J., & Swilling, M. (2019). The guiding logics and principles for designing emergent transdisciplinary research processes: Learning experiences and reflections from a transdisciplinary urban case study in Enkanini informal settlement, South Africa. *Sustainability Science, 14*, 823–841.

Van Breda, J., Musango, J., & Brent, A. (2016). Undertaking individual transdisciplinary PhD research for sustainable development: Case studies from South Africa. *International Journal of Sustainability in Higher Education, 17*(2), 150–166.

Vandenbussche, L., Edelenbos, J., & Eshuis, J. (2024). Collaboration in action: How micro-level relational dynamics are connected with issue frame convergence in collaborative governance networks. *Public Management Review*, 1–19.

Van De Poel, I., Asveld, L., Flipse, S., Klaassen, P., Kwee, Z., Maia, M., Mantovani, M., Nathan, C., Porcari, A., & Yaghmaei, E. (2020). Learning to do responsible innovation in industry: Six lessons. *Journal of Responsible Innovation, 7*(3), 697–707.

Van der Meij, M. G., Heltzel, A. A., Broerse, J. E., & Kupper, F. (2018). Frame reflection lab: A playful method for frame reflection on synthetic biology. *NanoEthics, 12*, 155–172.

Van der Velden, M., & Sjåfjell, B. (2022). Thinking with care: Exploring interdisciplinarity in a global research project. In *Interdisciplinary research for sustainable business: Perspectives of women business scholars* (pp. 23–40). Springer International Publishing.

Van der Wilt, G. J. & Reuzel, R. P. B. (2012) A transdisciplinary approach to the evaluation of medical technology: The case of cochlear implants for prelingually deaf children. In J. E. W. Broerse & J. F. G. Bunders (Eds.), *Transitions in health systems: Dealing with persistent problems* (pp. 115–128). VU University Press.

Van Mierlo, B. C., Regeer, B., van Amstel, M., Arkesteijn, M. C. M., Beekman, V., Bunders, J. F. G., Leeuwis, C., et al. (2010). *Reflexive monitoring in action. A guide for monitoring system innovation projects*. Communication and Innovation Studies, WUR; Athena Institute, VU.

Van Veen, S. C., de Wildt-Liesveld, R., Bunders, J. F., & Regeer, B. J. (2014). Supporting reflective practices in social change processes with the dynamic learning agenda: An example of learning about the process towards disability inclusive development. *International Journal of Learning and Change, 7*(3/4), 211–233.

Verran, H., Spencer, M., & Christie, M. (2022). Ground Up Inquiry: Questions and Answers About the Emergence and Development of a Northern Australian Tradition of Situated Research. *Learning Communities: International Journal of Learning in Social contexts*, *27*, 3–14. Article 1. https://digitalcollections.cdu.edu.au/nodes/view/5037

Vermeulen, W. J., & Witjes, S. (2020). History and mapping of transdisciplinary research on sustainable development issues: Dealing with complex problems in times of urgency. In M. M. Keitsch & W. J. V. Vermeulen (Eds.), *Transdisciplinarity for sustainability: Aligning diverse practices* (pp. 6–26). Routledge.

Verwoerd, L., Klaassen, P., & Regeer, B. J. (2021). How to normalize reflexive evaluation? Navigating between Legitimacy and Integrity. *Evaluation, 27*(2), 229–250.

Vinke-de Kruijf, J., Verbrugge, L., Schröter, B., den Haan, R.-J., Cortes Arevalo, J., Fliervoet, J., Henze, J., & Albert, C. (2022). Knowledge co-production and researcher roles intransdisciplinary environmental management projects. *Sustainable Development, 30*, 393–405. https://doi-org.vu-nl.idm.oclc.org/10.1002/sd.2281

Voss, J. P., Bauknecht, D., & Kemp, R. (Eds.). (2006). *Reflexive governance for sustainable development*. Edward Elgar Publishing.

Wenger, E. (1999). *Communities of practice: Learning, meaning, and identity*. Cambridge University Press.

Whitley, R. (1982). The Establishment and structure of the sciences as reputational organizations. In N. Elias, H. Martins, &R. Whitley (Eds.), *Scientific establishments and hierarchies. Sociology of the sciences a yearbook* (Vol. 6). Springer.

Whitney, D., & Cooperrider, D. L. (1998). The appreciative inquiry summit: Overview and applications. *Employment Relations Today, 25*(2), 17–28.

Wittgenstein, L. (1953). *Philosophical investigations*. In G. E. M. Anscombe & R. Rhees (Eds.) (G. E. M. Anscombe, Trans.). Blackwell.

Wyborn, C., Datta, A., Montana, J., Ryan, M., Leith, P., van Kerkoff, L., et al. (2019). Co-producing sustainability: Reordering the governance of science, policy, and practice. *Annual Review of Environment and Resources, 44*, 319–346.

Zachariah, B., Bunders-Aelen, J., & Regeer, B. (2023). Listening as a tool for transformative change in families and neighbourhoods: The case of SALT. In G. D. Bodie, D. L. Worthington, & Z. Beyene (Eds.), *Listening, community engagement, and peacebuilding: International perspectives* (pp. 55–78). Taylor & Francis.

Zuiderent-Jerak, T., Forland, F., & Macbeth, F. (2012). Guidelines should reflect all knowledge, not just clinical trials. *BMJ, 345*.

Zuiderent-Jerak, T. (2015). *Situated intervention: Sociological experiments in health care*. MIT Press.

CHAPTER 2

Trans-, Inter-, and Monodisciplinarity: Some Historical Considerations

Geert Somsen and Frans van Lunteren

Like most future-oriented movements, transdisciplinarity is very conscious of its past. Several transdisciplinary researchers have written histories of the approach that, taken together, give a good idea of the diversity of its origins and present state (Bernstein, 2015; Klein, 2015). But there is also a more elusive historical picture that is often implied or suggested and only occasionally spelled out (cf. Etzkowitz, 2002). This argues that transdisciplinarity was the natural 'next step' after interdisciplinarity, which itself was the logical response to monodisciplinarity. The story is a neat narrative in three stages, where each inevitably gave rise to the next generation. It also presents a clear starting point in everything that transdisciplinarity is *not*: a primordial state of single disciplines, as

G. Somsen (✉) · F. van Lunteren
Vrije Universiteit Amsterdam, Amsterdam, The Netherlands
e-mail: g.j.somsen@vu.nl

F. van Lunteren
e-mail: f.h.van.lunteren@vu.nl

F. van Lunteren
Leiden, The Netherlands

B. J. Regeer et al. (eds.), *Transdisciplinarity for Transformation*,
https://doi.org/10.1007/978-3-031-60974-9_2

isolated from each other as they were aloof from the needs of society. This Monodisciplinary Ivory Tower was the building to be razed to the ground.

As an advertisement the narrative has definite merits. But as historical description, we suggest, it is too simplistic. The story is as linear as it is predetermined. And it has a beginning that itself lacks history: a set of disciplines that appear always to have been there. Consequently, the narrative not only caricatures the past, but also fails to help understand the present conditions of transdisciplinary research. If the rise of transdisciplinarity is so straightforward and inevitable, then why are there still debates and struggles about its nature and position? In this chapter, we want to give an alternative historical account, not a complete counternarrative, which space and resources do not permit, but a few general points that arise from historical evidence. We hope these will help shed light on both the past and the current condition of transdisciplinarity.

2.1 Pre-disciplinarity

Let us start where the standard narrative begins: at the stage of single disciplines. Where did they come from and how did they develop? The historian and philosopher Bernadette Bensaude-Vincent (2016) asked this very question and concluded that disciplines should not be envisaged as being like trees, growing from their own trunk of pure core concepts and practices and branching out to ever wider applications. Disciplines, she argues, do not have and have never had a core. Rather, there has always been a whole range of approaches to reality (conceptually as well as experimentally), entangled with each other with no clear-cut distinctions. Rather than a tree, Bensaude-Vincent proposes the image of the rhizome to capture this continuum. A picture that also captures the situation is that of the physical map of the Earth—showing different kinds of landscapes blending into one another, sometimes with ruptures, and here and there a boundary, but nothing like the sharp and hard borders between uniformly coloured nation-states that a political world map presents. That map would be like the map of modern disciplines: clearly distinguished, each with their own, identifying colour.

The latter should not imply that disciplines are mere fictions (we will discuss their reality below), just that they are not naturally given, not determined by the complex variety of theories and practices that form the landscape of what scientists actually think and do. Just like nation-states, disciplines are superimposed on these practices, and the ways in which they cut up the field are deeply contingent.

The image of physical and political maps helps to make sense of the rise of disciplines. Just as nothing predetermined that the region of Limburg would end up in the Netherlands (rather than Germany or Belgium), there is little essential or decisive about what kind of scientific activity would ever belong to what discipline. The study of heat, for example, used to be a part of chemistry before it became a pillar of physics, sometime in the nineteenth century. Optics was long a branch of mathematics, as was the design of fortifications. And until the early twentieth century, Dutch students of the natural sciences self-identified not as scientists, but as philosophers (*philosoophen*). Labels have shifted while others have come and gone. What today is 'nanoscience' was 'colloid chemistry' a century ago. In fact, what we recognize as modern disciplines, fields like biology and physics, emerged as institutionalized entities only in the second half of the nineteenth century. Before that, there were no university programmes in psychology, history, or physics, while professors taught across the board, on subjects ranging from astronomy to medicine or from pharmacology to zoology. The map showed only a few broad territories, such as medicine and philosophy, with uncertain and flexible boundaries.

2.2 The Rise of the Disciplines

Modern disciplines as we know them are a product of the research university that emerged (at least in Europe, Japan, Canada, and the United States) in the second half of the twentieth century. They were not defined by the terrain of knowledge and research practices, but by the organization and institutionalization of the latter. Modern disciplines clustered contingently chosen sets of theories and research approaches into specialized training programmes resulting in disciplinary degrees. Hence, they created communities of experts, sharing specific vocabularies, outlooks, skills, and values. Disciplines materialized around a whole range of institutions, such as professorial chairs, university departments, specialized journals, research institutes, and disciplinary societies and their conferences (Van Lunteren, 2013). Hence modern disciplines became isolated, self-contained worlds, characterized by shared assumptions, practices, and mechanisms of inclusion and exclusion (Turner, 2017). They provided a major locus of belonging that gave their members a strong sense of identity. International conferences especially became occasions where disciplinary affiliation was celebrated, and participants were reminded that their field was 'The Goddess that We Serve' (Somsen, 2023).

Leading in the process of discipline formation were the sciences and humanities. These branches of knowledge had been liberated from their subordinate positions in the European universities through the elevation of the former faculty of arts to the level of the higher faculties of law, medicine, and theology. Increasing specialization in these fields, coupled with a new research-oriented pedagogy, resulted in a gradual transformation of the university. By the end of the nineteenth century, modern disciplines had become the main engines of knowledge production, and in the early twentieth century the disciplinary system expanded through the emancipation of the social sciences. New disciplines such as anthropology, psychology, political science, and sociology gained an independent academic status, at least in the rapidly expanding US universities, where the new disciplines were marked by the emergence of new departments (Ross, 1991). Although, strictly speaking, psychology is not a social science, it was increasingly recognized as such, as were economics and, in many cases, history.

Seen as such, transdisciplinarity's Other, 'monodisciplinarity', is not 'science as it always used to be' but a relatively recent phenomenon (Stichweh, 1984; Turner, 2000). While universities go back to the Middle Ages and Academies of Science to the seventeenth century, most disciplines that we are familiar with are hardly a century old. During many more preceding centuries, science and scholarship existed, but their cartography was different. They were organized more broadly, much less tightly institutionalized, and in ever-changing ways.

Nor can the Ivory Tower component of transdisciplinarity's Other be maintained. The Ivory Tower, as Steven Shapin has argued, never existed, except as a rhetorical figure used to mark unwanted academic practices as airy or irrelevant (Shapin, 2012). As for the modern disciplines, their institutionalization in fact relied strongly on their social relevance. In order to sustain themselves they needed society's support as well as a job market for their graduates. If research in the natural sciences was closely aligned with the needs of emerging chemical and electrical industries, and modern agriculture, the social sciences were expected to address the problems of the emerging industrial-urban societies. In that sense, the isolation of disciplines was always partial at best, and ties to the wider society have never been absent.

2.3 The Origins of Interdisciplinarity

It is against this late nineteenth-century background that we need to understand the rise of interdisciplinary approaches. If interdisciplinarity designates research involving different disciplines that are not just juxtaposed but interact in meaningful ways to solve certain problems, then its emergence may be said to have followed closely upon—or even been concomitant with—that of disciplinarity. In fact, it was precisely the creation of hard disciplinary boundaries that made certain combinations of approaches *look* like border-crossing—In the blurry, pre-disciplinary landscape, they would hardly have been recognizable as such. Hence, the new disciplines' new borderlands gave rise to hybrid specialties. These often involved the use of methods and concepts from one field to solve problems in the other. In several cases, members of both disciplines contributed to this process. This phenomenon became highly visible in around 1900 when fields like as astrophysics, physical chemistry, genetics, and biochemistry saw the light of day. Experimental psychology, combining physiology and philosophy, even became a full-fledged discipline of its own. More typically, however, the new fields were incorporated in one of the mother disciplines as new sub-specialties.

This process contributed to the fragmentation of the disciplines, along with a general tendency to specialization. Such fragmentation was a constant source of concern, which was voiced ever louder and more frequently after the turn of the century. While lamented, most scientists also agreed that specialization was an inevitable consequence of scientific progress, that could not be reversed. Moreover, fragmentation hardly affected the strength of the disciplinary system. University departments, societies, conferences, and journals guarded the integrity of the disciplines, as did an academic job market increasingly tuned to disciplinary distinctions (Somsen, 2023). Partial solutions were also sought in general overviews or 'popular' journals aimed at a broad scholarly public (Van Lunteren & Hollestelle, 2013). Interdisciplinarity was hence another belated proposal for overcoming the predicament. In most cases, external problems or needs—socio-political, economic, military, or healthcare-related—functioned as a trigger, as they were seen to require solutions that crossed disciplinary boundaries.

The major trigger was World War I. The mobilization of scientists for goal-oriented research, often through cross-disciplinary collaboration, resulted in a new way of thinking about the way science was organized. A well-known example is Fritz Haber's Kaiser Wilhelm Institute for Physical Chemistry, in Berlin. With a 50-fold increased budget, the institute set up interdisciplinary teams focusing on subjects such as gas masks, *Ersatz* explosives, and so on (Szöllösi-Janze, 2017, pp. 16–17). Similarly, the development of sound-ranging devices for artillery, resulting in human–machine combinations, involved a teaming up of theoretical physicists, acoustics experts, and experimental psychologists. These were not just cases of looking at a problem from different disciplinary angles, as all these approaches needed to be integrated in order to yield a solution. Nor do they present examples of 'applied science' as the problems to be solved required all kinds of basic research. And even if such cross-disciplinary collaborations may not have been new per se, the vast scale of such projects during wartime certainly was (Ash, 2019).

Whereas the main concern of disciplinary organizations remained the integrity and preservation of the disciplines, new post-WWI umbrella organizations, such as the US National Research Council (NRC), tended to pay more attention to the relationship *between* different disciplines and to the borderland areas (Cochrane, 1978, p. 176). Most outspoken in these forums was the astronomer George Ellery Hale, who had been instrumental in the foundation of the NRC during World War I. Even before then he had proposed that the National Academy of Science foster interest in 'subjects lying between the old-established divisions of science: for example, in physical chemistry, astrophysics, geophysics, etc.' (Ibid., p.327). He found an ally in the physicist Joseph Ames, who, upon his appointment as Chair of the Academy's Physical Sciences Division in 1924, established a 'Committee on Borderland Fields'. These campaigns for what soon was termed interdisciplinary research were continued by NRC President Isaiah Bowman in the 1930s (ibid., pp. 328–331).

A similar trend was visible in the Social Science Research Council (SSRC), which was founded in 1923 to promote empirical and policy-oriented research and to foster closer integration of social sciences, such as political science, economics, anthropology, and sociology (Worcester, 2001, pp. 15–33). It was at one of the early SSRC meetings that the term 'interdisciplinary' seems to have emerged. For many years, 'co-operative research' continued to be a more widely used label, but by the mid-1930s, the term started to take wing (Frank, 1988, p. 141). Its application was no longer reserved for research, but now also included education. Even more than in the natural sciences, interdisciplinary approaches were seen to be required for the study and solution of many social and economic problems confronting modern societies. To this end, the SSRC created special committees with representatives of several disciplines dedicated to specific issues such as the effects of the Prohibition or the Great Depression. The great challenge of the future was deemed to be the integration of the natural and the social sciences (Worcester, 2001, p. 6).

Interdisciplinarity was grafted onto disciplinarity, and this point is further illustrated by a contemporary counterexample. In interwar Europe, the social sciences were far less institutionalized than in the United States, and, as a consequence, European social scientists tended to have a much broader outlook. In both France and Germany, leading social scientists like Claude Lévi-Strauss and Max Horkheimer did not see themselves as representing a single well-defined field, but moved across sociology, anthropology, philosophy, and political science. The *Institut für Socialforschung* was founded in Frankfurt in 1923 to promote a critical understanding of the conditions of modern capitalist societies, and, to pursue this goal, mingled classical Marxism with a healthy dose of psychoanalytic theory and existentialist philosophy (Ross, 1991, pp. 224–225). However, none of this work was considered 'interdisciplinary', as that term presupposes the existence of well-established disciplines, which European social science lacked.

Interdisciplinarity only occurred when there were two or more disciplines present, yet it happened outside the places where they were most firmly institutionalized: universities and disciplinary societies. Early hotspots of interdisciplinary research were umbrella organizations, like the NRC. But even more effective were new research funding agencies, like the Carnegie and Rockefeller Foundations in the United States, the Kaiser Wilhelm Society in Germany, and the Medical Research Council in Britain (Ash, 2019, p. 628). It was the MRC that boosted the new field of biochemistry through the foundation of new research laboratories for nutritional chemistry and bacterial chemistry in Cambridge. The Kaiser Wilhelm Gesellschaft funded comparable extramural research institutes, some of which were clearly intended to meet industrial demands, like the Kaiser Wilhelm Institutes for Metals, Fibre, and Coal Research and the KWI for Anthropology, Human Heredity, and Eugenics (Ash, 2019, p. 628). The last of these was co-funded by the Rockefeller Foundation, as was the above-mentioned SSRC. Under the aegis of Warren Weaver, director of Natural Science Division of the Rockefeller Foundation, the division shifted its grants from physics to research in the life sciences, using physical and chemical methods. This deliberately interdisciplinary move resulted in the new field of'molecular biology', a term famously coined by Weaver in 1938.

2.4 Disciplines and Interdisciplines After World War II

World War II gave an even greater boost to interdisciplinarity than World War I had done. The reasons were the same: this war also produced a host of pressing problems as well as a new string of research institutes outside the universities. To harness scientists to the war effort the US government established the National Defense Research Committee followed by the even larger Office of Scientific Research and Development. Well-known examples of interdisciplinary wartime projects involving teams of physicists, chemists, mathematicians, and engineers are the Manhattan Project, the 'Rad Lab' (developing radar), and the work on proximity fuses. A German variety was the rocket programme. The war also created new interdisciplines such as operations research. This involved a mixed team of experts analysing particular wartime operations in order to improve tactical planning and decision making. To this end, all kinds of relevant data were assembled and, because of their secretive nature, made available in an operations room. The first operations research team, led by the British physicist P.M.S. Blackett, included physiologists, mathematicians, mathematical physicists, an experimental physicist, an astrophysicist, an army officer, and a surveyor. One of its tasks was to advise on the optimal use of radar in the defence of London against German bombing raids. Another was to develop a strategy that would prevent German U-boats from cutting off maritime supply-lines (Fortun & Schweber, 1993). Similar practices were also introduced in the United States, where the term 'operations research' was coined (Miser, 1980).

Such war-related efforts involving interdisciplinary teams were not limited to the natural sciences, but also involved the social sciences and the humanities. An example is the German *Kriegseinsatz der Geisteswissenschaften*, meant to provide a scholarly justification for the Nazis' new order in Europe. In the United States, research by social scientists on the engineering of public opinion, on stress resistance among American soldiers, and on psychological warfare had long-lasting effects on the sciences involved (Pooley, 2023). As in the previous world war, research on man–machine interactions coupled social scientists with engineers (Schweber, 2002). The Research and Analysis branch of the Office of Strategic Services (the forerunner of the CIA), established in 1942, recruited scholars from both the humanities and the social sciences to support intelligence work—among them seven future presidents of the American Historical Association and five of the American Economic Association. As Barry Katz has argued, their wartime experiences turned many of these participants into advocates of interdisciplinarity (Katz, 1989).

Another important new interdisciplinary field was systems analysis, which emerged in the immediate post-war period. It focused on future weapons systems and rational decision making on a quantitative basis amidst many uncertainties. An early adopter was the think tank RAND (short for Research ANd Development), created by Douglas Aircraft Company but later becoming a non-profit, the RAND Corporation. Although systems analysis soon widened its scope to all kinds of policy-related complex problems, most of its research during the Cold War concerned military issues. Most generally, it implied integrating several research techniques into a coherent framework to enable a balance between goals and their costs. This always implied collaboration between various kinds of experts, including economists, engineers, managers, and military officers (Fortun & Schweber, 1993).

One more major interdiscipline was cybernetics, the term coined by Norbert Wiener in 1947 for the new science of control mechanisms based on an exchange of information. It combined Wiener's experience with communication technologies with his interest in feedback mechanisms. Although likewise rooted in war-related problems (in this case anti-aircraft fire control) cybernetics became an all-encompassing worldview where the boundaries between living systems and machines were fully blurred, both being part of complex 'servomechanisms'. As Wiener stressed, from a cybernetic perspective there was little difference between a living creature and a machine: human purposeful behaviour was not different from that of self-regulating machines. As a way of understanding and doing it crossed disciplinary boundaries just as much as systems analysis did. Indeed, already in 1946, several meetings were held where natural scientists, mathematicians, and social scientists discussed circular causal systems and feedback mechanisms in the life sciences and social sciences (Galison, 1994).

A final example of a new interdisciplinary field, and likewise a Cold War product, was Area Studies. It emerged in the United States immediately after World War II in response to a widespread concern about the lack of knowledge about new global rivals, such as the Soviet Union and

China, and about political events in parts of Asia and Africa in the wake of the processes of decolonization. Already in 1946, the SSRC founded a Committee on World Area Research. New research institutes, such as Columbia's Russian Institute (1946) and Harvard's Russian Research Centre, could count on lavish funding by the Carnegie, Ford, and Rockefeller Foundations, as well as the US government. They also had close ties with the US intelligence agencies (Cumings, 1997).

And so interdisciplinarity had accompanied the disciplines almost as soon as they started—just as the notion of 'international' followed that of nation-states. The more disciplinary science was strengthened, the more crossovers became self-consciously interdisciplinary. But by the 1970s, the heyday of basic disciplinary science came to an end. Trust in the linear model, which saw innovation as a straight development from basic research to applied research to development, was starting to dwindle. Interdisciplinarity now appeared as an escape. The National Science Foundation, whose mission had been to support basic research, established a programme for applied science and called it Interdisciplinary Research Relevant to Problems of Our Society (Belanger, 1998).

Interdisciplinarity now also started to enter university education. Programmes such as Women's Studies, Cultural Studies, Area Studies, and Science and Technology Studies combined multiple approaches from the social sciences and humanities towards a particular problem area. These were followed in the natural sciences by programmes in, for example, environmental science, medical biology, bioinformatics, and—perhaps most recently—circular engineering. Today, there are even wider combinations, such as medical humanities and cultural heritage management. Still, none of these developments has meant the end of disciplines. Student numbers have dropped in some programmes, but it seems unlikely that chemistry, sociology, or history will disappear. In a sense, the rise of interdisciplinary research and teaching has as much corroded the disciplines as it has propped them up. Interdisciplinarity needs to draw on disciplinary approaches, or it ceases to be.

2.5 Transdisciplinarity and the Unity of Science

Transdisciplinarity, in many ways a product of the post-Cold War era, developed as seamlessly from interdisciplinarity as the latter did from disciplinarity. Thus, advocates of most schools of transdisciplinarity echoed the familiar arguments for interdisciplinarity: above all, the inability of discipline-oriented science to deal with complex social challenges. Moreover, in historical overviews the emergence of transdisciplinarity is usually dated to the International Conference on *Inter*disciplinary Research and Education held at Nice in 1970 (Bernstein, 2015; Klein, 2015). This conference was organized by the Centre for Educational Research and Innovation which had just been established by the Organisation for Economic Co-operation and Development (OECD) with the help of a grant from the Ford Foundation. It was the Swiss psychologist Jean Piaget who allegedly first used the word 'transdisciplinarity' in his talk on 'The Epistemology of Interdisciplinary Relationships'. His largely philosophical discourse addressed ways to integrate various disciplinary approaches in non-reductionist, non-hierarchical structuralist manner. At the end of his presentation, he expressed the hope of reaching a higher synthesis expressing the unity of science: 'This would be "transdisciplinarity", which would not only cover interactions or reciprocities between specialized research projects but would place these relationships within a total system without any firm boundaries between disciplines' (Piaget, 1972, p. 138).

The striving for epistemological unification in science had a long heritage, going back to late nineteenth-century philosopher-scientists like Ernst Mach and Karl Pearson. After World War I the unity of science became an overriding theme among the logical empiricists, resulting in a series of international conferences on the unity of science and the foundation by Otto Neurath of the Institute for the Unity of Science in The Hague (Cat, 2021; Kamminga & Somsen, 2016). Neurath and his Vienna companions coupled their philosophical programme explicitly to higher social and political goals, i.e. 'endeavors toward a new organization of economic and social relations, toward the unification of mankind, toward a reform of school and education' (Carnap, Hahn & Neurath (1928) cited in Uebel, 2020, p. 37).

After the war this tradition was continued in the United States, especially at Cambridge, Massachusetts, where Philipp Franck, also formerly a Vienna Circle member, founded another Institute for the Unity of Science. The new view of unity that emerged here was closely connected to the US-post-war perception of the cross-connections between disciplines and the rapid rise of interdisciplines like cybernetics and operations research (Galison, 1998).

In the 1960s, the Hungarian-British polymath Michael Polanyi, likewise a scientist-philosopher, teamed up with US colleagues to organize a Study Group on the Unity of Knowledge. Casting his net wider than his US predecessors, Polanyi not only sought to integrate different disciplinary approaches in a higher non-reductionist synthesis, but also aimed to include other forms of knowing such as art and religion—hence unity of *knowledge*, rather than *science*. In contrast to the logical empiricists' creed that a scientific worldview would help to solve humanity's problems, Polanyi considered the rampant scientism and positivism of his time to be responsible for the recent crises. The Study Group organized several international interdisciplinary conferences with financial support from the Ford Foundation. Leading scientists and scholars and, occasionally, artists attended these meetings to discuss a wide variety of problems and find common ground. One of the gatherings discussed the psychological theory of Piaget, whose anti-reductionist ideas about the relationship between the sciences went back to 1918, and who also participated (Breytspraak & Mullins, 2020).

The point of this little exposé is to show that neither the 1970 conference, nor the ideas presented there, were unprecedented. They were part of a long-standing debate that flared up again in the late 1960s. Indeed, even Polanyi's call for a higher synthesis that would incorporate non-academic, more spiritual kinds of knowledge, was hardly new. Such calls had been rampant in the early twentieth century, and even more so in the interwar period, and they generally overshadowed the more scientistic approaches of the logical empiricists (Baneke, 2008).

Piaget's linking of the term 'transdisciplinarity' to the long-standing concerns about the fragmentation of science and the resulting quest for a shared conceptual framework has been consolidated by several later advocates of transdisciplinarity, such as the US philosopher Kockelmans (1979), and more recently the Romanian particle physicist Basarab Nicolescu. The latter managed to institutionalize his views on transdisciplinarity in multiple ways. In 1987, he founded the International Center for Transdisciplinary Research and Studies in Paris. He was also the co-founder, with the Swiss philosopher and art historian René Berger, of the Study Group on Transdisciplinarity at UNESCO (1992), as well as being the main author of the *Charter on Transdisciplinarity* (1994) that resulted from the First World Congress on Transdisciplinarity held in Portugal in 1994. His co-authors were the French philosopher Edgar Morin and the Portuguese artist Lima de Freitas. Nicolescu also helped organize the 1997 Locarno International Congress of Transdisciplinarity as well as the 2005 Second World Congress of Transdisciplinarity in Brazil (Bernstein, 2015; McGregor, 2015).

The tenor of Nicolescu's writings was above all deeply humanistic, echoing several of Polanyi's concerns. Indeed, what was beyond the disciplines was, above all, the Human Subject. An improved methodology should include the subject, i.e. experiences, meanings, values, and emotions. Tellingly, the 1994 Charter related the need for a 'synthesis 'across, between and beyond' disciplines' to 'the complexity of our world and the present challenge of the spiritual and material self-destruction of the human species', 'a techno-science that obeys only the terrible logic of efficacy of efficacy's sake', and 'the present rupture between increasingly quantitative knowledge and increasingly impoverished inner identity' (De Freitas et al., 1994).

2.6 Transdisciplinarity and Mode 2 Knowledge Production

Another, more practical, research-oriented school of self-proclaimed transdisciplinarians likewise emerged in a smooth way from interdisciplinary foundations. Its main aim was to redirect scientific and technological research to the solution of the increasingly complex real-world problems confronting humanity, using different perspectives and methodologies. These may, but need not, derive from existing disciplines. Rather than striving for an overarching method based on an underlying metaphysics, as in the Nicolescu school, here the aim was to generate conceptualizations and methods in the context of application. As we have seen, many twentieth-century advocates of interdisciplinarity had shared this application-oriented outlook. However, several characteristics were added to the primacy of problem-oriented research.

An important source of inspiration for this school of transdisciplinarity was the publication of *The New Production of Knowledge* by the writers' collective Michael Gibbons, Camille Limoges, Helga Nowotny, Simon Schwartzman, Peter Scott, and Martin Trow (Gibbons et al., 1994). In contrast to the scientist-philosophers whom we encountered in the previous, more philosophical school, these authors were mostly rooted in the social sciences. In their book, they highlighted the supposedly new trend of 'Mode 2 knowledge production', which they contrasted with a long-standing traditional approach, named 'Mode 1'. The book was both a description of, and a plea for, this new approach.

Whereas Mode 1 was characterized by the focus on problems shaped by the interests of disciplinary communities, i.e. basic science, Mode 2 was seen to be application-oriented, transdisciplinary, socially distributed, and reflexive. Here transdisciplinarity referred to the transient networks of researchers with differing backgrounds and the methodological opportunism that transcended disciplinary boundaries, the social distribution to non-local collaborations, and the possible involvement of non-academics, such as representatives from industry, government, or non-governmental organizations (NGOs), and adopting reflexivity to replace the objective 'view from nowhere' by multiple situated views. At the organizational level, this latter aspect amounted to a shift from a 'culture of autonomy' to a 'culture of accountability' (Gibbons et al., 1994; Nowotny et al., 2001).

These ideas were in line with the research practices of Swiss and German researchers on environmental problems in the 1990s and, partly for that reason, they were central to the 2000 International Congress in Zurich on *Transdisciplinarity: joint problem-solving among science, technology, and society*. In the vision that emerged at this conference, reflexivity and, even more so, stakeholder participation, were seen as essential characteristics of transdisciplinarity, rather than Mode 2 add-ons. Ideally, stakeholders and 'end-users' should be involved in both the design of the transdisciplinary research projects as well as in their execution. The Congress gave rise to the foundation in 2002 of the Swiss-based Network for Transdisciplinary Research, which has been instrumental in spreading the new view of transdisciplinarity. Meanwhile, this view has been widely adopted by institutes and researchers all over the world (Bernstein, 2015; Klein, 2015; McGregor, 2015).

Although there is common ground between both schools—for instance, a joint concern about the complex problems facing today's world—there are also clear distinctions. Whereas the Nicolesu school strives to understand the world from a unitary viewpoint, the Zurich school aims to do science in a better and more useful way. As a result, current reviews of the literature on transdisciplinarity tend to distinguish the Nicolescu school from the Zurich or Swiss-German school. However, their approaches—one theoretical, the other more practical—are seen as complementary rather than as being in opposition (McGregor, 2015).

2.7 Conclusion

The development of disciplinary, interdisciplinary, and transdisciplinary science cannot, therefore, be seen as a process of three consecutive stages. On the contrary, interdisciplinarity emerged almost in tandem with the disciplines themselves, and transdisciplinarity repeated important orientations and justifications from interdisciplinary initiatives. Besides, the disciplinary matrix onto which the other forms were grafted cannot itself be seen as a preordained starting point. The modern disciplines are recent phenomena and, moreover, themselves contingent clusters of research practices around themes and social demands that presented themselves at particular moments. The actual, physical map of knowledge-making shows a rough and unruly landscape with no absolute borders—even what counts as scientific versus lay, experiential, or amateur knowledge is far from predetermined. What counted as science and disciplines were once projected onto this map, and interdisciplinary and transdisciplinary revisions are rearrangements of such projections.

References

Ash, M. G. (2019). Interdisciplinarity in historical perspective. *Perspectives on Science, 27*, 619–642.

Baneke, D. (2008). *Synthetisch denken. Wetenschappers over hun rol in de moderne maatschappij 1900–1940*. Verloren.

Belanger, D.O. (1998). *Enabling American innovation: Engineering and the National Science Foundation*. Purdue University Press.

Bensaude-Vincent, B. (2016). Building multidisciplinary research fields: The cases of materials science, nanotechnology and synthetic biology. In M. Merz & P. Sormani (Eds.), *The local configuration of new research fields* (pp. 45–60). Springer.

Bernstein, J. H. (2015). Transdisciplinarity: A review of its origins, development, and current issues. *Journal of Research Practice, 11*(1), Article R1.

Breytspraak, G., & Mullins, P. (2020). Polanyi and the study group for the unity of knowledge. *Tradition & Discovery: The Journal of the Polanyi Society, 46*(3), 4–27.

Cat, J. (2021). The unity of science. In Th. Uebel & Ch. Limbeck-Lilienau (Eds.), *The Routledge handbook of logical empiricism* (pp. 176–184). Routledge.

Cochrane, R. (1978). *The national academy of sciences: The first 100 years, 1863–1963*. National Academies Press.

Cumings, B. (1997). Boundary displacement: Area Studies and International Studies during and after the Cold War. *Bulletin of Concerned Asian Scholars, 29*(1), 6–26.

De Freitas, L., Morin, E., & Nicolescu, B. (1994). *The charter of transdisciplinarity*. In International Encyclopedia of Religion & Science. http://inters.org/Freitas-Morin-Nicolescu-Transdisciplinarity

Etzkowitz, H. (2002). The triple helix and the rise of the entrepreneurial university. In K. Grandin, N. Wormbs, & S. Widmalm (Eds.), *The science-industry nexus: History, policy, implications* (pp. 69–91). Science History Publications.

Fortun, M., & Schweber, S. S. (1993). Scientists and the legacy of World War II: The case of Operations Research (OR). *Social Studies of Science, 23*(4), 595–642.

Frank, R. (1988). 'Interdisciplinary': The first half century. *Issues in Integrative Studies, 6*, 139–151.

Galison, P. (1994). The ontology of the enemy: Norbert Wiener and the cybernetic vision. *Critical Inquiry, 21*, 228–266.

Galison, P. (1998). The Americanization of unity. *Daedalus, 127*(1), 45–71.

Gibbons, M., Limoges, C., Nowotny, H., Schwartzman, S., Scott, P., & Trow, M. (1994). *The new production of knowledge: The dynamics of science and research in contemporary societies*. Sage.

Kamminga, H., & Somsen, G. (2016). *Pursuing the unity of science: Ideology and scientific practice from the Great War to the Cold War*. Routledge.

Katz, B. M. (1989). *Foreign intelligence: Research and analysis in the Office of Strategic Services, 1942–1945*. Harvard University Press.

Klein, J. T. (2015). Reprint of 'Discourses of transdisciplinarity: Looking back to the future.' *Futures, 65*, 10–16.

Kockelmans, J. J. (1979). Why interdisciplinarity? In J. J. Kockelmans (Ed.), *Interdisciplinarity and higher education* (pp. 123–160). Pennsylvania State University Press.

McGregor, S. L. T. (2015). The Nicolescuian and Zurich Approaches to transdisciplinarity. *Integral Leadership Review, 15*(2), 6–16.

Miser, H. J. (1980). Operations research and systems analysis. *Science, 209*, 139–146.

Nowotny, H., Scott, P., & Gibbons, M. (2001). *Rethinking science: Knowledge and the public in an age of uncertainty*. Polity.

Piaget, J. (1972). The epistemology of interdisciplinary relationships. In: *Centre for Educational Research and Innovation (CERI), Interdisciplinarity: Problems of Teaching and Research in Universities* (pp. 127–139). Organisation for Economic Co-operation and Development.

Pooley, J. (2023). The plasticity of social knowledge: Paul F. Lazarsfeld and U.S. Communication Research, 1937–1952. *Journal for the History of Knowledge, 4*, 267–288.

Ross, D. (1991). *The origins of American social science*. Cambridge University Press.

Schweber, L. (2002). Wartime research and the quantification of American sociology. The view from 'the American soldier.' *Revue D'histoire Des Sciences Humaines, 6*, 65–94.

Shapin, S. (2012). The ivory tower: The history of a figure of speech and its cultural uses. *British Journal for the History of Science, 45*(1), 1–27.

Somsen, G. J. (2023). 'The goddess that we serve': Projecting international community at the First Serial Chemistry Conferences, 1893–1914. *British Journal for the History of Science, 54*(4), 453–467.

Stichweh, R. (1984). *Zur Entstehung des modernen Systems wissenschaftlicher Disziplinen: Physik in Deutschland* 1740–1890. Suhrkamp.

Szöllösi-Janze, M. (2017). The scientist as expert: Fritz Haber and German Chemical Warfare during the First World War and beyond. In B. Friedrich, D. Hoffmann, J. Renn, F. Schmaltz, & M. Wolf (Eds.), *One hundred years of chemical warfare: Research, deployment, consequences* (pp. 11–23). Springer.

Turner, S. (2000). What are Disciplines? And How is Interdisciplinary Different? In N. Stehr & P. Weingart (Eds.), *Practising interdisciplinarity* (pp. 46–65). University of Toronto Press.

Turner, S. (2017). Knowledge formations: An analytic framework. In R. Frodeman (Ed.), *The Oxford handbook of interdisciplinarity* (2nd ed. Online). Oxford Academic.

Uebel, Th. (2020). Intersubjective accountability: Politics and philosophy in the Left Vienna Circle. *Perspectives on Science, 28*(1), 35–62.

Van Lunteren, F. (2013). Het Ontstaan van het Systeem van Bètadisciplines: de Natuurkunde. *Studium: Revue d'histoire des sciences et des universités, 6*(2), 91–112.

Van Lunteren, F., & Hollestelle, M. (2013). Paul Ehrenfest and the Dilemmas of Modernity. *Isis, 104*(3), 504–536.

Worcester, K. W. (2001). *Social Science Research Council, 1923–1998*. Social Science Research Council.

PART I

Design and Evaluation

CHAPTER 3

Structuring Design & Evaluation in Transdisciplinarity for Transformation

Barbara J. Regeer, Kristiaan P. W. Kok, Alexandra Lux, Daniel J. Lang, and Barbara van Mierlo

3.1 Introduction

While transdisciplinary knowledge production is increasingly gaining traction in academic and policy environments, initiating and guiding such approaches is not straightforward and comes with challenges. These challenges concern, among other things, methodological and practical difficulties that arise in the 'fuzzy reality' of doing transdisciplinary

B. J. Regeer (✉) · K. P. W. Kok
Athena Institute, Faculty of Science, Vrije Universiteit Amsterdam, Amsterdam, The Netherlands
e-mail: b.j.regeer@vu.nl

K. P. W. Kok
e-mail: k.p.w.kok@vu.nl

A. Lux
ISOE—Institute for Social-Ecological Research, Frankfurt am Main, Germany
e-mail: alexandra.lux@isoe.de

D. J. Lang
Karlsruhe Institute of Technology (KIT), Karlsruhe, Germany
e-mail: daniel.lang@kit.edu

B. J. Regeer et al. (eds.), *Transdisciplinarity for Transformation*,
https://doi.org/10.1007/978-3-031-60974-9_3

research (Brandt et al., 2013; Lang et al., 2012). An overarching challenge for transdisciplinary research design and evaluation relates to the emergent nature of transdisciplinary efforts. Because such approaches aim to bridge the gap between knowledge and action, they should be designed and enacted to accommodate flexibility for, adaptation to and anticipation of emergent local needs and contextual developments (Fazey et al., 2018; Lux et al., 2019; Van Veen et al., 2014). Or, as Defila and Di Giulio eloquently phrase it in their chapter in this volume (Chapter 5, p. 140): there is the 'inevitability of the non-plannability' but also, from, among others, a funders' perspective, 'the necessity of having a reliable and robust research plan' (see also, e.g., Dahl Gjefsen et al., Chapter 4, this volume).

Given that transdisciplinarity can be characterized as deeply practice-based (West et al., 2019), situated (Zuiderent-Jerak, 2015), and highly contextualized (Caniglia et al., 2021; Norström et al., 2020), it is hard to imagine standardized procedures or a fixed body of methods, since these would very much depend upon the goals of the project and the background of those involved (see also Defila & Di Giulio, Chapter 5, this volume). The plurality of normative frameworks and the diverse ways of knowing and doing embodied by those involved pose challenges one would not encounter in more homogeneous collaborative (research) projects. Moreover, scholars emphasize that transdisciplinary co-production processes can also be understood as involving political practices (Kok et al., 2021; Turnhout et al., 2020), requiring those managing and facilitating transdisciplinarity to reflexively engage with, navigate and steer the political dynamics of co-production, while recognizing that it is challenging to do this from a neutral stance.

In the context of transdisciplinary research, such a manifestation of certainty is what transdisciplinary scholars, early-career researchers and novel transdisciplinary practitioners have advocated for, and developed, in the form of (methodological) guidelines, standards or frameworks supporting co-production processes (cf. Lang et al., 2012; Lux et al., 2019). Such guidelines or standards support teams in 'doing transdisciplinarity', and are particularly helpful at the planning and design stages of

---

B. van Mierlo
Knowledge, Technology and Innovation, Wageningen University & Research, Wageningen, The Netherlands
e-mail: barbara.vanmierlo@wur.nl

transdisciplinary projects, for instance when research proposals need to be submitted to PhD supervisors or to funders. Similarly, funding agencies are increasingly tasked with assessing or evaluating (the transformative impacts of) transdisciplinary research projects; the increasingly detailed evaluation frameworks or models (e.g. Kok et al., 2023; Lawrence et al., 2022; Luederitz et al., 2017; Lux et al., 2019; Schäfer et al., 2021; Schneider et al., 2019a; Walter et al., 2007) bear witness to this. While frameworks are of paramount importance in guiding transdisciplinary practice, they require necessary simplification and condensing of complex processes, thereby risking to not do full justice to the hugely entangled and messy character of transdisciplinarity. Simplification and reification are amplified by the schematic representation of transdisciplinary research design by means of models depicting a phased process with clearly delineated steps, or evaluation frameworks with a comprehensive set of criteria with which transdisciplinary research projects should comply (e.g. Belcher et al., 2016).

As the examples of learning questions in Chapter 1 show, this is where tensions arise: How to do justice to the emergent and situated nature of transdisciplinary practices, while there is also a need to plan and budget research and other activities? And subsequently, how can guidelines or standards be designed in a way that they provide guidance, while simultaneously embracing uncertainty and the open-ended nature and the fuzzy and political practice of 'doing' transdisciplinarity?

In Part I of this volume, we introduce chapters that variously seek to provide answers to the questions above. They aim to contribute to providing 'just enough structure' to those working in transdisciplinary research, from the level of 'micro-scale' interactions in inclusive spaces (Chapter 7), to supporting transformation-oriented research processes in the course of managing projects *ex durante* (Chapter 6) and designing (Chapter 4) and evaluating (Chapter 5) larger transdisciplinary projects. In this introductory chapter, we set the scene by elaborating on the different ways that the literature has dealt with design and evaluation of transdisciplinarity. We highlight a number of challenges and 'non-negotiables' in doing design and evaluation. Before exploring the challenges in applying frameworks in the 'real' fuzzy work of transdisciplinarity, let us look into a number of ideal–typical conceptualizations of transdisciplinary research.

3.2 Design and Evaluation Frameworks: Ideal–Typical Depictions of Messy Practices

Frameworks are deliberate simplifications of a phenomenon, process or situation; they are often constructed by decontextualizing situated complexities into more generic overviews. Frameworks are crucially important in transdisciplinarity, as they help researchers and practitioners to design, monitor and evaluate transdisciplinary processes. For instance, the much-cited 'conceptual model of an ideal–typical transdisciplinary process' (e.g. Jahn et al., 2012; Lang et al., 2012) is built up of years of experience with transdisciplinary research processes e.g. in the context of the Institute for Social-Ecological Research and the Leuphana University in Germany. It is qualified as 'ideal–typical' as it amplifies certain characteristics that are considered common in the pluriform practice of transdisciplinary research. The common characteristics are, first, the three phases that have formed the core of the model since its inception (Jahn & Keil, 2006); and, second, a set of generic design principles, linked to the three phases (see Table 3.1). Phase A focuses on framing the problem and team building; phase B on the co-creation of solution-oriented transferable knowledge and phase C on the (re-)integration and application of the created knowledge. Throughout the process, the emphasis is on bringing together actors from social and scientific practice and ensuring that what happens and is produced in the process is relevant for all involved. Feedback loops from both practices are seen as informing the transdisciplinary research process, which is therefore problem-oriented, integrative, and context-situated (Jahn et al., 2012). Together, these characteristics, collated into a model (see Fig. 3.1), aim to provide experience-based guidelines to practitioners and researchers alike.

A second example of a framework for transdisciplinary research is the Interactive Learning and Action (ILA)[1] approach, which was developed 30 years ago (Bunders, 1994) in the context of involving small-scale farmers in decision-making on biotechnological innovations, agricultural research and development in low- and middle-income countries.[2] This

[1] Note that in the area of biomedical and health (policy) research, the approach is more often referred to as the Dialogue Model (Abma & Broerse, 2010).

[2] Over the past decades, the approach was tested and evaluated in various different fields, including agricultural research (Broerse & Bunders, 2000; Swaans et al., 2009; Zweekhorst, 2004), biomedical and health (policy) research (Caron-Flinterman et al.,

Table 3.1 Design principles for transdisciplinary research in sustainability science

Phase	*Design principles*
Phase A	• Build a collaborative research team • Create joint understanding and definition of the sustainability problem to be addressed • Collaboratively define the boundary/ research object, research objectives as well as specific research questions, and success criteria • Design a methodological framework for collaborative knowledge production and integration
Phase B	• Assign and support appropriate roles for practitioners and researchers • Apply and adjust integrative research methods and transdisciplinary settings for knowledge generation and integration
Phase C	• Realize two-dimensional integration • Generate targeted products for both parties • Evaluate scientific and societal impact
General Design Principles (cutting across the three phases)	• Facilitate continuous formative evaluation • Mitigate conflict constellations • Enhance capabilities for and interest in participation

Note in the original article, each design principle is accompanied by a guiding question
Adapted from Lang et al. (2012)

process of prototyping, evaluating and adjusting the approach in a large number of fields over many decades has resulted in a robust framework for a multi-stakeholder, multi-phased, dialogical process. It revolves around four key interrelated factors (Betten et al., 2013):

2006; Harmsen et al., 2022) as well as in the area of emerging technologies such as synthetic biology (Betten et al.,2013) and neuroimaging (Arentshorst et al., 2015).

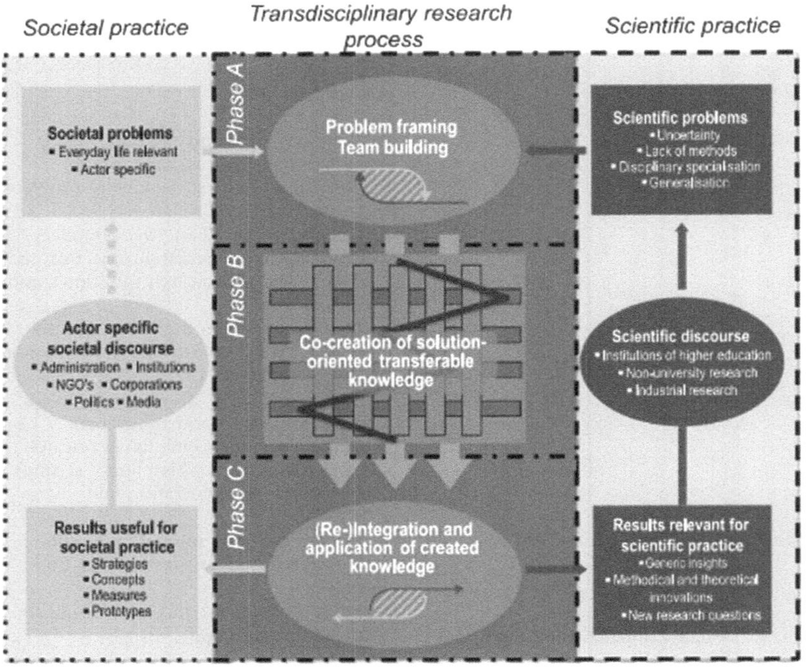

Fig. 3.1 Conceptual model of an ideal–typical transdisciplinary research process (Lang et al., 2012, p. 28, with reference to Bergmann et al., 2005; Jahn, 2008; Keil, 2009; Bunders et al., 2010)

- *Articulation of experiential knowledge*, especially of groups normally not engaged in research and innovation practices, such as citizens, patients or end-users;
- *Knowledge co-creation,* integrating social stakeholders' experiential and professional knowledge with researchers' scientific knowledge, taking into account real-life complexities and the myriad views, perspectives, needs and wishes that come with this complexity;
- *Embedding* or anchoring new ways of thinking about, organizing and doing research in the research system through realizing 'quick wins', and creating a support network of people with key positions in the 'system' for advice and support (Broerse, 1998); and

- *Process facilitation* to facilitate the above, while working to increase levels of trust between stakeholders with typically high-power differentials.

While the approach is characterized by an interactive and iterative emergent action-learning process, it is often described as roughly following five phases (see Box 3.1).

Box 3.1: The phases of the ILA approach (based on Swaans et al. [2009] and Betten et al. [2013]).

Phase 1: Exploratory phase

Establishing a research team, obtaining a preliminary overview of developments in the problem context through literature review and exploratory interviews, engaging with the local community and reaching agreement between stakeholders on general issues and procedures for collaboration.

Phase 2: In-depth phase

Identifying and analysing the problem perceptions, opinions, needs and ideas of the different stakeholders, including researchers. Because of asymmetry in power and knowledge between different stakeholder groups, in this phase stakeholders are consulted separately.

Phase 3: Integration phase

The perspectives of the different stakeholders are compared and, as much as possible, integrated by means of multi-stakeholder dialogue.

Phase 4: Priority setting and action planning

Stakeholders address conflicts and seek consensus on priority issues, common goals and plan of action.

Phase 5: Implementation phase

Participants determine and take action, monitor progress and evaluate results through continuous multi-stakeholder learning-action spirals.

After the first three phases, a spiral of activities keeps recurring: plan-act-observe-reflect-(re)plan, etc. (see Fig. 3.2).

Fig. 3.2 Action-learning spiral (Regeer et al., 2011)

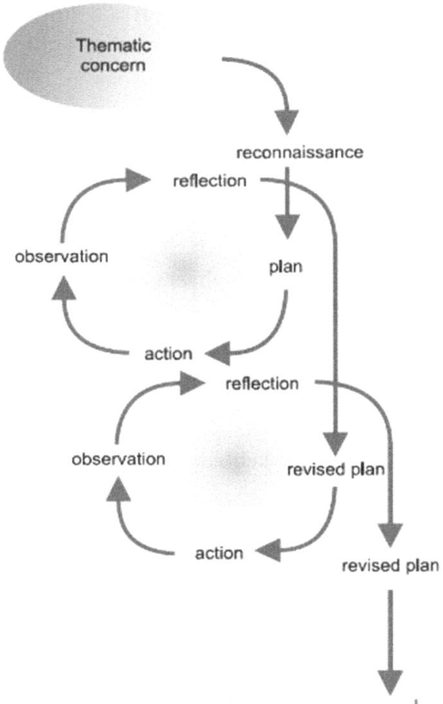

A third example we highlight here originates from an in-depth analysis of 16 transdisciplinary research projects, with the aim of identifying elements in approaches to transdisciplinary research that systemically strengthen the potential for social effectiveness (Lux et al., 2019). The elements identified together constitute the *TransImpact approach*, which provides guidance for the adaptive shaping of transdisciplinary research processes. They are not structured according to specific phases, but rather according to areas of prime importance in transdisciplinary research for transformation (see Fig. 3.3). It first states that awareness of the context in which the project takes place is key (left-hand side): this is about recognizing and understanding (a) the *history* of the given problem including the causes and dynamics behind it, and previous relationships between actors; (b) the *environmental* context more broadly; (c) the *heterogeneity* between and among different actor groups, in terms of their interests, expectations, institutional mindsets and organizational settings; and (d) the *funding* conditions. It secondly provides recommendations to clarify, observe, assess and adapt (a) all aspects that facilitate a better understanding of the *problem* situation and the application context; (b) *connectivity* to action contexts to enable uptake of results; (c) *roles and responsibilities* of each of the partners, in particular regarding knowledge integration and supporting knowledge transfer; (d) the *plural* interests, concerns, normative frameworks, hidden agendas or unshared objectives of those involved; and (e) a positive and inspiring *collaboration culture*.

All three example frameworks have value by condensing and structuring hugely entangled and messy processes. They guide the building of transdisciplinary research as a profession (Hoffmann et al., 2022) and provide credibility and legitimacy to different modes of research (Verwoerd et al., 2020). It is interesting, in this regard, that early-career researchers, in the context of a transdisciplinary doctoral programme, noted that developing attachments to transdisciplinarity at an epistemic

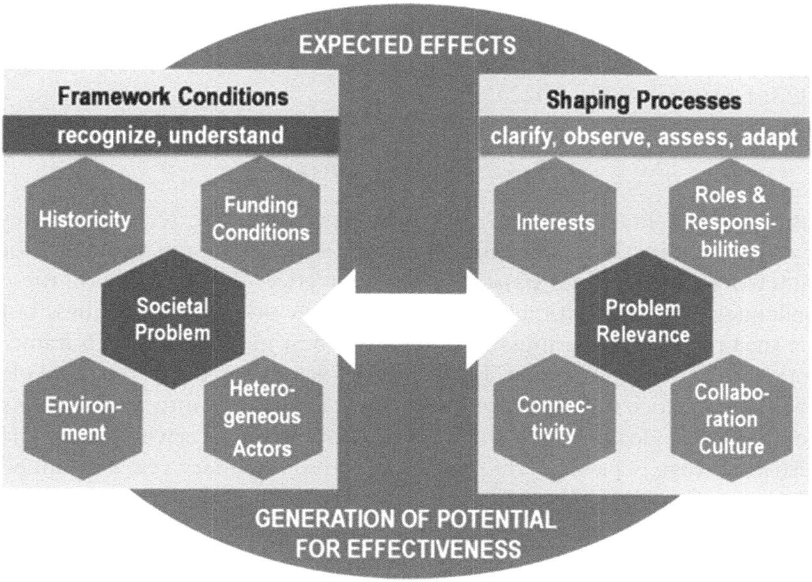

Fig. 3.3 TransImpact approach to foster potential for effectiveness in TDR (from Lux et al., 2019)

level was experienced as difficult, particularly because of a lack of adequate methodological or theoretical repertoire (Felt et al., 2013). As such, frameworks can become a 'point of reference' (Timmermans & Berg, 2010, p. 24, cited in Erisman, 2024) for those practising or evaluating transdisciplinary research projects.

While some authors express caution regarding the straightforward application of condensed models or frameworks (e.g. Erisman et al., 2024; Lawrence et al., 2022), others value exactly that. By definition, ideal types are not meant to mirror 'reality', but rather by highlighting certain common characteristics they tend to become prescriptive (Pohl et al., 2021). Belcher and colleagues (2016), for instance, constructed a comprehensive framework setting out principles and criteria for assessing the quality of transdisciplinary research, based on a systematic review, and stated that such a framework should be 'versatile': 'it should be useful to researchers and collaborators as a *guide* [emphasis added] to research design and management, and to internal and external reviews and assessors' (2016, p. 8). It is also in view of this need for *guidance* in conducting, but also for evaluating the impact of, transdisciplinary research, that scholars articulate challenges that find their origin in the tension between systematization and open-endedness (e.g. Kok et al., 2023; Lawrence et al., 2022).

3.3 Challenges

Based on the literature and our own experience, we see three types of challenges in the application of design and evaluation principles or frameworks as guidelines transdisciplinary practice. We elaborate on these challenges, not as a critique on existing frameworks and guidelines, but as a starting point to formulate strategies to best make use of such frameworks in practice. The first is that principles as such do not provide practical guidance if you do not know how to put them into practice. The second is that dealing with many principles can be overwhelming. The third one, that is particularly present in transdisciplinary research aimed at transformation, is that there is a tendency in the frameworks to focus mostly on process-oriented aspects of transdisciplinary research.

With regard to the first challenge, working according to the principles requires experience and acquired sensitivities. It is one thing to say that there are power imbalances of which the researcher should be aware and enhance trust and equitable relations. It is another to facilitate processes of empowerment and balance power differences (see also Part II of this volume). Further, it is one thing to say that historicity is important, and another to be sensitive to the difference between spending too much time and resources, or conversely too little, on preliminary investigation of path-dependency and the causes of lock-in. This also applies to principles that might seem contradictory. For instance, it is widely acknowledged that in transdisciplinary research we need to explicitly recognize the multiple ways of knowledge and doing (Norström et al., 2020), while we need to ensure that the process is goal-oriented by 'articulating clearly defined, shared and meaningful goals that are related to the challenge at hand' (ibid., p. 5). We can then yet again formulate a principle to that effect, such as balancing diversity and directionality (Kok et al., 2021) or balancing opening up and closing down (van Mierlo et al., 2020). This would, however, still require an idea of how to select participatory, inclusive and integrative methods and concepts and whom to include in what way in the project. How could one be sure that the selected methods are applied in an effective manner?

Second, many principles could be perceived to be too ambitious and thus do not seem realistic. We often read that transdisciplinary research means including *all relevant* stakeholders in *all phases* of the process. In most cases it is quite impossible to involve all stakeholders and various studies have indeed shown different kinds of involvement of a selection of stakeholders across phases (e.g. Enengel et al., 2012; Pohl et al., 2021). Do we demarcate the scope of stakeholders' involvement through the roles they may take (e.g. Defila & Di Giulio, this volume)? Or can we think about having some unusual relationships between different types of actors? Another example, to which we alluded above, is the emphasis on contextualization. Lux and colleagues (2019), for instance, emphasize that the environment of a problem shapes the possibilities and limits for projects that achieve an impact—and without understanding this context, projects may 'fail' or fall short of their potential. But what does that mean? There are so many different levels, so many developments that may seem relevant. The question is then, what would constitute smart or targeted contextualization? A final example of a principle that is often mentioned as essential but seems out of reach to most, even experienced, transdisciplinary practitioners is the emphasis on reflexivity. Whether understood as critical reflection of underlying assumptions and contextual conditions with the aid of various reflective exercises and tools (e.g. Van der Meij et al., 2018) or the ability of a transdisciplinary project to change both internal and external interdependencies, guiding rules and discourse (Beers & van Mierlo, 2017), the common idea is that the deeper the level of reflection, the more effective the (learning for) change will be (see also Part III). Moreover, reflection is often phrased as having to take place *continuously*. It may be clear that endless and continuous reflection to increase reflexivity is not only unrealistic, but also has its downsides, such as a show of confessional virtues, or strong constructivism leaving nothing 'real' upon which to act (Pels, 2000).

And third, frameworks tend to be oriented more to processes, while putting less emphasis on the contents and realization of desired future visions, ideas about solutions and the like: this can be suggesting that many frameworks might be more about 'transdisciplinarity' than about 'transformation'. Though this is understandable, as many frameworks are developed to guide transdisciplinary research, for those aspiring to contribute to transformation this could pose a challenge. Frameworks place an emphasis on the need for knowledge co-creation, for mutual learning, for addressing power dynamics, and for reflection. However, it is difficult, if not impossible, to guarantee that even if all these elements are in place, practices are changing, social situations are even improved in the experience of those affected, let alone that we are able to make tangible the transformative effects of the transdisciplinary intervention (Erisman et al., 2024; Kok et al., 2023; Lux et al., 2019; Schneider et al., 2019b). With a growing body of literature studying the dynamics of (and approaches to) transdisciplinary research practices comes the risk of understanding processes of collaboration, co-creation, social learning and reflexivity becoming ends in themselves, rather than a means to an end (see also the concluding Chapter 19 of this volume). Process criteria and outcome criteria are often treated as both important, but also separate (e.g. Swaans et al., 2009; Walter et al., 2007). Conceptualizing transdisciplinary research designs and evaluation schemes in such a way that a relationship with the aspired (short-, midterm- or long-term) transformation is integrated throughout, is still in its infancy (Lux et al., 2019; Williams & Robinson, 2020; see also Regeer et al., Chapter 1, this volume).

3.4 Outlook: Towards Just Enough Structure

In a recent overview of the field, and after highlighting challenges in transdisciplinary research, Lawrence and colleagues (2022, p. 58) state that 'evaluation frameworks such as those described previously need to be flexible enough for researchers to adapt them to their own context, while nevertheless being rigorous enough and retaining sufficient structure to allow a thorough analysis of the results and especially for comparing results across cases'. In our view, this recognition of inherent challenges associated with transdisciplinary research, and the fact that transdisciplinary practices, by definition, do not take place in splendid isolation, therefore calls for reflecting upon how frameworks can best provide guidance to practice. More precisely, we believe it calls for research design and evaluation frameworks that provide 'just enough structure'.[3]

One avenue to advance the debate is to further specify design and evaluation frameworks by connecting them to emerging research on different modes of transdisciplinary research. One could argue that every transdisciplinary practice is unique and contains so much variation, that a straightforward design or blueprint is not possible. A more nuanced take stresses that transdisciplinary comes in a number of forms and shapes, or in fact patterns or 'modes': Chambers and colleagues (2021), based on an analysis of 32 co-production initiatives to address complex sustainability challenges, identify six modes of co-production: (1) researching solutions; (2) empowering voices; (3) brokering power; (4) reframing power; (5) navigating differences; and (6) reframing agency. Similarly, Jahn and colleagues (2022) on the basis of 59 sustainability-oriented projects identify five (transdisciplinary) research modes: (1) purely academic research; (2) practice consultation; (3) selective practitioner involvement; (4) ideal–typical transdisciplinary research; and (5) practice-oriented research. One may be able to identify specific, or more tailored, process characteristics and guiding principles for each of the modes of co-production that may reduce the discrepancies between framework and actual practice.

Where further specification and focusing on different modes of transdisciplinary research is one appropriate strategy, another option is to zoom

[3] Being aware of the many interpretations of 'structure' in the social sciences, and studies on social transformation (cf. Kok, 2023), in this chapter we consider structure as both *enabling* actors in 'doing' and their transdisciplinary practice through guidance, while not *constraining* them to act otherwise, thus to allow for flexibility and open-endedness.

out and identify non-negotiable principles to which one must always adhere, regardless of the mode of transdisciplinarity, the setting, context or empirical domain. Such non-negotiables can complement existing frameworks, and help to reflect upon how these can provide guidance in practice. We—the authors—followed this strategy in a panel discussion on the topic at the Interdisciplinary Transdisciplinarity Conference 2021. In the panel, we explored how 'just enough structure' could be provided for transdisciplinary practices. Building on our shared exploration, in this section, we thus articulate four non-negotiables that we consider provide just enough structure for the practice of transdisciplinarity.

3.4.1 *Non-negotiable 1: Acknowledging Situatedness of Transdisciplinary Practice*

One important remedy in addressing the uncomfortable relationship between the messiness of practices and the neatness of ideal–typical depictions that we can recognize in all three examples above is the emphasis each approach places on the *situatedness* of transdisciplinary research (cf. Norström et al., 2020; Zuiderent-Jerak, 2015). This is also extensively recognized by the authors of the discussed frameworks themselves. For instance, Lux and colleagues (2019) speak about awareness and understanding of the historicity of the problem field, and Lang and colleagues (2012) emphasize the need for a phase of joint understanding and definition of the sustainability problem to be addressed. This is also articulated in work by Horcea-Milcu et al. (2022), who elaborate on the use of a 'phase 0' in transdisciplinarity. In the ILA approach, the first three phases are dedicated to the same issue, delving into both separate consultations of problem understandings of stakeholder groups, because of asymmetry in power (phase 2) and multi-stakeholder dialogues about these different understandings (phase 3). Furthermore, while Lux and colleagues (2019) discuss the importance of understanding and connecting to the context of action to ensure uptake, Lang and colleagues (2012) emphasize two-dimensional integration through targeted products and deliverables for the realms of society and of science, and the ILA approach integrates the notion of embedding or anchoring. In terms of including situatedness in the way the frameworks are presented, we see that Lang and colleagues (2012) attempt to 'breathe life into the principles through illustrative examples of challenges to comply with them [...] as encountered in transdisciplinary projects' across the globe (Lang et al., 2012,

p. 27) and that articles on the ILA such as Swaans and colleagues (2009) on promoting food security and well-being among HIV/AIDS-affected households in South Africa and Broerse and colleagues (2010) on including burn survivors in setting research agendas, are appreciated because they provide worked examples of transdisciplinary research in action.

3.4.2 Non-negotiable 2: Acknowledging Pluralities of Knowing—A Mindset of Curiosity

The second non-negotiable is the acknowledgement of plurality of knowledges, understandings and normative frameworks. And this is not a matter of just knowing, or writing about it. It goes much deeper and is far more a matter of being, or a mindset. It involves putting ourselves into other people's shoes (see also Schön & Rein, 1994) and trying to understand and value different ways of knowing. When we acknowledge the depth of the intertwinement between knowing and being (following Wittgenstein, 1953), it becomes evident how devious a recommendation this is. And, as outlined in Chapter 1 of this volume, our *shared* social and professional practices constitute our ways of being and knowing. So, while transdisciplinarity might be partly about creating transformative spaces, offering room to experiment without having to be accountable, in those spaces, to the rules of our professional or disciplinary homes, at the same time this poses tensions, because of other attachments, which might frustrate participants if unaddressed (Regeer & Bunders, 2009; Regeer et al., 2011).

Furthermore, some of us are considered, or consider themselves, as experts, which adds to the inherent epistemological challenge of being able to put ourselves into other people's shoes and view the world through their perspectives. Box 3.2 presents an illustrative example that we take from a transdisciplinary approach that has been referred to as Human Capacity for Response (HCfR), or the Community Life Competence Process (CLCP) (Zachariah et al., 2023). To acknowledge the plural ways of knowing and understanding, curiosity from a position of humility seems a good point to start. This might be harder for those who need to unlearn being an expert than it is for those new to the work (Zachariah et al., 2023).

Box 3.2: On the challenge of being an expert
The transdisciplinary Human Capacity for Response (HCfR) approach was first developed as a working model for AIDS competence by Jean-Louis Lamboray and colleagues in the 1990s (Campbell & Rader, 1995; Lamboray & Skevington, 2001), when HIV/AIDS was still a leading cause for death in many countries across the globe. It radically rejects earlier responses, whereby peer educators would go to a community and teach. In *What Makes Us Human*, in which Jean-Louis Lamboray reflects on decades of experience with the approach, he cites Toussaint, one of the facilitators in the Democratic Republic of Congo, who remembers: 'On any given day, I would go to a community, where I was invited as an expert. Installed at the high table, I would unpack my stuff and start my speech: "Pan, pan, pan, this is how you catch HIV. Pan, pan, pan, this is how you do not catch it. Pan, pan, pan, this is what you must do to avoid it." Then I would invite people to ask questions, and I'd answer them. If there were no more questions, I would pack my things and leave until the next meeting' (Lamboray, 2016, chapter 3). The HCfR framework reflects the belief that people have the capacity to *care, change, hope, lead, and belong as a community*, and that communities can harness these capacities to collectively address challenges (Lamboray, 2016). It uses the wisdom generated from people's experiences rather than from experts' knowledge and opinions. Toussaint continues: "Now [...] I come as a friend, I sit, and I ask questions. I let people talk about what they have done since my last visit, and I listen. And what do the communities do? They get the information they need to take action, they go en masse to get tested for HIV, and they visit families affected by AIDS" (Lamboray, 2016, chapter 3). Lamboray speaks of another facilitator, Antoine: 'Antoine likes to remind us that old habits die hard, and it is easy to resume the role of an expert. "The old man is asleep in us. He can wake up at any moment!" But when we have tasted the joy of sharing, and when we choose to appreciate the strengths of each person, of each family, of each community, then we progressively lose the desire to resume that role' (Lamboray, 2016, chapter 3).

3.4.3 Non-negotiable 3: Keeping Aspired Transformation Centre Stage

The third non-negotiable in the context of transdisciplinary project focusing on sustainability transformations is the need to not only focus on process criteria or process characteristics but to ensure a continual focus on the aspired change itself. There are legitimate reasons to concentrate primarily on input, process criteria and direct outputs (rather than aspired outcomes and impacts) in designing and evaluating multi-stakeholder approaches to complex societal problems. For one, there is the so-called ceiling of accountability (Bemme, 2019), which draws a line at what professionals can be held accountable for when conducting a programme or project. Due to the complex interplay of a multitude of factors in the context of 'wicked' problems—there are so many unforeseen reasons why a programme might or might not contribute to an aspired change and longer timeframes (sometimes decades) are needed to start seeing aspired outcomes—that it only seems fair to draw this line. However, with the ceiling of accountability having moved in the direction of increasingly tangible activities and outputs, aided by efforts to break down the complexity of transdisciplinary research into seemingly manageable activities (e.g. Belcher et al., 2016 define 27 criteria, or pointers for action, divided over four themes), the connection with the intended impacts risks of getting lost in the actual practice of transdisciplinarity (Schäfer et al., 2021). At the same time, there is an increasing awareness that moving the aspired change from beyond the ceiling of accountability, back to centre stage, may change the dynamics within the entire process, and between actors and organizations, and hence create more effective and lasting impacts.

3.4.4 Non-negotiable 4: Stimulate Action-Learning Spirals

Each of the approaches emphasizes the importance of iterative formative evaluation (Lang et al., 2012), action-learning spirals (Betten et al., 2013) and increasing reflexivity through a process of clarification, observation, assessment and adaption (Lux et al., 2019) in order to support the *doing in action* rather than the *designing* of transdisciplinary research. There are several promising developments that can aid in this process, because, in parallel to the (re)emergence and development of the idea of transdisciplinary research since the beginning of the century, the idea of accompanying this challenging practice with research to support its conduct has also emanated. It has gone under different names, ranging from the Interactive Learning and Action (ILA) monitoring approach (Regeer et al., 2009), transition monitoring (Taanman, 2012), Reflexive Monitoring in Action (RMA) (Van Mierlo et al., 2010) and accompanying research (Defila & Di Giulio, 2018; Schäpke, this volume).

The reasoning behind ILA monitoring was that 'if [transdisciplinary] approaches to persistent problems are so difficult to conduct, and if scholars of [these] approaches have indeed acquired relevant knowledge about these processes, then how can we contribute to accommodating these difficulties through our research? Where does theory meet practices?' (Regeer, 2010, p. 30). Or, as Taanman puts it, transition monitoring 'functions as a boundary object in the ongoing social learning and agenda setting between transition management in research and practice' (2012, p. 251).

In Reflexive Monitoring in Action (Arkesteijn et al., 2015; van Mierlo et al., 2010), for instance, one or several reflexive monitors are dedicated to supporting a diverse group of actors that aims to work on a collectively articulated aspired system change, which serves as a frame for reference for everyone involved and informs the kind of support and interventions of the monitor. Support is provided by stimulating recurring collective reflection on the results of actions in relation to the aspired system change as well as developments in the context that provide unexpected hindrances and also opportunities. While facing the everyday struggles of an ongoing transformative change process, these groups are stimulated to identify and experiment with solutions and ultimately change their practices, relationships and rules. Depending on the challenge at a specific moment, this happens by sparring in informal conversations, in interviews or with the aid of specific tools like the Dynamic Learning Agenda, a Collective System Analysis, or the Learning Mirror (de Wildt-Liesveld et al., 2015; van Mierlo, 2015; van Mierlo et al., 2010) that have been developed for keeping the focus on system change; stimulating learning, agenda setting and adaptation of activities or chosen direction of change; providing innovative forms for reporting; and allowing the group to carry it out collectively with the support of a reflexive monitor. The complexity and uncertainty associated with working on system change is fully acknowledged but also bounded because the locus of group action is supposed to be at the boundaries of the group and its direct social and institutional surroundings (Beers & van Mierlo, 2017). Similarly, it was found that questions on the learning agendas of specific cases reveal relevant boundaries (in terms of constraining conditions in the environment of the system) and help identify possible courses for action (Regeer et al., 2009).

3.4.5 Engaging with Non-negotiables in Practice

How can we engage with these non-negotiables in practice? First, through taking up the additional role of 'reflexive monitor' or 'accompanying researcher', besides project leaders and project participants in transdisciplinary research projects, building the capacity of these practitioners is supported. Training for prospective and upcoming monitors starts with urging them to start doing reflexive monitoring right away in a relevant project, programme or initiative in order to gain experiences of articulating aspired system change, and defining what monitoring activity would fit with a challenge at a specific moment in time. The provision of generic input about the foundations of, and principles for, the practice of reflexive monitoring as well as the abundant amount of possible monitoring activities are thus closely connected to training participants' own actual experiences as well as their earlier experiences and developed competences, while this training set-up also stimulates comparing and learning from each other's experiences.

This type of learning-by-doing is also supported in emerging structures such as Real-world Laboratories (RwLs) (Bergmann et al., 2021; Schäpke et al., 2018) or Living Labs (Erisman et al., 2024; Kok et al., 2023), which allow for open-ended processes within some kind of structure, and may provide an example of 'just enough structure' to implement and learn about conducting transdisciplinary research at the same time. Besides learning and reflecting on the job, small capacity-building workshops, for transdisciplinary researchers as well as partner collaborators, city administrators and civil society actors, could be part of this process.

At the same time, we believe that a small number of non-negotiables, combined with a spirit of action-learning, might speak to early-career researchers—especially those working on their own, or in small teams, on relatively small transdisciplinary research projects—who find it hard to identify with the more comprehensive frameworks because they suggest a rather large set-up, with a longer time horizon, and a large number of activities to generate and factors to take into account (see also Enengel et al., 2012; Van Breda et al., 2016).

Non-negotiables can easily also be applicable to micro-moments or single encounters between researchers and societal partners. For instance, when conducting an in-depth interview, the non-negotiables invite the researcher to:

- tailor the encounter to fit the specificities of the project, but also to invest in understanding the 'context of action', the historicity, and the multiple associations (Hallin et al., 2021, following Latour, 1986), through empiricizing work (Grijseels et al., 2024b);
- practise and enact a mindset of curiosity and openness, by also reflecting on, and putting to the test, their own assumptions, values and positionality, and supporting frame reflection by bringing insights from one interview into the conversation in the next interview;
- wonder 'what's in it for them' in order to refrain from extractive thinking and practice 'being alongside' (Grijseels et al., 2024a; Latimer, 2013), but also to solicit the interviewees' wisdom in co-creating responses to identified challenges or innovative options that help realize aspired changes.

All of which is supported by

- a spirit and process of action-learning, placing the interview and interviewee in the context of a larger, emerging, collaborative process of inquiry.

Transdisciplinary doctoral students in South Africa reflected on their process and observed that 'it was not so much the methods per se, but the philosophy and guiding principles underpinning the transdisciplinary approach which were most useful in navigating their individual research processes' (Van Breda et al., 2016, pp. 160–161). Incorporating the non-negotiables into regular research methods—such as interviews, participant observation, focus group discussions, document analysis—breathes into them the underlying philosophy of transdisciplinarity for transformation, and turns them into what we could call 'transformative methods'. In the same vein, non-negotiables can also be applicable to one-off events, such as a focus group discussion, a multi-stakeholder event or a dialogue session. Chapter 7 in this volume shows examples of ways in which non-negotiables were enacted in dialogical space making.

3.5 Concluding Remarks

We started this chapter by asking 'How can guidelines or standards be designed and used in a way that they do justice to open-ended nature of transdisciplinary processes, the fuzzy and political practice of doing transdisciplinarity, and provide guidance, while simultaneously embracing uncertainty?'. We have seen that guidelines or frameworks cannot be separated from the context in which they are employed. Scholars have argued for practice-based approaches or taking a 'praxeological perspective' (e.g. Lang et al., 2012, p. 27). Hence, paraphrasing Schön: 'rather than asking how those practising transdisciplinary research might make better use of frameworks for such research, or how scholars of transdisciplinary research might make their theories and models more palatable to those practising it, we can consider these practitioners as causal inquirers in their own right and ask how a different kind of research might enhance the types of causal inquiry they conduct in their efforts to support transformation' (1995, p. 96, cited in Laws & Hajer, 2008, p. 419). This has become a very relevant question for the practice of transdisciplinary research, especially since those new to the practice cannot be expected to have the required set of competences. And, more importantly, experience has shown that 'sensitivity and experience are at least as important as methodological skills and competences' (Regeer et al., 2011, p. 161): sensitivity to the fuzzy and cyclical nature of the transdisciplinary research, to the intangible aspects of the process, to the surrounding ecosystems and communities, to the chances and the obstacles that are specific to the situation, and to the project partners' viewpoints.

We have seen that the ability of standards, or frameworks, to reduce uncertainties, create transparency and travel across different (academic or transdisciplinary) spaces is valued, but at the same time they are criticized for their simplification of reality. Used in a prescriptive manner, a linearity is presumed whereby transdisciplinary practices are designed according to the model, executed according to plan and hence evaluated. At the same time, we have illustrated that scholars have tried to incorporate the messiness of reality and the emergent nature of transdisciplinary approaches in multiple ways in design and evaluation frameworks. We have stressed that this is not enough; there is a need for *ex durante* reflexive governance of transdisciplinary practice providing in situ guidance to practitioners. This, we believe, also means that *ex ante* design frameworks and *ex post* evaluation frameworks, and the assumptions and values inscribed in them, themselves become an actor in the conversation (Martinell Barfoed, 2018, cited by Erisman, 2024); they are a materiality and as such are part of the messy entanglements that characterize transdisciplinary research practices.

In this chapter, we have introduced four non-negotiables that could help researchers by providing 'just enough structure': (1) acknowledging situatedness of transdisciplinary practice; (2) acknowledging pluralities of knowing: a mindset of curiosity; (3) keeping aspired transformation centre stage; and (4) stimulating action-learning spirals. We argued that engaging with these non-negotiables in practice requires capacity building, and we stressed that these non-negotiables can be deployed both in large consortia and demarcated spaces (such as Labs), as well as in one-off events and smaller research projects. We hope our work can provide guidance to researchers, and stimulate reflection on the role of design and evaluation frameworks.

3.5.1 Outline of Part I

Each of the chapters in this part, from the authors' own experience, addresses one or more of the challenges outlined above. One common factor is that regardless of the 'scale' of the transdisciplinarity that is being practised, it is crucial to reflexively navigate the frameworks, so that these 'structures' help to realize the transformative and inclusive ambitions at play. All of these chapters also unravel the relationships between design and evaluation frameworks, and the intricate challenges and balancing acts transdisciplinary work poses for practitioners (see also Part II), as well as the roles and competences involved (see also Part III).

As we will see in all chapters, transdisciplinary practices are demarcated in one way or another, either by being defined as fundable projects (Gjefsen et al., this volume Chapter 4, Defila & Digiulio, this volume Chapter 5), or as real-world laboratories (Schäpke, this volume Chapter 6), or as communication spaces (Bruhn et al., this volume Chapter 7). Transdisciplinary practices can thus be seen as space making, which goes hand in hand with boundary setting. This boundary setting is amplified by funding requirements. With research endeavours (transdisciplinary or otherwise) being increasingly dependent on obtaining external, competitive funding, especially in Western Europe, much of transdisciplinary work takes place in the context of funded projects. While some funding requirements are becoming more favourable towards transdisciplinary research (Schneider et al., 2019a), the challenge of 'predicting' which activities to conduct during the duration of the project at the proposal stage is inherent to project funding, and specifically challenging for transdisciplinary research endeavours which are characterized by non-linearity and hence benefit from an emergent design approach. Moreover, project funding may lead to the so-called projectification of transformation efforts (e.g. Luger et al., 2023; Torrens & Von Wirth, 2021), leading to short-termism and undermining of transformative potential.

In Chapter 4, Gjefsen and colleagues reflect on the question of the *plannability* of transdisciplinary research processes from the perspective of participating in projects on research and innovation landscapes in relation to food and agriculture. They describe how considerations about *fundability* shaped project formulations and created path dependencies within the projects, affecting the transformation trajectories that were, or were not, pursued. The question then arises of how to plan ahead (at the project proposal stage) for transformative ambitions. They explore opportunities for a more 'authentic and honest engagement with "transdisciplinarity for transformation" within the structures afforded by project-based funding' and make the case for carving out more unstructured spaces for transdisciplinarity. From Chapter 4, it becomes clear that funding bodies play significant and powerful roles in driving the formation and 'societal relevance' of transdisciplinary projects (see also Fritz & Binder, 2020). Gjefsen and colleagues conclude that, transdisciplinary projects with transformative ambitions might not be 'a matter of "planning then doing", but rather a matter of "planning by doing"'. They make several recommendations to project coordinators and funders, as well as researchers and graduate students involved in transdisciplinary projects.

The observations in Chapter 4 beg more insight into *funding practices*, particularly the practices of setting assessment criteria and the conduct of review panels. This is exactly what Defila and Di Giulio set out to do in Chapter 5, where they present results of three case studies of the process of evaluating transdisciplinary research from a funders' perspective (in particular Federal State Funding, and a corporate foundation). They make a plea to the scholarly community of transdisciplinary researchers to move from debating and designing increasingly sophisticated design/evaluation frameworks and focus instead on the process of evaluation itself and how this process could be designed. As 'accompanying researchers' they were tasked with contributing to developing quality criteria and improving the evaluation process.

In Chapter 6, Schäpke further elaborates on the role of accompanying research: research that researches and supports other research, for instance in Real-world Labs. Based on work of Defila and Di Giulio (2018), Schäpke sets out the different types of contributions that accompanying them can provide to other researchers, such as generating and integrating knowledge generation, as well as process-oriented contributions, for instance supporting research teams through counselling. If accompanying research seeks to contribute to transformative knowledge production, it subsequently also requires a 'dynamically balanced, appropriately related and reflexive design' (Schäpke, this volume). In Chapter 6, the complex relation between accompanying research and 'evaluation' also becomes apparent: though accompanying research supports reflexivity and self-assessment of those being supported through accompanying research, it does not provide a traditional evaluation, but it could serve as a 'critical friend'. Through an empirical example on involvement in a co-creative reflection and dialogue space, this chapter highlights several balancing acts for accompanying researchers, and stresses the need for reflexivity in designing and doing accompanying research.

This relates, finally, to Chapter 7, in which Bruhn and colleagues reflect on their experiences with hosting Co-Creative Reflection & Dialogue Spaces at the 26th United Nations Framework Convention on Climate Change (UNFCCC) Conference of the Parties (COP26) in Glasgow, inviting all COP26 participants to (spontaneously) engage in joint reflection and mutual learning on a range of topics. We learn about designing and hosting spaces for dialogical exchange and trust-building in a tense environment in which it is difficult for participants to openly engage with each other, which privileges one-way dissemination of knowledge and which is experienced as cold and unsafe. Chapter 7 provides an overview of specific challenges that hosts encountered when trying to provide and maintain a 'safe enough' atmosphere; that is, an atmosphere that allows participants to engage in conversation outside their usual comfort zone and disclose potential learning edges or vulnerabilities. Principles that were enacted include 'listening to each other with compassion and curiosity'; 'staying aware of the impact of our contributions to the circle'; and 'suspending judgements, assumptions, and certainties'. Based on these hosting experiences and reflections, Chapter 7 gives concrete recommendations on how to design and host inclusive, safe enough spaces in not (yet) conducive contexts.

References

Abma, T. A., & Broerse, J. E. (2010). Patient participation as dialogue: Setting research agendas. *Health Expectations, 13*(2), 160–173. https://doi-org.vu-nl.idm.oclc.org/10.1111/j.1369-7625.2009.00549.x

Arentshorst, M. E., de Cock Buning, T., Boon, W. P., & Broerse, J. E. (2015). Prospecting responsible technology paths: Management options for an appropriate societal embedding of medical neuroimaging. *Science and Public Policy, 42*(6), 775–788.

Arkesteijn, M., Van Mierlo, B., & Leeuwis, C. (2015). The need for reflexive evaluation approaches in development cooperation. *Evaluation, 21*(1), 99–115. https://doi-org.vu-nl.idm.oclc.org/10.1177/1356389014564719

Beers, P. J., & Van Mierlo, B. (2017). Reflexivity and learning in system innovation processes. *Sociologia Ruralis, 57*(3), 415–436. https://doi-org.vu-nl.idm.oclc.org/10.1111/soru.12179

Belcher, B. M., Rasmussen, K. E., Kemshaw, M. R., & Zornes, D. A. (2016). Defining and assessing research quality in a transdisciplinary context. *Research Evaluation, 25*(1), 1–17. https://doi-org.vu-nl.idm.oclc.org/10.1093/reseval/rvv025

Bemme, D. (2019). Finding 'what works': "Theory of change, contingent universals, and virtuous failure in global mental health. *Culture, Medicine, and Psychiatry, 43*(4), 574–595.

Bergmann, M., Schäpke, N., Marg, O., Stelzer, F., Lang, D. J., Bossert, M., Sußmann, N., et al. (2021). Transdisciplinary sustainability research in real-world labs: Success factors and methods for change. *Sustainability Science, 16*, 541–564. https://doi-org.vu-nl.idm.oclc.org/10.1007/s11625-020-00886-8

Betten, A. W., Roelofsen, A., & Broerse, J. E. (2013). Interactive learning and action: Realizing the promise of synthetic biology for global health. *Systems and Synthetic Biology, 7*, 127–138. https://doi-org.vu-nl.idm.oclc.org/10.1007/s11693-013-9113-7

Brandt, P., Ernst, A., Gralla, F., Luederitz, C., Lang, D. J., Newig, J., Von Wehrden, H., et al. (2013). A review of transdisciplinary research in sustainability science. *Ecological economics, 92*, 1–15. https://doi-org.vu-nl.idm.oclc.org/10.1016/j.ecolecon.2013.04.008

Broerse, J. E. W. (1998). *Towards a new development strategy. How to include small-scale farmers in the biotechnological innovation process*. Eburon.

Broerse, J.E.W., & Bunders, J.F.G. (2000). Requirements for biotechnology development: The necessity for an interactive and participatory innovation process. *International Journal of Biotechnology, 2*(4), 275–296. https://doi-org.vu-nl.idm.oclc.org/10.1504/IJBT.2000.000140

Broerse, J. E., Zweekhorst, M. B., van Rensen, A. J., & de Haan, M. J. (2010). Involving burn survivors in agenda setting on burn research: an added value? *Burns, 36*(2), 217–231.

Bunders, J. F. G. (1994). *Participative Strategies for Science-Based Innovations*. VU University Press

Caron-Flinterman, J. F., Broerse, J. E., Teerling, J., Van Alst, M. L., Klaasen, S., Swart, L. E., & Bunders, J. F. (2006). Stakeholder participation in health research agenda setting: The case of asthma and COPD research in the Netherlands. *Science and Public Policy, 33*(4), 291–304. https://doi-org.vu-nl.idm.oclc.org/10.3152/147154306781778993

Campbell, I. D., & Rader, A. D. (1995). HIV counselling in developing countries: The link from individual to community counselling for support and change. *British Journal of Guidance & Counselling, 23*(1), 33–43. https://doi-org.vu-nl.idm.oclc.org/10.1080/03069889508258058

Caniglia, G., Luederitz, C., von Wirth, T., Fazey, I., Martín-López, B., Hondrila, K., Lang, D. J., et al. (2021). A pluralistic and integrated approach to action-oriented knowledge for sustainability. *Nature Sustainability, 4*(2), 93–100. https://doi-org.vu-nl.idm.oclc.org/10.1038/s41893-020-00616-z

Chambers, J. M., Wyborn, C., Ryan, M. E., Reid, R. S., Riechers, M., Serban, A., Pickering, T., et al. (2021). Six modes of co-production for sustainability. *Nature Sustainability, 4*(11), 983–996. https://doi-org.vu-nl.idm.oclc.org/10.1038/s41893-021-00755-x

Defila, R., & Di Giulio, A. (2018). What is it good for? Reflecting and systematizing accompanying research to research programs. *GAIA-Ecological Perspectives for Science and Society, 27*(1), 97–104.

Enengel, B., Muhar, A., Penker, M., Freyer, B., Drlik, S., & Ritter, F. (2012). Co-production of knowledge in transdisciplinary doctoral theses on landscape development—an analysis of actor roles and knowledge types in different research phases. *Landscape and Urban Planning, 105*(1–2), 106–117.

Erisman, J. C. (2024). *Navigating infrastructural mazes: An exploration of embodied practices, accountability and standardisation in a changing Dutch social care system* (Unpublished doctoral thesis). Vrije Universiteit Amsterdam.

Erisman, J.C., Feenstra, L.D., Kok, K.P.W., et al. (2024). Labbing for sustainability transformations: Learning about challenges and strategies for impact. *GAIA-Ecological Perspectives for Science and Society,* 33(1), 64-71.

Fazey, I., Schäpke, N., Caniglia, G., Patterson, J., Hultman, J., Van Mierlo, B., Wyborn, C., et al. (2018). Ten essentials for action-oriented and second order energy transitions, transformations and climate change research. *Energy Research & Social Science, 40*, 54–70. https://doi-org.vu-nl.idm.oclc.org/10.1016/j.erss.2017.11.026

Felt, U., Igelsböck, J., Schikowitz, A., & Völker, T. (2013). Growing into what? The (un-) disciplined socialisation of early stage researchers in transdisciplinary research. *Higher Education, 65*(4), 511–524.

Fritz, L., & Binder, C. R. (2020). Whose knowledge, whose values? An empirical analysis of power in transdisciplinary sustainability research. *European Journal of Futures Research, 8*(1), 3.

Grijseels, M., Zuiderent-Jerak, T., & Regeer, B. J. (2024a). Fostering Inclusive Technologies: Being Alongside Care in the Workplace. In D. Lydahl & L. C. Mossfeldt Nickelsen (Eds.), *Ethical and Methodological Dilemmas in Social Science Interventions: Careful Engagements in Healthcare, Museums, Design and Beyond* (pp. 99-113). Springer International Publishing.

Grijseels, M., Regeer, B., & Zuiderent-Jerak, T. (2024b). Empiricizing transitions: From empirical detail to transformation. https://doi-org.vu-nl.idm.oclc.org/10.2139/ssrn.4693130

Hallin, A., Karrbom-Gustavsson, T., & Dobers, P. (2021). Transition towards and of sustainability—Understanding sustainability as performative. *Business Strategy and the Environment, 30*(4), 1948–1957.

Harmsen, S., Pittens, C. A., Vroonland, E., van Rensen, A. J., & Broerse, J. E. (2022). Supporting health researchers to realize meaningful patient involvement in research: Exploring researchers' experiences and needs. *Science and Public Policy, 49*(5), 751–764. https://doi-org.vu-nl.idm.oclc.org/10.1093/scipol/scac024

Hoffmann, S., Deutsch, L., Klein, J. T., & O'Rourke, M. (2022). Integrate the integrators! A call for establishing academic careers for integration experts. *Humanities and Social Sciences Communications, 9*(1), 1–10. https://doi-org.vu-nl.idm.oclc.org/10.1057/s41599-022-01138-z

Horcea-Milcu, A. I., Leventon, J., & Lang, D. J. (2022). Making transdisciplinarity happen: Phase 0, or before the beginning. *Environmental Science & Policy, 136*, 187–197. https://doi-org.vu-nl.idm.oclc.org/10.1016/j.envsci.2022.05.019

Jahn, T., Bergmann, M., & Keil, F. (2012). Transdisciplinarity: Between mainstreaming and marginalization. *Ecological Economics, 79*, 1–10. https://doi-org.vu-nl.idm.oclc.org/10.1016/j.ecolecon.2012.04.017

Jahn, T., & Keil, F. (2006). Transdisziplinärer Forschungsprozess. *Soziale Ökologie. Grundzüge einer Wissenschaft von den gesellschaftlichen Naturverhältnissen*, 319–329.

Jahn, S., Newig, J., Lang, D. J., Kahle, J., & Bergmann, M. (2022). Demarcating transdisciplinary research in sustainability science—Five clusters of research modes based on evidence from 59 research projects. *Sustainable Development, 30*(2), 343–357. https://doi-org.vu-nl.idm.oclc.org/10.1002/sd.2278

Kok, K. P. W., Gjefsen, M. D., Regeer, B. J., & Broerse, J. E. (2021). Unraveling the politics of 'doing inclusion' in transdisciplinarity for sustainable transformation. *Sustainability Science, 16*, 1811–1826. https://doi-org.vu-nl.idm.oclc.org/10.1007/s11625-021-01033-7

Kok, K. P. W., van der Meij, M. G., Wagner, P., Cesuroglu, T., Broerse, J. E., & Regeer, B. J. (2023). Exploring the practice of Labs for sustainable transformation: The challenge of 'creating impact.' *Journal of Cleaner Production, 388*, 135994. https://doi-org.vu-nl.idm.oclc.org/10.1016/j.jclepro.2023.135994

Kok, K. P. W. (2023). Politics beyond agency? Pluralizing structure (s) in sustainability transitions. *Energy Research & Social Science, 100*, 103120. https://doi-org.vu-nl.idm.oclc.org/10.1016/j.erss.2023.103120

Lang, D. J., Wiek, A., Bergmann, M., Stauffacher, M., Martens, P., Moll, P., Thomas, C. J., et al. (2012). Transdisciplinary research in sustainability science: practice, principles, and challenges. *Sustainability Science, 7*, 25–43. https://doi-org.vu-nl.idm.oclc.org/10.1007/s11625-011-0149-x

Lamboray, J. L. (2016). *What makes us human?: The story of a shared dream.* Balboa Press.

Lamboray, J.-L., & Skevington, S. M. (2001). Defining AIDS competence: A working model for practical purposes. *Journal of International Development, 13*, 513–521. https://doi-org.vu-nl.idm.oclc.org/10.1002/jid.800

Latimer, J. (2013). Being alongside: Rethinking relations amongst different kinds. *Theory, Culture & Society, 30*(7–8), 77–104. https://doi-org.vu-nl.idm.oclc.org/10.1177/0263276413500078

Lawrence, M. G., Williams, S., Nanz, P., & Renn, O. (2022). Characteristics, potentials, and challenges of transdisciplinary research. *One Earth, 5*(1), 44–61. https://doi-org.vu-nl.idm.oclc.org/10.1016/j.oneear.2021.12.010

Laws, D., & Hajer, M. (2008). Policy in Practice. In M. Moran, M. Rein & R. Goodin (Eds.), *The Oxford handbook of public policy.* Oxford University Press.

Luederitz, C., Schäpke, N., Wiek, A., Lang, D. J., Bergmann, M., Bos, J. J., Westley, F. R., et al. (2017). Learning through evaluation—A tentative evaluative scheme for sustainability transition experiments. *Journal of Cleaner Production, 169*, 61–76. https://doi-org.vu-nl.idm.oclc.org/10.1016/j.jclepro.2016.09.005

Luger, J., Kotsila, P., & Anguelovski, I. (2023). The notion of justice in funded research on urban sustainability: performing on a postpolitical stage or staging the political? *Local Environment, 28*(1), 8–30.

Lux, A., Schäfer, M., Bergmann, M., Jahn, T., Marg, O., Nagy, E., Theiler, L., et al. (2019). Societal effects of transdisciplinary sustainability research—How can they be strengthened during the research process? *Environmental Science & Policy, 101*, 183–191. https://doi-org.vu-nl.idm.oclc.org/10.1016/j.envsci.2019.08.012

Norström, A. V., Cvitanovic, C., Löf, M. F., West, S., Wyborn, C., Balvanera, P., Österblom, H., et al. (2020). Principles for knowledge co-production in sustainability research. *Nature Susstainability, 3*(3), 182–190. https://doi-org.vu-nl.idm.oclc.org/10.1038/s41893-019-0448-2

Pels, D. (2000). Reflexivity: One step up. *Theory, Culture & Society, 17*(3), 1–25. https://doi-org.vu-nl.idm.oclc.org/10.1177/02632760022051194

Pohl, C., Klein, J. T., Hoffmann, S., Mitchell, C., & Fam, D. (2021). Conceptualising transdisciplinary integration as a multidimensional interactive process. *Environmental Science & Policy, 118*, 18–26. https://doi-org.vu-nl.idm.oclc.org/10.1016/j.envsci.2020.12.005

Regeer, B. J., & Bunders, J. F. (2009). *Knowledge co-creation: Interaction between science and society. A transdisciplinary approach to complex societal issues.* Advisory Council for Research on Spatial Planning, Nature and the Environment/ Consultative Committee of Sector Councils in the Netherlands (RMNO/ COS).

Regeer, B. J. (2010). *Making the invisible visible: analysing the development of strategies and changes in knowledge production to deal with persistent problems in sustainable development* (Unpublished doctoral thesis). Vrije Universiteit Amsterdam.

Regeer, B. J., Mager, S., & Van Orsouw, Y. (2011). *Licence to grow: Innovating sustainable development by connecting values.* VU University Press.

Regeer, B. J., Hoes, A. C., van Amstel-van Saane, M., Caron-Flinterman, F. F., & Bunders, J. F. (2009). Six guiding principles for evaluating mode-2 strategies for sustainable development. *American Journal of Evaluation, 30*(4), 515–537.

Schäfer, M., Bergmann, M., & Theiler, L. (2021). Systematizing societal effects of transdisciplinary research. *Research Evaluation, 30*(4), 484–499. https://doi-org.vu-nl.idm.oclc.org/10.1093/reseval/rvab019

Schäpke, N., Stelzer, F., Caniglia, G., Bergmann, M., Wanner, M., Singer-Brodowski, M., Lang, D. J., et al. (2018). Jointly experimenting for transformation? Shaping real-world laboratories by comparing them. *GAIA-Ecological Perspectives for Science and Society, 27*(1), 85–96. https://doi-org.vu-nl.idm.oclc.org/10.14512/gaia.27.S1.16

Schneider, F., Buser, T., Keller, R., Tribaldos, T., & Rist, S. (2019a). Research funding programmes aiming for societal transformations: Ten key stages. *Science and Public Policy, 46*(3), 463–478. https://doi-org.vu-nl.idm.oclc.org/10.1093/scipol/scy074

Schneider, F., Giger, M., Harari, N., Moser, S., Oberlack, C., Providoli, I., Zimmermann, A., et al. (2019b). Transdisciplinary co-production of knowledge and sustainability transformations: Three generic mechanisms of impact generation. *Environmental Science & Policy, 102*, 26–35. https://doi-org.vu-nl.idm.oclc.org/10.1016/j.envsci.2019.08.017

Schön, D., & Rein, M. (1994). *Frame reflection: Toward the resolution of intractable policy controversies*. Basic Books.

Swaans, K., Broerse, J., Meincke, M., Mudhara, M., & Bunders, J. (2009). Promoting food security and well-being among poor and HIV/AIDS affected households: Lessons from an interactive and integrated approach. *Evaluation and Program Planning, 32*(1), 31–42. https://doi-org.vu-nl.idm.oclc.org/10.1016/j.evalprogplan.2008.09.002

Taanman, M. (2012). Working in the science-policy interface: Transition monitoring in the Dutch energy transition program. In G. Verbong & D. Loorbach (Eds.), *Governing the energy transition: Reality, illusion or necessity?* (pp. 251–276). Routledge.

Torrens, J., & von Wirth, T. (2021). Experimentation or projectification of urban change? A critical appraisal and three steps forward. *Urban Transformations, 3*(1), 1–17. https://doi-org.vu-nl.idm.oclc.org/10.1186/s42854-021-00025-1

Turnhout, E., Metze, T., Wyborn, C., Klenk, N., & Louder, E. (2020). The politics of co-production: Participation, power, and transformation. *Current Opinion in Environmental Sustainability, 42*, 15–21. https://doi-org.vu-nl.idm.oclc.org/10.1016/j.cosust.2019.11.009

Van Breda, J., Musango, J., & Brent, A. (2016). Undertaking individual transdisciplinary PhD research for sustainable development: Case studies from South Africa. *International Journal of Sustainability in Higher Education*, *17*(2), 150–166.

Van der Meij, M. G., Heltzel, A. A., Broerse, J. E., & Kupper, F. (2018). Frame reflection lab: A playful method for frame reflection on synthetic biology. *NanoEthics, 12*, 155–172.

Van Mierlo, B. C., Regeer, B., van Amstel, M., Arkesteijn, M. C. M., Beekman, V., Bunders, J. F. G., Leeuwis, C., et al. (2010). *Reflexive monitoring in action. A guide for monitoring system innovation projects*. Communication and Innovation Studies, WUR; Athena Institute, VU.

Van Mierlo, B., Beers, P. J., & Hoes, A. C. (2020). Inclusion in responsible innovation: Revisiting the desirability of opening up. *Journal of Responsible Innovation, 7*(3), 361–383. https://doi-org.vu-nl.idm.oclc.org/10.1080/23299460.2020.1780409

Van Mierlo, B. (Ed.). (2015). *Learning mirror: Additional tool for reflexive monitoring in action—WUR*. Wageningen University, Knowledge, Technology and Innovation. https://www.wur.nl/fr/publications.htm?publicationId=publication-way-353039353234

Van Veen, S. C., de Wildt-Liesveld, R., Bunders, J. F., & Regeer, B. J. (2014). Supporting reflective practices in social change processes with the dynamic learning agenda: An example of learning about the process towards disability inclusive development. *International Journal of Learning and Change, 7*(3/4), 211–233.

Verwoerd, L., Klaassen, P., Van Veen, S. C., De Wildt-Liesveld, R., & Regeer, B. J. (2020). Combining the roles of evaluator and facilitator: Assessing societal impacts of transdisciplinary research while building capacities to improve its quality. *Environmental Science & Policy, 103*, 32–40. https://doi-org.vu-nl.idm.oclc.org/10.1016/j.envsci.2019.10.011

Walter, A. I., Helgenberger, S., Wiek, A., & Scholz, R. W. (2007). Measuring societal effects of transdisciplinary research projects: Design and application of an evaluation method. *Evaluation and Program Planning, 30*(4), 325–338. https://doi-org.vu-nl.idm.oclc.org/10.1016/j.evalprogplan.2007.08.002

West, S., van Kerkhoff, L., & Wagenaar, H. (2019). Beyond "linking knowledge and action": Towards a practice-based approach to transdisciplinary sustainability interventions. *Policy Studies, 40*(5), 534–555. https://doi-org.vu-nl.idm.oclc.org/10.1080/01442872.2019.1618810

de Wildt-Liesveld, R., Bunders, J. F., & Regeer, B. J. (2015). Governance strategies to enhance the adaptive capacity of niche experiments. *Environmental Innovation and Societal Transitions, 16*, 154–172. https://doi-org.vu-nl.idm.oclc.org/10.1016/j.eist.2015.04.001

Williams, S., & Robinson, J. (2020). Measuring sustainability: An evaluation framework for sustainability transition experiments. *Environmental Science & Policy, 103*, 58–66.

Wittgenstein, L. (1953). *Philosophical Investigations*. In G.E.M. Anscombe and R. Rhees (Eds.) [trans. G.E.M. Anscombe]. Blackwell.

Wolfram, M. (2016). Conceptualizing urban transformative capacity: A framework for research and policy. *Cities, 51*, 121–130. https://doi-org.vu-nl.idm.oclc.org/10.1016/j.cities.2015.11.011

Zachariah, B., Bunders-Aelen, J., & Regeer, B. (2023). *Listening as a Tool for Transformative Change in Families and Neighborhoods: The case of SALT*. In Listening, Community Engagement, and Peacebuilding (pp. 55–78). Routledge.

Zuiderent-Jerak, T. (2015). *Situated Intervention: Sociological Experiments in Health Care*. MIT Press.

Zweekhorst, M. B. M. (2004). *Institutionalising an interactive approach to technological innovation: the case of the Grameen Krishi Foundation* (PhD thesis). VU Amsterdam.

CHAPTER 4

Confronting the Projectification of Transdisciplinarity for Transformation

Mads Dahl Gjefsen, Kristiaan P. W. Kok, and Richard Helliwell

4.1 Introduction

While the potential of transdisciplinary approaches to stimulate transformation and face society's grand challenges has been the subject of much debate, a crucial question deserves attention: *How plannable are these processes, really?* This is worth asking in a knowledge economy where practising transdisciplinarity is fragmented both in relation to time (tied to specific projects or initiatives with fixed lifetimes and attendant [fundable] predefined plans and objectives), and to space (tending to include

M. D. Gjefsen (✉)
Studio Apertura, NTNU Social Research, Trondheim, Norway
e-mail: mads.gjefsen@samforsk.no

K. P. W. Kok
Athena Institute, VU Amsterdam, Amsterdam, The Netherlands
e-mail: k.p.w.kok@vu.nl

R. Helliwell
RURALIS Institute for Rural and Regional Research, Trondheim, Norway
e-mail: richard.helliwell@ruralis.no

B. J. Regeer et al. (eds.), *Transdisciplinarity for Transformation*,
https://doi.org/10.1007/978-3-031-60974-9_4

knowledge actors and stakeholders who are separate from each other, both in terms of disciplines and also physically and organizationally). Combined with probing the indeterminacy that stands at the heart of transformation efforts concerned with research and innovation (R&I) and its impacts—the acknowledgement in Science & Technology Studies and elsewhere that things could be otherwise having prompted descriptions and interventions into versions of this *otherwise* via knowledge-symmetrical, dialogical, and participatory approaches—much can be said for the ***unplannability*** of the situations where meaningful opportunities for change arise.

There might be a case for carving out more space for the unstructured and unexpected in transdisciplinarity. In this chapter, we take a modest step in that direction by problematizing the project-based nature of transdisciplinarity for transformation. We seek to identify ways in which initial funding, organizational, and other logics tend to set up cycles of promises that may direct transformative attention in less-than-constructive directions and discourage the seizing of opportunities for meaningful interventions that may occur along the way. On this basis, we propose some 'rules of thumb' to guide the development of promises in relation to impact, transdisciplinarity, and transformation with integrity, coupled with principles by which the delivery on such promises might be evaluated. At the same time, understanding how project 'planning' results in path dependencies for 'doing' might also open avenues for project designs that encourage and help facilitate promising and meaningful new directions during project practice.

The chapter is based on our collective reflections on our experiences of working in several projects. These past and ongoing projects in different ways operationalize transdisciplinarity and transformation. Thematically, the projects focus on systemic interventions pertaining to R&I landscapes on food and agriculture, areas where systemic approaches are widely used (Klerkx et al., 2010; Springmann et al., 2018), between which productive synergies are sought (Kok et al., 2019), and where insights are thought to be transferable to many other areas where social-technical complexities lie at the heart of efforts to produce positive change—such as climate and energy, mobility, public health, and digitalization, to name a few. While, in keeping with a long tradition of neat reporting on scientific activity that excludes the seemingly mundane and un-methodologically standardized practices (Latour & Woolgar, 1979), such project accomplishments are generally published without much critical attention to their own funding context. Breaking from this tradition in this chapter, we describe how considerations relating to fundability, together with the realities imposed by bureaucratic accountability structures and project-based research funding, influenced project formulations and path dependencies, ultimately affecting the transformation trajectories that were, or were not, pursued.

The projects we report on were funded by the European Union (EU) and national funding agencies. Most prominently they include FIT4FOOD2030[1] (2017–2020), an EU coordination and support action project (a project category in Horizon 2020 which emphasizes stakeholder integration and dissemination over goals for disciplinary research outputs) coordinated by Vrije Universiteit Amsterdam (VU) with partners across Europe. The project worked towards EU food system transformation by engaging a wide variety of stakeholders in 25 diverse transformation-oriented Labs, which were embedded in schools, universities, science centres, and national ministries, and supported and monitored by a project consortium intended to work in a transdisciplinary manner. SYNAGRI[2] (2021–2024) is a three-year research project funded by the Research Council of Norway coordinated by Ruralis—Institute for Rural and Regional Research in Norway. The project's point of departure is the acknowledgement that there are distinct promissory discourses around what can be respectively termed *eco-* and *bio-economic* food systems (Marsden & Farioli, 2015), with contrasting emphases on technology, localness, ecological principles, optimization, land use, consumption, and farming infrastructure. In addition, we also draw on experiences from other system- and stakeholder-focused projects.

There are many important differences between these projects, both in terms of funding stipulations, and in terms of researchers' choices about how to frame and operationalize key concepts. We do not claim to offer an exhaustive view of what is or is not transdisciplinary, but rather consider family resemblance to be implied when calls for projects include requirements for participatory co-creation by researchers and societal groups, or when a project consortium claims to integrate disciplinary and other forms of knowledge and experience. This means that

[1] FIT4FOOD2030, Fostering Integration and Transformation for FOOD 2030, Horizon 2020, grant agreement ID: 774088.

[2] SYNAGRI, Developing synergies between the bioeconomy and regional food systems for a sustainable future, Research Council of Norway grant agreement ID: 325403.

we admittedly compromise on clarity of conceptual distinctions between transdisciplinarity (moving beyond disciplinary knowledge), multidisciplinarity (combining several disciplines), and interdisciplinarity (integration and exchange across disciplines) in an attempt to describe elements of transdisciplinarity practice; this approach enables us to formulate practical recommendations at the end of the chapter.

We first introduce several theoretical considerations regarding the 'planning' and 'doing' transdisciplinarity for sustainable transformation, which help to place our analysis in context, and guide us in developing further insights from the cases. We then go on to describe our methods, followed by integrated empirical presentation and discussion, along with recommendations for researchers, project reviewers, and research-funding organizations.

4.2 Theoretical Background: Planning and Doing Transdisciplinarity

In recent decades, transdisciplinary R&I projects have rapidly emerged as means to contribute to resolving complex societal problems (Thompson Klein et al., 2001; Pohl & Hadorn, 2007) by seeking to include a wide range of societal actors, knowledge, and problem perceptions in reflexive and experimental R&I processes. These actors can be described as societal stakeholders with socio-political interests, practitioners who engage the topics discussed in practice, citizens, and other non-certified experts who have particular knowledge relevant to the transdisciplinary R&I project (see Defila & Di Giulio, 2015). In that light, transdisciplinary R&I is considered promising as it could contribute not only to normative and democratic ambitions of stakeholder inclusion in R&I (e.g. Kok et al., 2021; Nowotny et al., 2001) but also to create knowledge and innovations that are considered more legitimate and socially robust by those affected by them (e.g. Owen et al., 2012; Schmidt et al., 2020). The transformative potential of transdisciplinarity is highlighted in its potential to create 'societal impact' in the face of urgency (e.g. Fazey et al., 2018; Lang et al., 2012).

The 'planning' activities (which are continuously intertwined with and scarcely separable from 'doing') associated with transdisciplinary R&I projects have increasingly been the object of scholarly scrutiny (e.g. Lang et al., 2012; Lux et al., 2019; Schneider et al., 2019). 'Projectification' of sustainability and transformation-focused efforts are an emerging research area in their own right (cf. Ika & Munro, 2022). Questions regarding the 'how and why' of design and agenda-setting of R&I projects that aim to create societal 'impact' and 'engagement' have also increasingly become an object of study (see Reed et al., 2021 for a recent review), and in Responsible Research and Innovation (RRI, see for instance, Stahl et al., 2021). Planning for transdisciplinarity is challenging given its need for flexibility (such as anticipation and adaptation) based on context-specific and often unexpected developments, stakeholder needs or emerging insights that evolve in the practice of seeking to solve complex societal problems (cf. Lang et al., 2012). How does one plan for what kinds of strategies, activities, and impacts need to be evaluated at the end of a fuzzy, complex, and non-linear multi-year project (Reed et al., 2021; Schäfer et al., 2021)?

To aid the development of both design and evaluation, transdisciplinary projects increasingly adopt so-called Theories of Change (ToCs) (e.g. Deutsch et al., 2021; Schneider et al., 2019; cf. Earl, 2001), offering '*guiding framework[s] for all stages of thinking, action and sense-making when a project or a program intervenes in processes of social change*' (Deutsch et al., 2021, p. 29, drawing on van Es et al., 2015, p. 12). Deutsch and colleagues (2021) argue that ToCs can be useful in 'planning' the activities of the project and their relations to aspired (short-term) outputs (such as concrete products) and (longer-term) societal impacts. They stress that ToCs are both an important process and an output in transdisciplinarity but also observe that the pre-set 'planned' impacts could become path-reinforcing instruments during the project's 'doing', even when other directionalities might have been more fitting from policy or practice perspectives (Deutsch et al., 2021, p. 37).

Moreover, while planning for impact (for instance through ToCs) is an important part of 'planning' transdisciplinarity (alongside 'planning the process', and 'planning the results'), scholars have also stressed that the impact focus of 'planning' can lead to 'impact sensationalism' and over-promising societal impacts in grant applications (e.g. Chubb & Watermeyer, 2017), an understandable consequence of competitive funding processes and of effectiveness-oriented (transdisciplinary) R&I aimed at sustainable transformation (Musch & Von Streit, 2020). In addition, the 'projectification' of sustainability science (Torrens & Von Wirth, 2021) with many short-term (but often large) projects that strive for impact could further lead to closing down, rather than opening up pathways for reflexive and transformative experimentation in transformation-oriented projects, given a focus on quick and quantitative deliverable outputs (Torrens & Von Wirth, 2021). These developments not only risk the development of overconfident impact narratives in project design, but also of contributing to other questionable practices such as (un)intended negative effects for participating stakeholders (Musch & Von Streit, 2020) that could even lead to reinforcing existing power relations rather than unpacking or exploring them (Turnhout et al., 2020).

In their analysis of power dynamics in five transdisciplinary projects, Fritz and Binder (2020) elaborate on how actors (researchers, practitioners, and funders) influence the planning phase (or: agenda-setting phase) of transdisciplinarity. They argue that researchers are powerful[3] in shaping project design 'based on their authority as project leaders, their knowledge about proposal writing, and their financial/time resources' (Fritz & Binder, 2020, p. 12). Meanwhile, funding bodies 'exercise structural power by means of its material sources and discursively frames the nature of research agendas worthy of funding' (p. 10).

4.3 Methodological Approach and Limitations

This chapter builds on the authors' engagement with two projects—FIT4FOOD2030 and SYNAGRI—and reflections on these and other projects from their perspectives as permanent employees at research institutes reliant on project-based extramural funding (MDG, RH), and a fixed-term doctoral candidate at a major research university (KK).

The work, undertaken *within* the two main projects we consider here, was as follows. In FIT4FOOD2030, one co-author (MDG) contributed to a designated work package aimed at monitoring and facilitating learning for transformation, a variant of accompanying research (see Schäpke, *this volume*) where activities included documenting various forms of stakeholder engagement and impact and stimulating learning and reflection among the coordinators of 25 City, Food and Policy Labs, the main venues for participatory activities in the project. Another co-author

[3] Fritz and Binder (2020) draw upon a widely used three-dimensional conceptualization of power (cf. Arts & van Tatenhove, 2004) in which there is *instrumental power* (as 'who decides in decision-making', building on Dahl [1962]), *structural power* as setting preconditions influencing decision-making processes (Bachrach & Baratz, 1962) and *discursive power* as influencing norms and beliefs (Lukes, 2004).

(KK) supported the project coordination, including facilitating interactions between project activities. In SYNAGRI, one co-author (RH) led the application submission and currently serves as Principal Investigator, after also being involved in the preparation of an earlier unsuccessful grant application for a previous iteration of the project. Another co-author (MDG) temporarily led a work package on food systems and planning. Project documents and notes from scholarly and administrative meetings constitute a rich source of empirical data on which we draw to illustrate key points.

In addition to this work, and specifically for this chapter, we also draw on repeated conversations and reflections among the authors, using elements of autoethnography (Holman Jones, 2007) and of heuristic inquiry (Djuraskovic & Arthur, 2010). In this chapter, we are concerned less with our individual positionality in terms of dimensions such as race or class or with our own interpretative process, but more with what we take to be the normative ambitions of our academic fields and our own agency in realizing these from within the institutional structures that project-based research funding and implementation provide. These concerns were recurrent themes in our discussions, which followed Moustakas' stages of heuristic research (1990, cf. Wall, 2006).

Our methodological approach has limitations both in terms of partiality of perspective and in terms of potential selectiveness and bias in reporting and analysis of events. We seek to limit these weaknesses by corroborating our own experiences with other empirical data. Accordingly, four formal qualitative interviews and informal exchanges were undertaken with initiators and grant writers involved in securing funding. We also re-examined the qualitative interviews with the coordinators of 14 labs within FIT4FOOD2030 conducted as part of its reflexive monitoring in practice approach (cf. Svare et al., 2023).

4.4 Empirical Findings and Discussion

Our purpose here is to explore the consequences of 'project-basedness'—by which we mean the organization into discrete projects with fixed timeframes, resources and plans, goals and outputs for transdisciplinary transformation efforts. Specifically, we focus on the challenges for 'planning' and 'doing' transdisciplinarity within project-based work, both at the proposal stage and later, and the ways in which this 'project-basedness' enables and constrains the arena and processes of transdisciplinary working.

As stated above, the FIT4FOOD2030 project was funded by the EU as a so-called CSA-project, which in Horizon 2020 is understood as a project-type emphasizing dissemination, awareness-raising, network-building, learning, and policy dialogue across member states, and which expressly does *not* include strictly research-focused knowledge goals. FIT4FOOD2030's consortium partners included national research-funding agencies, food industry associations, and one municipality, plus a large number of additional non-research organizations recruited later to host and coordinate two dozen City and Policy Labs. These labs would contribute to generating relevant bottom-up content as well as to societal receptiveness to European food system transformation. The FIT4FOOD2030 project description presented a ToC where the establishment and operation of such Labs (together with an EU think tank), evolving into a self-sustained platform and coupled with systemic approaches to food and R&I, would increase the impacts of European and national R&I investments. This approach addressed the Horizon 2020 call for projects to support the policy framework FOOD 2030 and its aim of 'underpin[ning] the transformation of food systems in Europe so as to make them "future proof"'; projects were expected to be based on 'multi-actor engagement and awareness-raising in support of the initiative and its action plan' (European Commission, 2017).

The SYNAGRI project, by contrast, was originally submitted as a proposal for a Collaborative and Knowledge-building Project, where the emphasis is on usable knowledge and project outputs with relevance to non-research actors, who are obliged to contribute actively to projects via self-funded work. That submission was desk-rejected for insufficient involvement of non-research partners and revised into a research project the following year. The research project call was for projects concerned with sustainable food systems, and emphasized the need for food system transformation to address economic, social, health, climate, and environmental concerns. The call stressed that projects should adopt systems thinking. In response, the SYNAGRI project proposal argued that there are potential conflicts between regional and bio-economic food systems (each of which is a topic of policy support), and proposed to combine system mapping and modelling, using participatory methods to develop strategies for promoting integrated and sustainable food systems. While the emphasis on transdisciplinarity and transformation was less explicit in the proposal than in the call text, the project design sought to integrate stakeholders from across the value chain to inform project definitions, framings, strategies, and policy recommendations.

In what follows, we discuss instances where aspects of 'project-basedness' served to constrain or determine courses of action with respect to stakeholder integration and transdisciplinarity. We pay particular attention to instances where the project architecture derived from proposal writing determined courses of action, and on challenges to project coordination, before transitioning to lessons and to a conclusion that offers practical recommendations for funders and researchers in the field.

4.4.1 The Role of Funding Schemes and Structures

The two funding mechanisms defined certain parameters for the form that transdisciplinarity could take in either project. The EU's CSA call encompassed two potentially conflicting imperatives for how non-research actors should contribute to the project, as both a source of special insights and creative impulses for generating solutions and content, and as audiences to be targeted by project outputs intended to reach large and diverse societal groups. On the one hand, including stakeholders might yield unique insights, increase the chances of uptake, and lend necessary legitimacy to investigations. On the other hand, the co-existence of two different imperatives for non-researcher involvement also casts a certain ambiguity over stakeholder interactions during the project, where stakeholders may be approached (or interpreted) as either an audience, collaborators, or both, at different times, and where researchers' and stakeholders' expectations and interpretations of situation framings may not necessarily align. While the FIT4FOOD2030 proposal included participatory activities and monitoring routines that were intended to support these goals separately, there were also instances during the project where stakeholder interactions could be seen as serving both ends at the same time, or where project participants (especially Lab coordinators) found the imperatives to be pulling them in different directions, as when one coordinator reflected on the trade-offs they experienced between focus and diversity in stimulating stakeholder exchanges:

> ...if you want to have people learn from their ideas, so that they understand each other's [perspective] and how it iterates their own life, then I guess usually low diversity is good, as they will be more relatable to each other.

> But if you have high officials and grocery shop keepers, and waiters and researchers all at the same table, then… they might have their interest in bits and pieces to share with each other, but it could be very difficult for them to build upon each other's ideas to generate something… (Lab coordinator)

In SYNAGRI, conversely, the attempt to move from an initial highly transdisciplinary design towards a more multidisciplinary direction, retaining aspects of stakeholder engagement in the second, sparked reflections on the ramifications of the first call's self-financing requirement.

> …many organizations are potentially contractually committed to ongoing projects which shrinks the opportunity/desire for more spontaneous engagements and means that stakeholders are conscious that more 'promising' projects might come up in future which they want to be involved in. The contractual basis means that clearly defined activities at the application stage are more important. (Helliwell reflection notes)

With stakeholders contractually obliged to fulfil certain tasks, there is a desire for well-defined activities at the start so that they know what these obligations will be. This is to allow them to assess the opportunity cost of joining one project and not another. But the requirements for a clear definition of activities involving stakeholders at the proposal stage conflicts with the researchers' strategic desire to maintain flexibility and openness regarding the actual direction of the project. The researcher, who led the initial SYNAGRI application and took on a mentoring role at the coordinating institution in the revised submission, expressed concerns about funders' requirement that stakeholders' commitment to specific activities, timelines, and project designs should be secured at the application stage. At the time of the first application, applicants were required to quantify non-research partners' work in monetary terms and secure letters of intention from each before they could submit a project application, a mechanism intended to ensure that collaboration is substantial and meaningful. In the researchers' view, this requirement might dissuade non-research actors from participating, while also having methodological implications. Project designs that involve a high number of small-time engagements, including interviews, short meetings and one-on-one discussions with key stakeholders, are dissuaded in favour of a smaller number of choreographed all-day events involving the full group.

These are administratively, contractually, and practically easier to quantify and coordinate, both for stakeholders and the project researchers (SYNAGRI co-applicant, interview).

In FIT4FOOD2030, stakeholder engagement was integral to the project particularly via the Labs and Lab coordinators. Learning session notes reveal that the topic of stakeholder engagement remained a significant challenge for Lab facilitators well into the project. As one coordinator described it:

> ...it was difficult mainly because it was too general and they [stakeholders targeted for participation] didn't understand it, what we are doing or aiming, and even network building wasn't good enough for them because they say: 'Okay, we will become a part of the network, but why?'. [...] while working on the [educational] modules, it was much easier because they were able to have something tangible as a result. (Lab coordinator, interview)

In SYNAGRI, co-author RH recalled worrying about exhausting networks of relevant stakeholders or jeopardizing long-standing collaborations as the team applying for grant funding sought to recruit collaborators to meet the RCN's threshold for self-funded non-research collaborators. After SYNAGRIs initial desk rejection, the emphasis on stakeholders was curtailed as the project proposal was adapted to a new call that no longer stipulated a stakeholder participation threshold. Nevertheless, and somewhat to the coordinators' surprise, those stakeholders who had agreed to take part in the first version remained interested in the second version, since they recognized the urgency and necessity of the project (SYNAGRI co-applicant, interview). Paradoxically then, the project concept and societal interests appeared to push towards transdisciplinarity, even though the funding structure discouraged such a project architecture by emphasizing conventional disciplinary and interdisciplinary research plans over mechanisms for wider stakeholder involvement and inclusion.

4.4.2 Are We Trapped in Our Project Architectures?

Whereas projects in general find themselves confronted with emerging needs that could not be planned for in the 'planning phase', this is even more the case with transdisciplinary projects aimed at catalysing societal transformation. These projects in particular deal with large diversities of stakeholders who might see different needs emerging during the project in response to changing dynamics in local contexts. Thus, if they aim to create meaningful impact, projects need to find ways of planning for the unplanned in order to seize opportunities to create impact, or to mitigate challenges that have emerged, during project implementation, while also meeting funders' expectations for credible and accountable planning. The question is when and under which circumstances such expectations and their corresponding institutional and bureaucratic approaches and logics might conflict with the indeterminacies and unplannabilities of transdisciplinarity.

In FIT4FOOD2030, coordinators and project partners felt that the implementation of the project required collective reflection on its ambitions and discourses, requiring more frequent consortium-wide meetings than originally planned. In addition, to better share lessons and insights, the project developed 'impact narratives' and seized opportunities to disseminate Lab work in novel ways (such as in EC, 2021). Many steps were taken throughout and during certain parts of the project that could be instructive for other transdisciplinary projects. These include efforts to organize consortium-wide meetings around the ToC rather than around discrete work packages, task-, or milestone-specific updates, and establishing a 'taskforce for impact' with representation from partner institutions and work packages to seize on impact opportunities towards the end of the project. While sensible, the unplanned elements of these various efforts also needed the investment of resources and were at times tediously difficult to implement or to enrol the whole consortium in, especially when they conflicted with the carefully planned allocation of working hours across very specific tasks (a general feature of EU projects, which varies by funding body). When changes were instituted, they were made possible because partners found it sufficiently important to spend time on and incorporate the new activities.

Furthermore, in accordance with recommendations for reflexive monitoring, the FIT4FOOD2030 project included accompanying research in the form of a Dynamic Learning Agenda (DLA, van Mierlo et al., 2010), and a training programme addressing certain predefined topics while remaining adaptive to incorporate training according on changing needs. Thus, the project was able to create a structure for supporting a degree of 'planning while doing' during the project lifetime. As DLA records show, this space was useful partly for containing a range of different approaches adopted by the various Labs, and for enabling coordinators to learn about contrasting ways of addressing their shared overarching objectives (Svare et al., 2023). While affording coordinators flexibility and autonomy, there were also limits to the project's ability to support diverging approaches:

> I remember sensing that certain labs went above and beyond in terms of the scope and ambition of their activities, and that they did so because of what they understood to be the overall 'spirit' of the project, namely to stimulate new and improved interactions between the R&I and the food system. But at the same time the project design did not expect the labs to do so, or to do it quite so ambitiously, and the consortium was not able to support or stimulate the more pioneering labs quite as much as we would have liked, because we were more focused on helping the labs who were struggling. The pioneering labs might have experienced it as the project holding them back somewhat. So I had the feeling we did not do as much as we could have, there. (Gjefsen reflection notes)

This begs reflection from researchers and evaluators on when and where activities need to be 'planned' and where there might be space for more serendipitous engagements that can be signalled in a project design/proposal, and about the need to maintain flexibility in distributing (or re-distributing) resources in the face of different strands of stakeholder-driven work proving more or less ambitious and/or impactful over the course of a project.

Designing for flexibility is especially important as projects that aim to both serve emerging needs of the stakeholders involved, as well as being relevant in the policy context, need to deal with an ever-changing world, in which unforeseen opportunities and threats emerge along the way. During FIT4FOOD2030, the European Commission slowly changed its policy focus from four R&I priority areas towards 10 R&I pathways, in turn requiring the project to adapt its activities (for instance by designing new multi-stakeholder workshops based on these pathways) to retain relevance for the EU policy processes that were unfolding. While FIT4FOOD2030 included stakeholder integration via City and Policy Labs from the beginning of the project, additional stakeholders were also enrolled about halfway into the project period in the form of new labs. This allowed the project to build and improve on initial lab experiences, drawing lessons and exploring upscaling of promising efforts, while partly also addressing changes in the policy landscape. From the perspective of fundability, the inclusion of initial labs at the application stage also served to demonstrate stakeholder commitment from the very beginning.

4.4.3 What Challenges Arise in Project Coordination?

One particular dynamic we observed is the challenge in (collectively) deciding which priorities matter most, especially when trade-offs became evident between planned ambitions (such as key performance indicators in the project proposal), (new) needs of funders during the project, and emerging needs of stakeholders. Expectation management can be challenging both internally, among partners, and externally, among funders and audiences. In FIT4FOOD2030, a considerable amount of time and energy was spent on drafting the project's mid-term review. From a project management perspective, it was desirable to receive positive reviews from reviewers and funders, and while the mid-term reporting process helped sharpen internal alignment and project coherence, it also placed considerable pressure on partners and Labs to provide *both new and existing* qualitative and quantitative indicators, fulfilling initial application promises, *and* demonstrating the project's seizing of unplanned opportunities to achieve greater impact. The pressures were further exacerbated by the European Commission's ad hoc requests for concrete and generalizable lessons and findings from the 'ground' (the Labs and stakeholders). Another related challenge involved diverging views within the consortium on whether the primacy regarding impact accountability should be put on emphasizing KPI attainment or less tangible but substantive qualitative impact indicators.

Funded as a Researcher Project (a project category where the Research Council of Norway places the main emphasis on research content and outputs, and which does not include specific requirements for stakeholder participation), the prioritization of impact in SYNAGRI is tilted towards quantifiable scientific outputs, the number of which is stipulated in the proposal. The initial challenge facing SYNAGRI was in establishing the necessary groundwork and alignment required for realizing these outputs in the context of its multidisciplinary and integrative aspirations, while stakeholder impact accountability is much more vaguely defined. The absence of planned engagement activities in the project proposal in favour of looser aspirations has raised recurrent questions about when and how to engage most constructively with stakeholders, and about what it even means to demonstrate qualitative, project-linked impacts in the context of saturated food system research with numerous projects simultaneously vying for and engaging with regional and national stakeholders' attention.

As such, diverging interpretations on which indicators 'matter most' remain an intrinsically political, yet important and ongoing, effort during project implementation. To avoid the trap of focusing only on numerical outputs, and neglecting capacities for change that are built through project activities, projects might benefit from mechanisms to foster a focus on substantive ambitions, such as activities to align work package leaders about substantive goals or prioritization.

As experienced by both projects, there was a constant churn of people. People take maternity and sick leave, are promoted into new roles, or change jobs completely. The obvious problem is that adjusting to this churn is time-consuming and disruptive. Finding the right expertise to replace losses can be challenging or simply impossible. Research is an embodied process. Its aims, objectives, and project-specific ways of working are collectively developed and negotiated over time and 'onboarding' someone into an ongoing and emergent process of research is challenging. 'Handovers' of ongoing work, if they can happen at all, are a poor attempt to bridge the gap and embed someone in a network of already established relationships and collaborative and concurrent research activities. In practice, dealing with turnover might mean redefining the research questions and focus of a particular work package to meet different expertise and interests of new personnel—a need that arose early on in SYNAGRI. In FIT4FOOD2030, coordinators were struck by how long it took to integrate new staff members into the project, attributing this to its unusually high degree of complexity as well as to the project's emergent design, which rendered much of the wording in the initial proposal insufficiently specific about subsequent project choices to fully serve as a reference guide for incoming staff (FIT4FOOD2030 co-applicant A, interview).

Related questions also concerned project coordination and administration more broadly. One of the people who contributed to drafting the FIT4FOOD2030 proposal and subsequently to the coordination of the project stated that FIT4FOOD2030 frequently needed to deviate from the project description to adapt to changing conditions, and that there was greater need for facilitating consortium discussions and clarifications than usually seen in EU projects. These were consequences, in their view, of the project's conceptual complexity and process focus, as well as of the Commission's framing of its own call. While project management in Horizon 2020 generally seeks to avoid change because they require formal change requests to the European Commission and increase the administrative burden on the project, in a project like FIT4FOOD2030, they said: 'You can't avoid it [change]. You just have to live with constant changes, because that is what the Commission asked for. [...] They said we want you to be adaptive' (FIT4FOOD2030 co-applicant A, interview).

Co-author KK recalls the experience of clarifying project proposal intentions at a project meeting:

> We were trying to figure out what was 'expected from us' with regard to identifying drivers and barriers to potential R&I breakthroughs. Which meant we were trying to figure out what the 'DoA' [*Description of the Action*, the project proposal] had in mind [...]. Of course, it was not entirely clear because it had been written ages ago, and the situation had changed by then. I remember that one partner referred to the DoA as the 'bible', and because I came from the coordinating institution, I also felt that all eyes were directed at me whenever the 'bible' was not immediately clear on what to do. In short, I felt like a priest who was to give meaning to what was written. (Kok, reflection notes)

While it is hardly unusual for the coordinating institution to be called on to clarify intentions from project descriptions, given the strict length requirements and need for brevity and thus also a degree of abstractness in all project proposals, for KK the experience was rooted partly in this project's ambitions for system transformation, which necessitated continuous re-interpretation and clarifications within the consortium.

4.4.4 What Can We Learn from This?

What lessons can we draw from the above experiences? As stated in the methodology, our account is based on project immersion and reflections, where events and interactions appeared to either live up to or fall short of the ambitious and societally relevant language we had considered necessary in order to attain funding. We can formulate observations about recurring features and tensions and provide advice on how to anticipate and address these (as we do below) but must also acknowledge that more systematic research may result in other verdicts regarding the ability of current funding regimes to stimulate transdisciplinary efforts that match the normative ambitions highlighted in the literature. There is a need for more research, but also more acceptance and recognition among researchers and funding bodies alike of the risks of sensationalism and overpromising in highly competitive funding regimes.

Our modest exploration also calls into question the capacity of incumbent design and evaluation practices (e.g. the use of key performance indicators, planning the whole project in advance, and allocating resources very specifically, the absence of mechanisms to easily adapt project plans during implementation) in the context of transdisciplinary projects aimed at contributing to sustainable transformation (cf. Kok et al., 2023; Lawrence et al., 2022). In the reality of 'doing' transdisciplinarity with its reflexive re-evaluation of project activities; additional (un)planned activities based on Theory of Change thinking; adaptation of project ambitions based on emerging needs of stakeholders, funders and policy contexts, the 'planning' never really seems to end. It does seem, however, that these types of projects are not a matter of 'planning then doing', but rather a matter of 'planning by doing'. The question of what one should or could do at the proposal stage, then, is not just one for transdisciplinary researchers to grapple with, but one that funders and evaluators also need to take seriously.

The shift in relationships between science and society suggested by the advent of transdisciplinary research has brought with it heady claims of a new transformative potential of research to address societal challenges. But some of the limits to its applicability and capacity to enact meaningful change might be observed precisely in the project-basedness current knowledge-economic regimes impose. As transdisciplinarity has become another funding requirement, it becomes something else to be accommodated in the research process. Necessity supersedes applicability and suitability. This raises questions about the ability of projects to resist transdisciplinary ways of working where they might not be suitable. Key stakeholders might be fundamentally committed to resisting any transformative research framing that they see as threatening to their core interests. Alternatively, stakeholders are not always already mobilized or willing to acknowledge an issue of concern.

4.5 Conclusion

It has been over 20 years since scholars began talking about the potential of transdisciplinary working to address pressing and escalating societal challenges. But what is the capacity of all this transdisciplinary research and what does it add up to? As the world seemingly lurches ever more rapidly towards ecological collapse, the window for action shrinking, and political institutions actively abetting, or unwilling to control the excesses of capitalist extraction and plunder, researchers working in transdisciplinarity for transformation may increasingly expect to be asked for evidence of this hopeful, transformative, transdisciplinary science leading change *in* the world.

These questions are important to researchers and funding agencies alike. Therefore, we seek to offer advice as a modest first step towards a more authentic and honest engagement with 'transdisciplinarity for transformation' within the structures afforded by project-based funding, and how to support this type of research. The advice here is limited to pragmatic adjustments to project design, engagement with stakeholders, and guidance to project funders and evaluators, and we do not seek to grapple with the more foundational conceptual underpinnings of the research fields involved, which presents a much larger task. Far greater challenges will remain in employing these lofty terms with integrity. Funders' implicit expectation that something approximating 'transformation' can really occur through short-term projects—or that the ethos of 'transdisciplinarity' can be meaningfully realized in project after project—by and with researchers—should itself be critically questioned; even more so as tendencies within peer-reviewed research and competitive features of the academic job market also encourage researchers to envelop their various workshopping methodologies in such terms.

Based on the above experiences we suggest that the following features might be expected to arise in projects employing transdisciplinarity in pursuit of transformation, and that project coordinators and others would do well to anticipate and incorporate these issues in their work:

- **Project changes are the rule rather than the exception.** Higher project management burdens and increased needs for coordination should be expected—this may result in seemingly less efficient or competitive proposals if allocated to project management work streams but might alternatively be integrated across all project activities.
- **The world also changes.** While initial funding might depend on claims to social and political relevance, subsequent change in these domains may also warrant project changes; this is an argument for

carving out spaces for the unstructured and unexpected in proposal writing. The establishment of a taskforce for impact and the project's enrolment of a second round of stakeholders around the City and Policy Lab models to build on and make improvements to lessons in the initial stages of FIT4FOOD2030 offer instructive examples of this.
- **Stakeholder engagement is fraught with Catch 22s.** Goals of inclusivity, diversity, and establishing cohesion and dialogue, flexibility that allows for bottom-up deliberation and prioritization, as well as formal needs for project planning and commitment to quantified work effort, will often conflict. While there is no simple solution to these trade-offs, increased openness about them can help manage expectations and ultimately improve project outcomes.
- **Critical and diverging reflections on project process can be expected.** With stakeholder roles and process design being open to any number of interpretations at a given time, discussions and disagreements are to be expected; carving out spaces for dialogue not merely on topical questions, but on the very project structure and work process, can help accommodate such discussions. Efforts to focus consortium discussions in FIT4FOOD2030 on ToC ambitions rather than isolated tasks and work packages offer one way to address this.

Accordingly, we pose the following recommendations to research-funding organizations:

- **Beware of the consequences of forcing transdisciplinarity.** Transdisciplinary in its broadest sense implies a radical co-creative process and possible tensions with regimes and processes for administrative oversight. Funding bodies should consider when this level of co-creation is preferable to other forms of actor interactivity and be open to learning about and exploring their own roles in the governance process, for instance by instilling processes of self-reflection (Regeer et al., 2016).
- **Make project changes as easy as possible to allow room for the unplanned.** Support researchers in exercising judgement and seizing opportunities by reducing the administrative burdens of project changes.

- **Allow researchers to influence your expectations.** Be curious about the 'unplanned' project lessons and remain open to adjusting own expectations and requirements according to researchers' experiences.

Finally, we suggest the following recommendations to researchers and graduate students in the field:

- **Be prepared for turnover.** Often overlooked in grant writing and project planning, turnover can be a major factor in project management, especially in complex and relationship-based transdisciplinary transformation projects. The need for time and attention to support 'onboarding' of new partners and explain decisions and interpretations not clearly visible in project proposals and formal planning documents should not be underestimated.
- **Listen to your intuition and consider which outcomes matter most.** Project applications may require you to formulate a wide range of outputs, impacts, and performance indicators—being clear about what outcomes are the most important, to yourself and collaborators, and revisiting and repeating this throughout the project for continuous realignment across the consortium, can help maintain motivation and project substance.

References

Arts, B., & Tatenhove, J. V. (2004). Policy and power: A conceptual framework between the 'old' and 'new' policy idioms. *Policy Sciences, 37*(3), 339–356. https://doi-org.vu-nl.idm.oclc.org/10.1007/s11077-005-0156-9

Bachrach, P., & Baratz, M. S. (1962). Two faces of power. *The American Political Science Review, 56*(4), 947–952.

Chubb, J., & Watermeyer, R. (2017). Artifice or integrity in the marketization of research impact? Investigating the moral economy of (pathways to) impact statements within research funding proposals in the UK and Australia. *Studies in Higher Education, 42*(12), 2360–2372. https://doi-org.vu-nl.idm.oclc.org/10.1080/03075079.2016.1144182

Dahl, R. A. (1962). *Who governs? Democracy and power in an American city*. Yale University Press.

Defila, R., & Di Giulio, A. (2015). Integrating knowledge: Challenges raised by the 'inventory of synthesis.' *Futures, 65*, 123–135. https://doi-org.vu-nl.idm.oclc.org/10.1016/j.futures.2014.10.013

Deutsch, L., Belcher, B., Claus, R., & Hoffmann, S. (2021). Leading inter- and transdisciplinary research: Lessons from applying theories of change to a strategic research program. *Environmental Science & Policy, 120*, 29–41. https://doi-org.vu-nl.idm.oclc.org/10.1016/j.envsci.2021.02.009

Djuraskovic, I., & Arthur, N. (2010). Heuristic inquiry: A personal journey of acculturation and identity reconstruction. *Qualitative Report, 15*(6), 1569–1593.

Earl, S., Carden, F., & Smutylo, T. (2001). *Outcome mapping: Building learning and reflection into development programs*. IDRC.

European Commission. (2017). *Support to the development and implementation of FOOD 2030—A European research and innovation policy framework for food and nutrition security TOPIC ID: SFS-18-2017*. https://ec.europa.eu/info/funding-tenders/opportunities/portal/screen/opportunities/topic-details/sfs-18-2017

Fazey, I., Schäpke, N., Caniglia, G., Patterson, J., Hultman, J., Van Mierlo, B., Wyborn, C., et al. (2018). Ten essentials for action-oriented and second order energy transitions, transformations and climate change research. *Energy Research & Social Science, 40*, 54–70.https://doi-org.vu-nl.idm.oclc.org/10.1016/j.erss.2017.11.026

FIT4FOOD2030. (2017). *Proposal number: 774088 Proposal acronym: FIT4FOOD2030. Fostering Integration and Transformation for FOOD 2030*. Project description [unpublished].

Fritz, L., & Binder, C. R. (2020). Whose knowledge, whose values? An empirical analysis of power in transdisciplinary sustainability research. *European Journal of Futures Research, 8*(1), 1–21. https://doi-org.vu-nl.idm.oclc.org/10.1186/s40309-020-0161-4

Holman Jones, S. (2007). *Autoethnography*. Blackwell.

Ika, L. A., & Munro, L. T. (2022). Tackling grand challenges with projects: Five insights and a research agenda for project management theory and practice. *International Journal of Project Management, 40*(6), 601–607. https://doi-org.vu-nl.idm.oclc.org/10.1016/j.ijproman.2022.05.008

Klerkx, L., Aarts, N., & Leeuwis, C. (2010). Adaptive management in agricultural innovation systems: The interactions between innovation networks and their environment. *Agricultural Systems, 103*(6), 390–400. https://doi-org.vu-nl.idm.oclc.org/10.1016/j.agsy.2010.03.012

Kok, K. P. W., Den Boer, A. C. L., Cesuroglu, T., Van Der Meij, M. G., de Wildt-Liesveld, R., Regeer, B. J., & Broerse, J. E. W. (2019). Transforming research and innovation for sustainable food systems—A coupled-systems perspective.

Sustainability, 11(24), 7176. https://doi-org.vu-nl.idm.oclc.org/10.3390/su11247176

Kok, K. P. W., Gjefsen, M. D., Regeer, B. J., & Broerse, J. E. W. (2021). Unraveling the politics of 'doing inclusion' in transdisciplinarity for sustainable transformation. *Sustainability Science, 16*(6), 1811–1826. https://doi-org.vu-nl.idm.oclc.org/10.1007/s11625-021-01033-7

Kok, K. P. W., van der Meij, M. G., Wagner, P., Cesuroglu, T., Broerse, J. E., & Regeer, B. J. (2023). Exploring the practice of Labs for sustainable transformation: The challenge of 'creating impact'. *Journal of Cleaner Production, 388*, 135994. https://doi-org.vu-nl.idm.oclc.org/10.1016/j.jclepro.2023.135994

Lang, D. J., Wiek, A., Bergmann, M., Stauffacher, M., Martens, P., Moll, P., Thomas, C. J., et al. (2012). Transdisciplinary research in sustainability science: Practice, principles, and challenges. *Sustainability Science, 7*(1), 25–43. https://doi-org.vu-nl.idm.oclc.org/10.1007/s11625-011-0149-x

Latour, B., & Woolgar, S. (1979). *Laboratory life*. Princeton University Press.

Lawrence, M. G., Williams, S., Nanz, P., & Renn, O. (2022). Characteristics, potentials, and challenges of transdisciplinary research. *One Earth, 5*(1), 44–61. https://doi-org.vu-nl.idm.oclc.org/10.1016/j.oneear.2021.12.010

Lukes, S. (2004). *Power: A radical view*. Macmillan International Higher Education.

Lux, A., Schäfer, M., Bergmann, M., Jahn, T., Marg, O., Nagy, E., Theiler, L., et al. (2019). Societal effects of transdisciplinary sustainability research—How can they be strengthened during the research process?. *Environmental Science & Policy, 101*, 183–191.https://doi-org.vu-nl.idm.oclc.org/10.1016/j.envsci.2019.08.012

Marsden, T., & Farioli, F. (2015). Natural powers: From the bio-economy to the eco-economy and sustainable place-making. *Sustainability Science, 10*, 331–344. https://doi-org.vu-nl.idm.oclc.org/10.1007/s11625-014-0287-z

Moustakas, C. (1990). *Heuristic research: Design, methodology, and applications*. Sage.

Musch, A. K., & von Streit, A. (2020). (Un) intended effects of participation in sustainability science: A criteria-guided comparative case study. *Environmental Science & Policy, 104*, 55–66. https://doi-org.vu-nl.idm.oclc.org/10.1016/j.envsci.2019.10.004

Nowotny, H., Scott, P., & Gibbons, M. (2001). *Re-thinking science: Knowledge and the public in an age of uncertainty* (p. 12). Polity.

Owen, R., Macnaghten, P., & Stilgoe, J. (2012). Responsible research and innovation: From science in society to science for society, with society. *Science and Public Policy, 39*(6), 751–760. https://doi-org.vu-nl.idm.oclc.org/10.1093/scipol/scs093

Pohl, C., & Hadorn, G. H. (2007). *Principles for designing transdisciplinary research*. oekom.

Reed, M. S., Ferré, M., Martin-Ortega, J., Blanche, R., Lawford-Rolfe, R., Dallimer, M., & Holden, J. (2021). Evaluating impact from research: A methodological framework. *Research Policy, 50*(4), 104147. https://doi-org.vu-nl.idm.oclc.org/10.1016/j.respol.2020.104147

Regeer, B. J., de Wildt-Liesveld, R., van Mierlo, B., & Bunders, J. F. G. (2016). Exploring ways to reconcile accountability and learning in the evaluation of niche experiments. *Evaluation, 22*(1), 6–28. https://doi-org.vu-nl.idm.oclc.org/10.1177/1356389015623659

Schäfer, M., Bergmann, M., & Theiler, L. (2021). Systematizing societal effects of transdisciplinary research. *Research Evaluation, 30*(4), 484–499. https://doi-org.vu-nl.idm.oclc.org/10.1093/reseval/rvab019

Schmidt, L., Falk, T., Siegmund-Schultze, M., & Spangenberg, J. H. (2020). The objectives of stakeholder involvement in transdisciplinary research. A conceptual framework for a reflective and reflexive practise. *Ecological Economics, 176*, 106751. https://doi-org.vu-nl.idm.oclc.org/10.1016/j.ecolecon.2020.106751

Schneider, F., Giger, M., Harari, N., Moser, S., Oberlack, C., Providoli, I., Zimmermann, A., et al. (2019). Transdisciplinary co-production of knowledge and sustainability transformations: Three generic mechanisms of impact generation. *Environmental Science & Policy, 102*, 26–35.https://doi-org.vu-nl.idm.oclc.org/10.1016/j.envsci.2019.08.017

Springmann, M., Clark, M., Mason-D'Croz, D., Wiebe, K., Bodirsky, B. L., Lassaletta, L., Willett, W., et al. (2018). Options for keeping the food system within environmental limits. *Nature, 562*(7728), 519–525.https://doi-org.vu-nl.idm.oclc.org/10.1038/s41586-018-0594-0

Stahl, B. C., Akintoye, S., Bitsch, L., Bringedal, B., Eke, D., Farisco, M., Ulnicane, I., et al. (2021). From responsible research and innovation to responsibility by design. *Journal of Responsible Innovation, 8*(2), 175–198. https://doi-org.vu-nl.idm.oclc.org/10.1080/23299460.2021.1955613

Svare, H., Gjefsen , M.D., den Boer, A.C.L., & Kok, K.P.W. (2023). Learning systems and learning paths in sustainability transitions. *Ecology and Society*, 28(1), article 22. https://doi-org.vu-nl.idm.oclc.org/10.5751/ES-13868-280122

Thompson Klein, J., Grossenbacher-Mansuy, W., Haberli, R., Bill, A., Scholz, R. W., & Welti, M. (Ed.). (2001). *Transdisciplinarity: Joint problem solving among science, technology, and society: An effective way for managing complexity*. Springer Science & Business Media.

Torrens, J., & von Wirth, T. (2021). Experimentation or projectification of urban change? A critical appraisal and three steps forward. *Urban Transformations, 3*(1), 1–17. https://doi-org.vu-nl.idm.oclc.org/10.1186/s42854-021-00025-1

Turnhout, E., Metze, T., Wyborn, C., Klenk, N., & Louder, E. (2020). The politics of co-production: Participation, power, and transformation. *Current Opinion in Environmental Sustainability, 42*, 15–21. https://doi-org.vu-nl.idm.oclc.org/10.1016/j.cosust.2019.11.009

Van Es, M., Guijt, I., & Vogel, I. (2015). *Theory of change thinking in practice: A stepwise approach*. Hivos.

Van Mierlo, B. C., Regeer, B., van Amstel, M., Arkesteijn, M. C. M., Beekman, V., Bunders, J. F. G., Leeuwis, C., et al. (2010). *Reflexive monitoring in action. A guide for monitoring system innovation projects*. Communication and Innovation Studies, WUR; Athena Institute, VU.

Wall, S. (2006). An autoethnography on learning about autoethnography. *International Journal of Qualitative Methods, 5*(2), 146–160. https://doi-org.vu-nl.idm.oclc.org/10.1177/160940690600500205

CHAPTER 5

Transdisciplinary Development of Quality Criteria for Transdisciplinary Research

Rico Defila and Antonietta Di Giulio

5.1 Introduction

The question of how to evaluate transdisciplinary research has been debated for some time. In this debate we can distinguish two dimensions of what evaluation might address, because they are informed by different, although complementary, goals that are both related to success (see also Lawrence et al., 2022). One dimension addresses impact: does the transdisciplinary project lead to the expected practical solutions and/or societal changes? The other dimension addresses quality: does the transdisciplinary project meet the specific quality requirements of transdisciplinary research? In this chapter, we[1] focus on the latter.

[1] In the entire chapter, 'we', 'our', 'us', and the like always denote the authors of the chapter.

R. Defila · A. Di Giulio (✉)
Department of Social Sciences, Social Transitions Research Group (STR), University of Basel, Basel, Switzerland
e-mail: antonietta.digiulio@unibas.ch

R. Defila
e-mail: rico.defila@unibas.ch

B. J. Regeer et al. (eds.), *Transdisciplinarity for Transformation*,
https://doi.org/10.1007/978-3-031-60974-9_5

In order to narrow down the topic more specifically, we distinguish between those who initiate an evaluation and those who conduct it (see Table 5.1). This chapter focuses on external evaluations that are conducted by third parties, which is an important issue on which to focus because it refers to funding structures and review processes that are decisive, for instance, with regard both to whether transdisciplinary research obtains funding and to which kind of transdisciplinary projects are funded. Current funding structures and review processes are still considered as among the major barriers to the scaling up transdisciplinary research (e.g. Koier & Horlings, 2015; Schneider et al., 2023). It therefore makes sense to learn from experiences and provide new avenues and guidelines for funding agencies and review panels dealing with transdisciplinary research evaluations.

In recent decades, the scholarly discourse on the evaluation of transdisciplinary research has yielded a considerable number of highly differentiated sets of criteria (some examples are Bergmann et al., 2005; Defila & Di Giulio, 1999; Jahn & Keil, 2015; a review on the subject is provided, for instance by Belcher et al. [2016], Boix Mansilla et al. [2006], Klein[2008], Pohl et al. [2011], Schuck-Zöller et al. [2017], and Steelman et al. [2021]). Most of these lists of criteria have been developed by scholars that investigate and are involved in inter- and transdisciplinary research. Accordingly, they are informed by the concerns, terminology, and theoretical approaches of this community. The concern

Table 5.1 Internal vs. external evaluation, self-evaluation vs. third-party evaluation

An evaluation can be …		… initiated from within a project/activity or from outside	
		Internal evaluation	*External evaluation*
… conducted by the participants themselves or by third parties	**Self-evaluation**	E.g. principal investigators	E.g. funding agency asking for self-report
	Third-party evaluation	E.g. advisory board selected by project	E.g. review panel selected by funder

Based on Defila et al. (2006)

that the special nature of transdisciplinary research is not appropriately captured by evaluations runs like a thread through this discourse as does the tacit assumption that providing elaborate lists of criteria is a remedy for transdisciplinary research not being sufficiently valued.

There is, as a result, no shortage of proposals on how evaluation processes for transdisciplinary projects should be conducted (see also, e.g. the results of the comprehensive review by Laursen et al. [2022]). But quite often they are not evidence-based. Too little is known about the evaluation processes actually taking place and about the dynamics that arise in these processes, one recent exception being an experience-based report by Gerhardus et al. (2016). Improving these processes depends on having more knowledge about the challenges faced by those involved in evaluating transdisciplinary research and about what they perceive to be supportive. Hence, what is missing in the discourse is a scholarly engagement with the actual evaluation practice of 'well-meaning and well-informed actors', that is, actors who value transdisciplinary research, are experienced in dealing with it, and are aware of the specific nature and requirements of such research. This could uncover promising paths both for review processes and for future research.

In this chapter, we concentrate on evaluation processes that are characterized by 'well-meaning and well-informed actors'.[2] In this first section, we explain our approach to transdisciplinary research, defining our understanding of transdisciplinary research and of its specific quality. We conclude this section by identifying the challenges of evaluating the quality of transdisciplinary research. Based on this, in the subsequent sections we present experiences and results from three case studies in which we have accompanied processes of evaluating transdisciplinary research before finally drawing conclusions for funders and for researchers.

We proceed from the following *definitions*. While in a *multidisciplinary approach* experts of different fields explore the same topic but do not relate their perspectives, in an *interdisciplinary approach*, scholars of at least two academic disciplines collaborate with the aim of producing integrated results, of producing a synthesis (e.g. Andersen & Wagenknecht, 2013; Holbrook, 2013; Zweekhorst et al., 2001; for the

[2] By this, we do not deny the problem of reviewers often not having the expertise to judge the specific quality of transdisciplinary research or that such research does not always gain academic acceptance.

scholarly discourse, e.g. Defila & Di Giulio, 1998; Hvidtfeldt, 2018; Klein, 2010; Vermeulen & Witjes, 2020). In *transdisciplinary research*, in addition to scholars from different academic disciplines, actors from outside academia participate in the research process (e.g. Bogner, 2012; Gibbons et al., 1994; Mielke et al., 2016; Pohl & Hirsch Hadorn, 2007; Regeer & Bunders, 2003). These actors contribute substantially to the research—they are not just a source of information, data, and/or feedback but are involved in co-designing the research and in co-producing the integrated knowledge. Such an actor-oriented understanding of transdisciplinary research is what Mobjörk (2010) refers to as 'participatory transdisciplinarity' (in contrast to 'consulting transdisciplinarity').

According to these definitions, the specific *quality of inter- and transdisciplinary research* can be described by three terms which all denote processes that must take place (based on Di Giulio & Defila, 2017; see also Bergmann & Schramm, 2008; Defila & Di Giulio, 1999; Huutoniemi, 2010; Jahn & Keil, 2015; Klein, 1990; Pohl et al., 2011; Röbbecke et al., 2004):

- *Consensus:* Those participating in the research have to arrive at a shared problem framing. They need to develop joint research goals they all equally want to reach and shared research questions they all equally want to answer. They have to reach a joint understanding about the theoretical and methodical approach for dealing with these questions, and to develop a common language. Consensus does not mean that individuals should abandon their different perspectives and replace them with a 'group perspective' or that their different perspectives should dissolve into just one perspective. Rather, they have to develop a shared point of view—which is not an identical point of view but one with which they can all identify to a certain extent and are prepared to proceed from and to relate their findings. A shared problem framing and the like have to be developed by applying methods of (cognitive) consensus-building.
- *Integration:* The research must lead to common outputs (results and products). In other words, those participating in the research have to develop common answers to their shared research questions by integrating, from the very start, the findings from the different disciplines and/or non-academic fields that are involved. To this end, findings and approaches have to be selected in terms of their contribution to the common answers, they have to be reprocessed,

related, and integrated. The common result is the integrated knowledge produced in this process, the so-called synthesis. The synthesis has to be achieved by applying methods of knowledge-integration.
- *Diffusion:* As a rule, the audience of inter- and transdisciplinary research is neither disciplinary nor purely scholarly, and nor are the users of the products (products can be publications, but tool kits, recommendations, technologies, materials, etc., are products as well). The research outputs (results and products) must feed into different academic and non-academic discourses and fields of practice. This means that the results must be translated in order to fit with the 'logic' of the targeted discourses and to be accessible to the different target audiences and their perspectives. This is not simply a matter of the language used nor is it only about disseminating results and products and promoting reception on the part of the audience.

Defining transdisciplinary research implies defining *who are the actors from outside academia* that are to be involved in the research. This should be done by considering the intended aim of involving them and, related to that, the contribution to the research expected from them. This can best be captured by referring to the concepts of credibility, salience, and legitimacy, which are part of the discourse on scientific policy advice (e.g. Cash et al., 2003; Hastie, 2007). 'Credibility' refers to the scientific legitimacy of the knowledge that is produced. It denotes the scientific adequacy of the evidence and arguments. 'Salience' refers to the practical legitimacy of the knowledge that is provided. It denotes the relevance to the needs of decision-makers. 'Legitimacy' refers to the political/societal legitimacy of the results. It denotes the perception that the knowledge production has been respectful of stakeholders' divergent values and beliefs, unbiased in its conduct, and fair in its treatment of opposing views and interests. Drawing on these concepts, we suggest distinguishing three goals of participation leading to three types of participating actors and three types of contributions:

- *Participation of uncertified experts to increase credibility:* Participation can serve the goal of broadening the knowledge that is considered in framing problems, in investigating problems, and in providing answers and solutions. That is, one goal of participation is to ensure that the relevant knowledge is considered and integrated in the research regardless of whether it is academic or non-academic. In this case, the participating actors are 'experience-based experts' (or 'uncertified experts') with regard to the topic being investigated (while the participating scholars are 'certified experts') (Collins & Evans, 2002). They contribute by providing expertise.

- *Participation of (future) users to increase salience:* Participation can serve the goal of including first-hand experiences about actual needs and usability. So, one goal of participation is to ensure that the outputs of research (knowledge and products) can in fact be used, that they answer practical needs, and that they are linked to users' options for action. In this case, the participating actors are (future) users (including those who have agency and/or practitioners) in the field that is being explored. They contribute by providing practical experience and knowledge about the practice.
- *Participation of stakeholders to increase societal legitimacy:* Participation can serve the goal of strengthening the societal legitimacy of the research and its outputs. In other words, one goal of participation is to ensure that the production of knowledge and its outputs are sensitive to socio-political interests, fair in the treatment of opposing views and interests, and that they consider and respect divergent values and beliefs. In this case, the participating actors are stakeholders and actors representing (affected) groups in civil society. They mirror the relevant socio-political interests in the field and contribute by providing their everyday experiences, feelings, and concerns.

One actor may of course belong to more than one of these groups. For instance, an actor may simultaneously be both an uncertified expert and a (future) user.

Taking the three process requirements (i.e. the quality of transdisciplinary research) and the differentiations with regard to participation (i.e. goals, criteria of involvement, contributions) seriously has the following *implications with regard to the tasks of evaluating*:

- The evaluation needs to assess whether the different perspectives covered by the participants are integrated, i.e. whether processes aimed at consensus and integration take place (and are conducted state of the art) and whether the research is informed by their results.

- The evaluation must assess whether the research has the potential to produce the intended impact by generating appropriate results and products and by conducting activities of dissemination that support the diffusion of the outputs.
- The evaluation has to assess whether the participation is expedient, i.e. whether the 'right' actors are involved and whether they contribute substantially (according to the specific goals) to the research.

The *challenges* that have to be mastered *in evaluating the quality of transdisciplinary research* present themselves as follows:

- Being unable to rely primarily on quantitative and indirect indicators for measuring quality: Whether a transdisciplinary project does or does not meet the specific quality requirements of transdisciplinary research may often not be judged simply by using quantitative criteria and indicators (e.g. Stokols et al., 2003). That is, not using qualitative criteria and indicators in evaluating transdisciplinary research would impair the quality of the evaluation. This is reinforced given that an evaluation of transdisciplinary research that relies only on indirect indicators and does not also include direct indicators that target the processes will be unable to judge the specific quality of such research (e.g. Love et al., 2022; Steelman et al., 2021; Wagner et al., 2011). This also means that an individual reviewer's judgment carries considerable weight and that individual reviewers thus bear a high responsibility.
- Doing justice to diversity and coping with the lack of common ground: Transdisciplinary projects do not have standardized procedures. The methods used depend not only on the disciplinary background of the scholars involved but also on the background (and possibly vulnerability) of the non-academic actors and on the different types of goals that are pursued by involving them. The methods must be appropriate in relation both to the goals and questions of a given project and to the people who are involved. That is, each project is much more unique than projects that take place in a disciplinary context while, at the same time, there is no shared and agreed on body of methods or state of the art approaches that

can be used as a common point of reference in evaluating transdisciplinary projects, making 'evaluation a custom task' (Koier & Horlings, 2015, p. 47; see also, e.g. Laursen et al., 2022, and the different contributions in Stoll-Kleemann & Pohl, 2007).

- Navigating between the need to have a reliable and robust research plan and the inevitability of the non-plannability: For transdisciplinary research processes to be open and informed by the results of the ongoing processes of consensus-building and of knowledge-integration makes it indispensable to have a flexible research plan (e.g. Defila et al., 2016; Verwoerd et al., 2023; see also Dahl Gjefsen et al., this volume). The more a transdisciplinary project's research plan is fixed from the very beginning, the lower its transdisciplinary quality is likely to be. But this in turn very often conflicts with the expectations and requirements of funding bodies that expect carefully and detailed worked-out research plans (e.g. Lawrence et al., 2022; Vermeulen & Witjes, 2020). And it adds to the difficulty of evaluating a project because the less a transdisciplinary project's research plan is fixed, the more demanding its evaluation.

The guiding question in our chapter is how evaluation processes can be improved with regard to supporting actors involved in coping with these challenges and with second-order challenges that might arise from managing these first-order challenges. One rather obvious way of approaching the first-order challenges is to arrange for review panels to take funding decisions as a group or to agree as a group on how to react to mid-term or final reports. This might lead to second-order challenges with regard to the dynamics and interaction in the review panel.

5.2 Experiential and Empirical Background

Our evidence is presented in three case studies. In all three, we supported the process of the external evaluation of transdisciplinary projects as certified experts (but without being involved in the evaluation of proposals/projects). All three processes were characterized by 'well-meaning and well-informed actors' (and in all three, some of the members of the review panel were certified experts of inter-/transdisciplinary research). The question of how to evaluate transdisciplinary research and of the criteria to use, as we have shown above, is not new. The novelty of our approach is that in all three case studies, we applied a transdisciplinary design to answer this question for a specific research program. In the following, we describe the three case studies by summarizing both our role and our methodical approach before going into the details of our experiences and results in the subsequent sections.

Case study 1 (CS-1) is the accompanying research project[3] to the funding program 'Research for sustainable development' (WfNE) in Lower Saxony, managed by the Volkswagen Foundation. WfNE had three rounds of funding (2014, 2015, 2017). The accompanying project was funded by the Ministry of Science and Culture of Lower Saxony. It had three principal investigators—the two of us and Claudia Binder. We were in charge of the research question which was devoted to the appropriate evaluation of transdisciplinary research. Working on this question covered not only investigating the topic, but also contributing to the development of quality criteria that were used in making funding decisions in WfNE. The practitioners with whom we collaborated in our part of the project were the Volkswagen Foundation, the Ministry of Science and Culture, and the interdisciplinary group of scholars responsible for reviewing the research proposals. We observed the discussions of the reviewers (tape recording), we interviewed the reviewers as well as members of the foundation and the ministry (qualitative interviews) in order to learn about their experiences in conducting the evaluation, and we asked the applicants how they experienced the process of submission and evaluation (qualitative interviews; online questionnaire). Subsequently, we discussed the empirical results with the members of the foundation and the ministry involved in the management of the research program and provided a collaboratively revised list of evaluation criteria. This new list was used in the second round of funding, and again we observed the discussions of the reviewers and asked the different actors about their experiences in using this list and about their judgment of the adequacy and applicability of each of the criteria. In CS-1 a transdisciplinary collaboration took place with uncertified experts and (future) users but not with stakeholders (those affected by the evaluation, the applicants).

Case study 2 (CS-2) is another accompanying research project we were in charge of. Over the 2015–2019 period, the Federal State of Baden-Württemberg funded projects running as real-world laboratories (two rounds of funding). During this time, the real-world laboratories

[3] Project 'Civil society and research for sustainable development: demanding and fostering transdisciplinarity' (ZiFoNE, 2014–2019).

program had two accompanying projects. We led one of them[4] (Defila & Di Giulio, 2018, see also Schäpke, Chapter 6, this volume). In this project, we contributed to the development of quality criteria used in the mid-term and in the final evaluation and to the improvement of the corresponding evaluation processes. The practitioners with whom we collaborated in this project were the interdisciplinary group of scholars responsible for reviewing the ongoing research, and the research teams conducting real-world laboratories. In 2016, the reviewers evaluated the mid-term reports of the research teams in the first round of funding. Both the reviewers and those being evaluated criticized the procedure and the result, which initiated a process of reflection and revisioning. We were in charge of designing and facilitating this process. We analyzed the evaluation reports the reviewers had produced as well as the critique the research teams had voiced with a view to the coherence and consistency of the evaluation and to how the reviewers had justified and interpreted the criteria. Based on the results of the analysis, we suggested how the evaluation process could be improved (criteria and procedure) for the second round of funding. Both the criteria and procedure suggested were subjected to a participatory process with the research teams (both rounds of funding) and a feedback process with the review panel. That is, the research teams participated in the development of the quality criteria that then were applied to their own projects. At a later stage, part of this process was repeated in order to develop the criteria for the final evaluation of the projects (both rounds of funding). In CS-2 a transdisciplinary collaboration took place with stakeholders (those affected by the evaluation, the research teams).

Case study 3 (CS-3) is situated in the same funding context as CS-2, the real-world laboratories program funded by the Federal State of Baden-Württemberg. A third round of funding real-world laboratories started in 2021; in 2023, the projects had the possibility of submitting a proposal for a two-year-extension (starting in 2024). We are mandated to provide methodical support and coaching for the teams that are conducting the projects. In addition, we had a time-limited mandate (2022) to support the process of setting in place the mid-term evaluation and the evaluation of the proposals for the two-year-extensions. The practitioners with whom we collaborated in fulfilling this mandate were the Ministry of

[4] Project 'Linking, understanding, continuing real-world laboratories' (2015–2019).

Science, Research and Arts Baden-Württemberg, the professional agency in charge of organizing the evaluation, and the research teams conducting real-world laboratories. In contrast to CS-2, we did not interact with the review panel. We provided a first input into the process by reminding the funder (including the professional evaluation agency in charge of organizing the evaluation) of the process that had taken place in the 2015–2019 period and by providing the materials that had been produced in this process. Based on this, the funders decided how they would like to proceed and what evaluation criteria they would like to apply. We provided feedback on their concepts and helped to design and facilitate an online workshop in which the research teams had the opportunity to discuss and comment on the criteria. Based on the research teams' feedback, the criteria were revised and handed over to the review panel for the final decision. That is, also in CS-3, the research teams participated in developing the quality criteria that then were applied to their own projects. In CS-3 a transdisciplinary collaboration took place with stakeholders (those affected by the evaluation, the research teams).

In the following section, we report on our learnings from these case studies, focusing on three topics: the practical requirements with regard to evaluation criteria, the interdisciplinary nature of the process of evaluating transdisciplinary research, and the benefits of a transdisciplinary approach to developing criteria and procedures for evaluating transdisciplinary research. We draw mainly on the experiences and results of CS-1 in which we collected empirical data but complement this by the experiences in CS-2 and/or CS-3.

5.3 Requirements for Practicable Criteria to Evaluate the Quality of Transdisciplinary Research

In 2014, WfNE was launched (CS-1). The aim of the program was to fund research looking into issues of sustainability without further limitations regarding the topics to address or the scientific fields invited for submission. In the first round of funding, two related but not identical sets of criteria were used (see Table 5.2). One of them was published in the call for projects, the other resulted from specifying these criteria for the review process. Both lists were provided by the funder (ministry and foundation).

Table 5.2 The two sets of criteria used in the first round of funding of WfnE in 2014. The list in the left-hand column was communicated to the applicants, the list in the right-hand column was communicated to the review panel

Criteria call for projects	*Operationalized criteria for the review process*
A comprehensible presentation of the state of research, encompassing both the disciplinary state of the art and relevant transdisciplinary studies	Locating the project work planned in the sustainability discourse Presentation of disciplinary state of research and how the work planned in the project is related to it Presentation of the transdisciplinary state of research and how the work planned in the project is related to it Expected contribution to the disciplinary advance of knowledge
An explicit reflection and explanation of the project's contribution to the goal of an ecologically, socially and economically sustainable and intergenerational just development	Consideration of the social, ecological and economic dimensions Contribution to the 'reproductive capacity' of current living conditions for future generations ('intergenerational just development')
A methodically and administratively regulated collaboration transcending disciplinary borders taking place on the basis of recognized disciplinary research of those involved in the project	Interdisciplinarity of the teams Organization of the collaboration Qualification of the applicants
A presentation of the potential to provide solutions to one or several virulent societal challenges	Topical question having a high societal relevance Presentation of the potential to provide solutions
The consideration of the perspective of affected citizens as well as social groups in the process of weighing up different specific suggestions developed for how to solve societal problems	Consideration of the perspective of actors from civil society (where appropriate by involving these actors)
	Originality of the research question

We asked the applicants (qualitative interviews, lasting about 15 minutes) what they believe are important criteria to evaluate sustainability research and whether the different criteria published in the call (Table 5.2, left-hand column) make sense to them. We asked the reviewers (qualitative interviews, lasting about an hour) whether the different criteria they were asked to use (Table 5.2, right-hand column) were adequate for the purpose of assessing and selecting proposals for funding. And we asked both groups whether they felt any criteria were lacking. The applicants voiced a number of difficulties with the criteria as they had been published in the call, and the reviewers also expressed difficulties as the criteria had been specified for evaluating the project proposals. The analysis of the data was informed by two questions: What criteria do the respondents use in judging the suitability of criteria? What do the respondents suggest in terms of revising the criteria?

The findings from the first question (see Table 5.3) show that the main criteria the applicants and the reviewers used to assess the criteria differ although showing some overlap and that they are informed by the respondents' respective roles.

In the interviews, both groups of respondents made suggestions for how to change the criteria. Based on these suggestions, we reprocessed the list of criteria. This was a collaborative process with the funder that resulted in a new list of criteria that was used in the second round of funding 2015 (see Table 5.4). Major changes covered the following: the new criteria did not predetermine a specific theoretical approach to sustainable development, actual or perceived redundancies were eliminated, and the criteria were presented and operationalized in the format of questions whereby the questions used to specify each criterion were not meant to be applied cumulatively but rather to define the conceptual space to consider in applying the criteria to an individual proposal.

Again, we asked both the applicants (online questionnaire this time) and the reviewers (qualitative interviews as before) about their experiences and how they judge the adequacy of the criteria. After that, we discussed the results and the criteria in a meeting with the funder and the review panel. Applicants, reviewers, and the funder were satisfied with the new set of criteria and judged it to be suitable (the funder took the final decision on the criteria). In the third round of funding in 2017, this list of criteria was published as a part of the call for projects.

The process in CS-2 (evaluation of mid-term reports, starting in 2016) was again initiated by the actors (reviewers, project teams) not being satisfied with a first list of nine criteria and resulted in a collaboratively agreed list of seven. The process was designed as follows: Based on our analyses of the evaluation reports, on the critique by the research teams, and on our experiences in CS-1, we provided a first list of revised criteria that were operationalized in the format of questions. In workshops, the research teams discussed and revised this list, and the result of this process

Table 5.3 The main criteria applicants and reviewers used in judging the suitability of the criteria

Main criteria used in judging the suitability of the criteria		
	By the applicants: They want …	*By the reviewers: They want …*
Clarity	… to know what exactly is meant by the different criteria in order to know how they should consider them in their proposal	… to know what exactly is meant with the different criteria in order to know how they have to be operationalized when they are applied to a proposal
Compliance, Applicability	… criteria they actually feel able to comply with in their research design and in their proposal	… criteria they can judge on the basis of the information and documents provided
Weighting	… to know how the different criteria are weighted in making funding decisions	–
Flexibility	–	… criteria covering all relevant dimensions of a project, but at the same time they want their operationalization to allow them to be adapted to the specific characteristics of individual projects
Compatibility	–	… criteria that are not tied to specific theories but can be applied to a broad range of approaches. In other words, criteria should not be phrased in ways that presuppose the projects adopt particular theories (such as a specific notion of sustainability), and technical language should be kept to a minimum
Manageability	–	… short lists of criteria, being as distinct as possible with each focusing on a different aspect of a project
Objectivity	–	… criteria that are not solely dependent on their individual and subjective opinion (such as whether something is societally relevant)

Table 5.4 The revised criteria used in the second round of funding of WfnE in 2015, showing the entire list (left-hand column) and how selected criteria were operationalized (right-hand column)

Revised criteria	*Operationalized criteria (selection)*
Does the project have the potential to make a relevant contribution to sustainable development?	Is the notion of sustainability used by the project convincing, i.e. does it proceed from a multidimensional notion of sustainability that also encompasses intergenerational justice? Is the subject addressed in the project topical and relevant in relation to sustainable development? Does the project have the potential to make a relevant contribution to societal transformation toward sustainability? Does the project have the potential to make a relevant contribution to sustainability studies? Are there findings and products to be expected from the project that can be practically implemented?
Is the scientific quality of the proposal convincing?	(...)
Are the applicants sufficiently qualified to run the project?	(...)
Is the project original and innovative?	(...)
Are the interdisciplinary team composition and the organization of the interdisciplinary collaboration convincing?	Is the composition of the consortium convincing in relation to the aims and questions of the project? Is it convincingly explained how the different work steps will be coordinated in terms of content? Is the methodical design to achieve a synthesis convincingly explained? Is it sufficiently apparent that joint products will be developed? Does the planned project structure appear to be suitable with a view to a synthesis-building process?

(continued)

Table 5.4 (continued)

Revised criteria	*Operationalized criteria (selection)*
Is the way in which civil society actors are to be integrated into the project convincing?	Is it convincingly explained for each of the individual civil society actors why they participate in the project? Is the contribution to be made by each civil society actor sufficiently clear, and is it convincingly shown that this is a substantial contribution to the project? Is it convincingly explained what methods will be applied in the collaboration with the civil society actors? Did the practice partners provide binding and sufficiently concrete commitments as to what they will have to contribute?

was subsequently discussed with and revised by the reviewers (the review panel took the final decision on the criteria).

At a meta-level, the lists of criteria resulting from these processes (CS-1, CS-2) can be characterized as follows: The lists differ with regard to the content of the criteria, that is, both lists of criteria are tailored to the individual funding program. They cannot be simply transferred to another funding context. Although the criteria in both lists are in line with core requirements of transdisciplinary research as they are formulated in the scholarly discourse, the language used in how they are formulated is not entirely the technical language this academic community uses. If, in our role as certified experts, we had provided the lists of criteria, these would have been, at least in part, formulated differently. Only one of the six criteria in CS-1 addresses the transdisciplinary process directly (two if we also count the criterion addressing the interdisciplinarity of the consortium). The list in CS-2 has two criteria that address the transdisciplinary process directly (knowledge-integration, participation) and two that address processes of diffusion and of generating societal impact. In both lists, all the criteria are qualitative, and their number is limited. In both funding contexts, the actors considered it useful to have criteria in the format of questions and to specify these by questions.

5.4 The Interdisciplinary Nature of the Process of Evaluating Transdisciplinary Research

The process of evaluation in CS-1 (research proposals) was rather complex. Roughly it worked as follows. First, applicants submitted a full proposal, which was evaluated by an interdisciplinary panel of reviewers. Each member of the panel was assigned several proposals, which they read against the criteria provided. They met for a one-day discussion to decide which of the proposals they deemed to be eligible for funding (roughly 30% of the submitted proposals). These applicants were invited to present their projects in a two-day colloquium open to the public. During and directly after this colloquium the reviewers met for several rounds of discussion and decided which of the projects should receive funding (approximately half of the eligible projects). In the interviews that took place some months after this process, the reviewers described in hindsight how they had experienced the process. The following is based on the main points that were voiced in these interviews by several of the reviewers.

Neither the broad spectrum of research fields nor the diversity of disciplines covered by the submitted proposals were represented in the review panel. In other words, the review process in CS-1 was of a multidisciplinary nature both with regard to the composition of the team of reviewers and with regard to the disciplinary background of the proposals each reviewer had to assess. This meant that the reviewers were constantly forced to move out of their individual comfort zone:

> Sometimes, I found it difficult to judge to what extent the disciplinary state of research was well presented and whether the research question was sufficiently related to it, which is also one of the criteria. And we always had only one statement on this topic [by a member of the panel] to draw on [...] and sometimes none. (Interview with reviewer)[5]

Review decisions in CS-1 had to be backed by the entire group, which meant that the reviewers had to integrate their different perspectives and reach a decision they all agreed with. Thus, the review process in CS-1 aimed to be interdisciplinary because it aimed to reach an integrated judgment. In this process, the reviewers experienced the problems

[5] The illustrative quotes are from different reviewers.

that tend to characterize any interdisciplinary collaboration—problems of bridging different disciplinary worldviews as well as the problem that the processes of consensus-building and of knowledge-integration are not always carefully designed and supported:

> When I think of the [...] engineers and the sociologists, these are two different worlds, aren't they? We did have diametrical sensitivities and perceptions and also judgements. (Interview with reviewer)

> And then it started, how can I say, a fundamental discussion about the relationship between certain sciences and the pecking order in sciences, and who's better now than the other and so on. (Interview with reviewer)

> But I also got the impression that in this interdisciplinary communication there might have been a little more exchange at some points. (Interview with reviewer)

Achieving an integrated judgment—that is, succeeding in the interdisciplinary integration of perspectives—was experienced to be individually rewarding and regarded as adding substantially to the quality of the results of the funding decisions:

> And I also experience it to be enriching, because one does learn from each other, that is, one learns how other people do actually look at the proposals with their disciplinary backgrounds, that is, what do they read in this proposal, which I read with a specific lens and perspective and with regard to which I have a specific perception and judgement. (Interview with reviewer)

> Of course, in one's own review, in the course of the individual preparation, there were always a few questions left unanswered, but the group served this purpose to discuss these questions in the group. And this always worked, that in the group these questions could be answered very quickly. Well, I was not left alone with anything. (Interview with reviewer)

> And, of course, there might always be projects, which, if I had been the only person to decide whether to fund them or not, might not have been funded. But that is the advantage of considering different perspectives in deciding and of deciding with different people. (Interview with reviewer)

> That there were always several people who were discussing and deciding on a proposal, that maybe the bias which one has or where one had to exceed the personal comfort zone or expertise, perhaps, then hopefully was compensated for. (Interview with reviewer)

One of the risks of not striving for and achieving an interdisciplinary—an integrated—judgment is ending up with unbalanced funding decisions that might systematically privilege some approaches and/or knowledge systems and/or topics. Another risk manifested itself in CS-2. In the first round of funding, the reviewers wrote individual comments on the project teams' mid-term reports, which were used to produce the mid-term reviews without them having first been subject to an in-depth discussion and an interdisciplinary integration of perspectives in the (interdisciplinary) group of reviewers. Our comparative analysis of these mid-term reviews found that the different evaluation criteria had been interpreted differently by the members of the review panel resulting in an inconsistent and sometimes even contradictory mid-term review.

Our case studies confirm that the process of evaluating transdisciplinary research is inevitably multidisciplinary. But they also show that this process is not always interdisciplinary, meaning that it is not always organized in such a way as to lead to integrated judgments although this does, if successful, improve the quality of the evaluation and the decision-making. Aiming at integrated judgments is time-consuming because it requires reviewers to engage in intensive interdisciplinary processes of consensus-building and of knowledge-integration, and it is demanding because these processes must be designed and moderated. In such a process reviewers learn from each other and broaden their horizons. This might strengthen what Misra et al. (2015) call a 'transdisciplinary orientation', because it provides them with a positive experience of interdisciplinary collaboration—and such experiences do possibly add to reviewers' willingness to engage in such time-consuming and cognitively challenging processes.

But the interdisciplinary interaction should not be limited to the group of reviewers as was emphasized by the applicants in CS-1. In a funding context that addresses any scientific field, it is quite a challenge to ensure comprehensibility for a broad spectrum of disciplines. In such a context the applicants cannot know what information they need to explain in their proposals and what information the reviewers will be able to infer. This problem can be eased by an oral exchange between reviewers and applicants. That is, ideally an interdisciplinary process of evaluating transdisciplinary research plans for such an exchange (and removes the review panels' anonymity). The value such an exchange could have can be illustrated by the experience in CS-3. Evaluation criteria might be interpreted differently by scholars from different disciplinary backgrounds. This became obvious in how the project teams discussed the evaluation criteria suggested by the funder (mid-term evaluation) in the online workshop. Discussing criteria with those who have to comply with them makes it possible to identify the criteria that need to be reformulated (or explained) in order to avoid misunderstandings.

5.5 The Potential of Adopting a Transdisciplinary Approach in Setting up the Evaluation of Transdisciplinary Research

In all three case studies, the process of how the external evaluation of transdisciplinary projects was developed (criteria and procedure) shows transdisciplinary elements by involving actors who play different roles. The three case studies differ in terms of the intensity of the transdisciplinary collaboration with the different actors involved (see Table 5.5).

An intensive collaboration of review panels and funders (uncertified experts, (future) users) in developing criteria and procedures seems an

Table 5.5 Intensity with which the uncertified experts, (future) users, stakeholders, and certified experts were involved in the development of the criteria and procedures for the external evaluation of transdisciplinary projects

Actors involved	*Case study 1*	*Case study 2*	*Case study 3*
Uncertified experts (review panel, funder)	Participated in development	Were asked for feedback	Results of participative process were handed over
(Future) Users (funder, review panel)	Participated in development	Were asked for feedback	Results of participative process were handed over
Stakeholders (applicants, project teams)	Were asked for feedback	Participated in development	Participated in development
Certified experts (authors of chapter)	Design, facilitation of process	Design, facilitation of process	Consulting on process

obvious thing to do. But in many cases, this is not done systematically. Rather, as a rule, the funder provides the criteria to be used and the review panel can modify these criteria to a certain extent. This was the case with CS-1 in which the review panel's deep dissatisfaction with the criteria led to the transdisciplinary process of revising these. The funder had not planned this, and it would not have happened systematically without the involvement of certified experts who designed and facilitated the process. In CS-1, the scholarly knowledge about transdisciplinarity, the practical needs and experiences of funders and reviewers as well as the experience-based expertise of the funders and reviewers fed into the process and its result in a transdisciplinary way. This increased the credibility and the salience of the evaluation. The stakeholders' perspectives were included by extractive methods and via feedback.

While in CS-1 the applicants (the stakeholders) did not actively participate in the collaborative development of the criteria, this was the case in CS-2 and CS-3. In CS-2, they were involved because they criticized their mid-term review. Again, such a process was not planned by the funder, and it would not have been possible without the involvement of certified experts who designed and facilitated the process. In CS-3, the stakeholders were involved because based on the previous experience (CS-2) the funder wanted this to happen in order to improve the quality and transparency of the evaluation. The certified experts were involved as consultants, and they reminded the funder about the previous achievements (CS-2) and thus guaranteed that the current process built on what had been learned and developed in the past. In CS-2 and CS-3, the scholarly knowledge about transdisciplinarity as well as the stakeholders' concerns, experiences, and interests fed into the process and its result in a transdisciplinary way. This increased the legitimacy of the evaluation. The funders' and the reviewers' perspectives also fed into the process, but there was no point in which all actors engaged in a direct discussion and collaboration with each other.

One might ask whether it is reasonable to involve applicants and project teams in developing criteria and procedures that will be applied to their own proposals and projects. In CS-2 and CS-3, this worked out well and lead to criteria on which all actors agreed. Furthermore, in CS-3, the project teams were asked what they expect from the mid-term evaluation, which was done in an online workshop setting. The answers were collected on a whiteboard cover four dimensions:

- Expectations of the evaluation's quality: fair; taking the individuality of the projects into account; efficient, transparent with regard to the criteria; considering both qualitative and quantitative dimensions.
- What should be taken into account in evaluating the projects: the specialty of the research format (real-world laboratories); external factors that influence progress but are out of reach of the projects; what can realistically be achieved at the mid-term; that activities aimed at including stakeholders or at achieving long-term impact should be acknowledged, even if not all of them are successful.
- What the evaluation should yield for the projects: opportunity to reflect and learn about the progress of the project; getting (constructive) feedback, food for thought, and suggestions with a view to the second phase; opportunity to question and revise the design and plan for the second phase; making visible the efforts of the first phase; a special focus on the methods used to implement participation.
- Expectations about how the evaluation should contribute to the broader discourse about the research format of real-world laboratories.

Based on our experiences in all three case studies, the benefits of including certified experts and stakeholders in the collaborative development of the external evaluation of transdisciplinary projects can be summarized as follows:

- The *certified experts* contribute expertise not only in relation to the topic of transdisciplinarity and of evaluating transdisciplinary research, but also with regard to how the inter- and transdisciplinary processes of consensus-building and knowledge-integration related to the evaluation processes could be designed and facilitated. Compared to the other actors involved in the process the certified experts are neutral on the set of criteria and the evaluation procedure. In procedures that extend over a period, the certified experts can serve as a measure of quality assurance for the process. They encourage self-reflection by questioning practices or by presenting results from the accompanying research.
- The *stakeholders* contribute, of course, their concerns, experiences, expectations, and interests. Considering the stakeholders' perspectives adds to their commitment to high-quality transdisciplinary

research. In addition, in most cases, being academics, the stakeholders amplify the spectrum of disciplinary perspectives that are considered in formulating the criteria. This in turn could help funders and reviewers in doing justice to the diversity of disciplines, non-academic actors, topics, and approaches that are represented in a transdisciplinary research program and to the individuality of transdisciplinary projects.

Finally, a transdisciplinary approach should not focus solely on developing the criteria, but target the entire process of evaluation as was emphasized by reviewers in CS-1:

> Because I think that actually the evaluation procedures should be carried out exactly in this way. That is: as transparent as possible and in compliance with comprehensible criteria, but also be willing to review and change the adopted process at any time, and where there is a need for change and the possibility to change to then actually do so. (Interview with reviewer)

In emphasizing the benefits of applying a transdisciplinary approach to the evaluation of transdisciplinary research we do not advocate a democratic approach—the final decisions about both criteria and procedures rest with the funder or the review panel.

5.6 Conclusion

The processes in interdisciplinary review panels that evaluate transdisciplinary research face the same problems, can yield the same added value, and need the same support as any form of interdisciplinary collaboration. Thus, such processes should meet the same quality requirements as any other interdisciplinary collaboration and must be carefully designed and facilitated in order to lead to shared problem framings and integrated results. A high-quality evaluation of transdisciplinary research requires time-consuming processes in which reviewers from different disciplines interact with each other and with the applicants. Reviewers must be willing to leave their individual comfort zone and to enter an interdisciplinary collaboration. But they must not be left alone to deal with the challenges they encounter while doing this. Taking decisions in a group supports the individual reviewers, eases their responsibility—and might facilitate their willingness to leave their comfort zone and to admit to personal uncertainties. The community of scholars doing research on the evaluation of transdisciplinary research should insist that the procedural quality of such processes is respected and examine the dynamics of these processes.

A transdisciplinary approach has the potential of adding to the credibility, salience, and legitimacy of how the quality of transdisciplinary research is evaluated. Funders should consider involving applicants and project teams (stakeholders) in developing quality criteria that are then applied to their own projects, and they should consider collecting and serving the expectations of those that are affected by mid-term evaluations in order to enhance the beneficial impact of mid-term evaluations (see also Defila & Di Giulio, 2020). This requires a reconsideration of the relationship between those that do an evaluation and those that are evaluated, such that the usually hierarchical relationship is replaced with one that is based on partnership. Funders should consider involving certified experts of inter- and transdisciplinarity, such as in the format of an accompanying research project, that examine and support the ongoing process(es). The certified experts in turn should be sensitive to their role when they engage in such processes without themselves being reviewers. A collaboration with funders, reviewers, and applicants is a transdisciplinary collaboration. Certified experts engaging in such transdisciplinary collaborations must be aware that they cannot impose their criteria and theories on the uncertified experts, (future) users, or stakeholders but need to enter a process of consensus-building and knowledge-integration with them.

The scholarly approach of certified experts to the topic of how to evaluate the quality of transdisciplinary research should be reconsidered. One question on which to reflect is whether the perspectives of uncertified experts, (future) users, and stakeholders are sufficiently considered. In developing suggestions for how to evaluate transdisciplinary research that shall actually be used in funding decisions and in mid-term or final evaluations, not only the scholarly knowledge about transdisciplinarity has to be taken into account, but also the expertise of the uncertified experts (funders and reviewers), what (future) users (funders, and reviewers) actually need from a practical perspective, and what bothers stakeholders (applicants, project teams). If these perspectives do not feed into the scholarly discussion, the suggestions emerging from this discussion will not be sufficiently linked to the practice of doing evaluations. This question also touches on the language used in the scholarly discourse about the evaluation of transdisciplinary research; it has to be asked whether the terminology that is used is to a too large extent loaden with the theories and terms of the academic community rather than linking to the language of uncertified experts, (future) users, and stakeholders. A second question arises from the fact that in an actual evaluation of transdisciplinary research it is impossible to use a large number of criteria that cover all aspects that are, according to the knowledge of certified experts, important to achieve high-quality transdisciplinary research. Furthermore, the sets of criteria that are used cannot be limited to mirroring transdisciplinary quality but also have to mirror the funding context and the funders' worldviews and policies. Against this background, the topic that has dominated the scholarly debate for quite some time now has to be questioned: All the highly differentiated lists of criteria that have been developed are useful as a source of inspiration for external evaluations (as well as those developed primarily for internal self-evaluation). But they cannot be more than that, because they are too differentiated and comprehensive for the purpose of an external evaluation. The question thus is whether it would be better to stop developing ever more elaborate sets of criteria and turn to other questions instead.

We might summarize our conclusions and learnings in the form of three general messages:

> Message 1: Criteria that can be used in taking funding decisions or in mid-term or final evaluations have themselves to meet criteria, and these criteria should be informed by the expertise and the practical needs of those applying them and consider the concerns of those affected by them.
>
> Message 2: There are enough suitable sets of criteria that can (and have to) be adapted for specific evaluations. The scholarly debate on inter- and transdisciplinarity should now move forward and focus on the process of evaluation itself and on how this process should be designed and supported.
>
> Message 3: It makes sense to adopt a transdisciplinary approach to develop evaluation criteria for transdisciplinary research and to improve the evaluation process. The evaluation process in turn must itself meet the same quality criteria as any other inter- or transdisciplinary process.

We are convinced that the quality of how transdisciplinary research is evaluated can be improved by developing criteria to assess the quality of evaluations as well as by transdisciplinary collaborations. What our case studies did not cover and thus remains to be investigated is what role non-academics can and should play in this.

Acknowledgements Our thanks go, first of all, to the funders that made our case studies possible by funding them and by collaborating with us. A special thanks also goes to the members of the review panels because they gave us their time and were willing to engage in discussions with us in addition to their task of reviewing proposals and project reports. In addition, we wish to thank Claudia Binder for many interesting discussions in the project team of ZiFoNE and Livia Fritz and Thorsten Schilling who were also members of this project team for their support in collecting data. A big thanks goes also to Alexandra Lux for providing suggestions for our chapter.

References

Andersen, H., & Wagenknecht, S. (2013). Epistemic dependence in interdisciplinary groups. *Synthese, 190*(11), 1881–1898. https://www-jstor-org.vu-nl.idm.oclc.org/stable/41932110

Belcher, B. M., Rasmussen, K. E., Kemshaw, M. R., & Zornes, D. A. (2016). Defining and assessing research quality in a transdisciplinary context. *Research Evaluation, 25*(1), 1–17. https://doi-org.vu-nl.idm.oclc.org/10.1093/reseval/rvv025

Bergmann, M., & Schramm, E. (Eds.). (2008). *Transdisziplinäre Forschung*. Campus Verlag.

Bergmann, M., Brohmann, B., Hoffmann, E., Loibl, M. C., Rehaag, R., Schramm, E., & Voss, J.-P. (2005). *Quality criteria of transdisciplinary research. A guide for the formative evaluation of research projects*. ISOE. http://www.isoe-publikationen.de/fileadmin/redaktion/ISOE-Reihen/st/st-13-isoe-2005.pdf

Bogner, A. (2012). Wissenschaft und Öffentlichkeit: Von Information zu Partizipation. In S. Maasen, M. Kaiser, M. Reinhart, & B. Sutter (Eds.), *Handbuch Wissenschaftssoziologie* (pp. 379–392). Springer VS. https://doi-org.vu-nl.idm.oclc.org/10.1007/978-3-531-18918-5_30

Boix Mansilla, V., Feller, I., & Gardner, H. (2006). Quality assessment in interdisciplinary research and education. *Research Evaluation, 15*(1), 69–74. https://doi-org.vu-nl.idm.oclc.org/10.3152/147154406781776057

Cash, D. W., Clark, W. C., Alcock, F., Dickson, N. M., Eckley, N., Guston, D. H., Jäger, J., & Mitchell, R. B. (2003). Knowledge systems for sustainable development. *PNAS, 100*(14), 8086–8091. https://doi-org.vu-nl.idm.oclc.org/10.1073/pnas.1231332100

Collins, H. M., & Evans, R. (2002). The third wave of science studies: Studies of expertise and experience. *Social Studies of Science, 32*(2), 235–296. https://doi-org.vu-nl.idm.oclc.org/10.1177/0306312702032002003

Defila, R., & Di Giulio, A. (1998). Interdisziplinarität und Disziplinarität. In J.-H. Olbertz (Ed.), *Zwischen den Fächern – über den Dingen? Universalisierung versus Spezialisierung akademischer Bildung* (pp. 111–137). Leske & Budrich. https://doi-org.vu-nl.idm.oclc.org/10.1007/978-3-322-90935-0_6

Defila, R., & Di Giulio, A. (1999). *Evaluating transdisciplinary research—Evaluation criteria for inter and transdisciplinary research.* Panorama, Special Issue 1/99. https://www.ikaoe.unibe.ch/forschung/ip/Specialissue.Pano.1.99.pdf. 18 March 2024.

Defila, R., & Di Giulio, A. (2018). What is it good for? Reflecting and systematizing accompanying research to research programs. *GAIA—Ecological Perspectives for Science and Society, 27*(S1), 97–104. https://doi-org.vu-nl.idm.oclc.org/10.14512/gaia.27.S1.17

Defila, R., & Di Giulio, A. (2020). Science policy recommendations for funding real-world laboratories and comparable formats. *GAIA—Ecological Perspectives for Science and Society, 29*(1), 63–65. https://doi-org.vu-nl.idm.oclc.org/10.14512/gaia.29.1.14

Defila, R., Di Giulio, A., Fischer, D., Gölz, S., Kaufmann-Hayoz, R., & Schäfer, M. (2016). Was wir noch zu sagen hätten – Briefe an unsere Leserinnen und Leser. In R. Defila & A. Di Giulio (Eds.), *Transdisziplinär forschen – zwischen Ideal und gelebter Praxis. Hotspots, Geschichten, Wirkungen* (pp. 332–341). Campus Verlag. https://doi-org.vu-nl.idm.oclc.org/10.12907/978-3-593-43381-3

Defila, R., Di Giulio, A., & Scheuermann, M. (2006). *Forschungsverbundmanagement. Handbuch für die Gestaltung inter- und transdisziplinärer Projekte.* vdf Hochschulverlag an der ETH Zürich.https://doi-org.vu-nl.idm.oclc.org/10.3218/4106-4

Di Giulio, A., & Defila, R. (2017). Enabling university educators to equip students with inter-and transdisciplinary competencies. *International Journal of Sustainability in Higher Education, 18*(5), 630–647. https://doi-org.vu-nl.idm.oclc.org/10.1108/IJSHE-02-2016-0030

Gerhardus, A., Becher, H., Groenewegen, P., Mansmann, U., Meyer, T., Pfaff, H., Puhan, M., et al. (2016). Applying for, reviewing and funding public health research in Germany and beyond. *Health Research Policy and Systems, 14*, 43. https://doi-org.vu-nl.idm.oclc.org/10.1186/s12961-016-0112-5

Gibbons, M., Limoges, C., Nowotny, H., Schwartzman, S., Scott, P., & Trow, M. (1994). *The new production of knowledge: The dynamics of science and research in contemporary societies*. Sage. https://doi-org.vu-nl.idm.oclc.org/10.4135/9781446221853

Hastie, J. (2007). The role of science and scientists in environmental policy. In J. Pretty, A. S. Ball, T. Benton, J. Guivant, D. R. Lee, D. Orr, M. J. Pfeffer, & H. Ward (Eds.), *The SAGE handbook of environment and society* (pp. 519–535). Sage. https://doi-org.vu-nl.idm.oclc.org/10.4135/9781848607873.n36

Holbrook, J. B. (2013). What is interdisciplinary communication? Reflections on the very idea of disciplinary integration. *Synthese, 190*(11), 1865–1879. https://www-jstor-org.vu-nl.idm.oclc.org/stable/41932109

Huutoniemi, K. (2010). Evaluating interdisciplinary research. In R. Frodeman, J. T. Klein, & C. Mitcham (Eds.), *The Oxford handbook of interdisciplinarity* (pp. 309–320). Oxford University Press.

Hvidtfeldt, R. (2018). *The structure of interdisciplinary science*. Palgrave Macmillan. https://doi-org.vu-nl.idm.oclc.org/10.1007/978-3-319-90872-4

Jahn, T., & Keil, F. (2015). An actor-specific guideline for quality assurance in transdisciplinary research. *Futures, 65*(Special Issue Advances in transdisciplinarity 2004–2014), 195–208. https://doi-org.vu-nl.idm.oclc.org/10.1016/j.futures.2014.10.015

Klein, J. T. (1990). *Interdisciplinarity. History, theory & practice*. Wayne State University Press.

Klein, J. T. (2008). Evaluation of interdisciplinary and transdisciplinary research: A literature review. *American Journal of Preventive Medicine, 35*(2), 116–123. https://doi-org.vu-nl.idm.oclc.org/10.1016/j.amepre.2008.05.010

Klein, J. T. (2010). A taxonomy of interdisciplinarity. In R. Frodeman, J. T. Klein, & C. Mitcham (Eds.), *The Oxford handbook of interdisciplinarity* (pp. 15–30). Oxford University Press.

Koier, E., & Horlings, E. (2015). How accurately does output reflect the nature and design of transdisciplinary research programmes? *Research Evaluation, 24*(1), 37–50. https://doi-org.vu-nl.idm.oclc.org/10.1093/reseval/rvu027

Laursen, B. K., Motzer, N., & Anderson, K. J. (2022). Pathways for assessing interdisciplinarity: A systematic review. *Research Evaluation, 31*(3), 326–343. https://doi-org.vu-nl.idm.oclc.org/10.1093/reseval/rvac013

Lawrence, M. G., Williams, S., Nanz, P., & Renn, O. (2022). Characteristics, potentials, and challenges of transdisciplinary research. *One Earth, 5*(1), 44–61. https://doi-org.vu-nl.idm.oclc.org/10.1016/j.oneear.2021.12.010

Love, H. B., Fosdick, B. K., Cross, J. E., Suter, M., Egan, D., Tofany, E., & Fisher, E. R. (2022). Towards understanding the characteristics of successful and unsuccessful collaborations: A case-based team science study. *Humanities & Social Science Communications, 9*, 371. https://doi-org.vu-nl.idm.oclc.org/10.1057/s41599-022-01388-x

Mielke, J., Vermaßen, H., Ellenbeck, S., Milan, B. F., & Jaeger, C. (2016). Stakeholder involvement in sustainability science—A critical view. *Energy Research & Social Science, 17*, 71–81. https://doi-org.vu-nl.idm.oclc.org/10.1016/j.erss.2016.04.001

Misra, S., Stokols, D., & Cheng, L. (2015). The transdisciplinary orientation scale: Factor structure and relation to the integrative quality and scope of scientific publications. *Journal of Translational Medicine & Epidemiology, 3*(2), 1042. https://www.researchgate.net/publication/290193780. 18 March 2024.

Mobjörk, M. (2010). Consulting versus participatory transdisciplinarity: A refined classification of transdisciplinary research. *Futures, 42*(8), 866–873. https://doi-org.vu-nl.idm.oclc.org/10.1016/j.futures.2010.03.003

Pohl, C., & Hirsch Hadorn, G. (2007). *Principles for designing transdisciplinary research*. Oekom Verlag. https://doi-org.vu-nl.idm.oclc.org/10.14512/9783962388638

Pohl, C., Perrig-Chiello, P., Butz, B., Hirsch Hadorn, G., Joye, D., Lawrence, R., Nentwich, M., et al. (2011). *Questions to evaluate inter- and transdisciplinary research proposals* (Working paper, td-net for Transdisciplinary Research, Berne). https://t1p.de/Pohl-etal-2011

Regeer, B. J., & Bunders, J. F. (2003). The epistemology of transdisciplinary research: From knowledge integration to communities of practice. *Interdisciplinary Environmental Review, 5*(2), 98–118. https://doi-org.vu-nl.idm.oclc.org/10.1504/IER.2003.053901

Röbbecke, M., Simon, D., Lengwiler, M., & Kraetsch, C. (2004). *Inter-Disziplinieren. Erfolgsbedingungen von Forschungskooperationen*. Edition Sigma.

Schuck-Zöller, S., Cortekar, J., & Jacob, D. (2017). Evaluating co-creation of knowledge: From quality criteria and indicators to methods. *Advances in Science and Research, 14*, 305–312. https://doi-org.vu-nl.idm.oclc.org/10.5194/asr-14-305-2017

Schneider, F., Patel, Z., Paulavets, K., Buser, T., Kado, J., & Burkhart, S. (2023). Fostering transdisciplinary research for sustainability in the Global South: Pathways to impact for funding programmes. *Humanities and Social Sciences Communications, 10*(1), 1–11. https://doi-org.vu-nl.idm.oclc.org/10.1057/s41599-023-02138-3

Steelman, T., Bogdan, A., Mantyka-Pringle, C., Bradford, L., Reeds, M. G., Baines, S., Fresque-Baxter, J., et al. (2021). Evaluating transdisciplinary research practices: insights from social network analysis. *Sustainability Science, 16*, 631–645. https://doi-org.vu-nl.idm.oclc.org/10.1007/s11625-020-00901-y

Stokols, D., Fuqua, J., Gress, J., Harvey, R., Phillips, K., Baezconde-Garbanati, L., Unger, J., et al. (2003). Evaluating transdisciplinary science. *Nicotine &*

Tobacco Research, 5(Suppl_1), S21–S39. https://doi-org.vu-nl.idm.oclc.org/10.1080/14622200310001625555

Stoll-Kleemann, S., & Pohl, C. (Eds.). (2007). *Evaluation inter- und transdisziplinärer Forschung*. Oekom Verlag.

Vermeulen, W. J. V., & Witjes, S. (2020). History and mapping of transdisciplinary research on sustainable development issues: Dealing with complex problems in times of urgency. In M. M. Keitsch & W. J. V. Vermeulen (Eds.), *Transdisciplinarity for sustainability. Aligning diverse practices* (pp. 6–26). Routledge. https://doi-org.vu-nl.idm.oclc.org/10.4324/9780429199127

Verwoerd, L., Brouwers, H., Kunseler, E., Regeer, B., & de Hoop, E. (2023). Negotiating space for knowledge co-production. *Science and Public Policy, 50*(1), 59–71. https://doi-org.vu-nl.idm.oclc.org/10.1093/scipol/scac045

Wagner, C. S., Roessner, J. D., Bobb, K., Klein, J. T., Boyack, K. W., Keyton, J., Rafols, I., & Börner, K. (2011). Approaches to understanding and measuring interdisciplinary scientific research (IDR): A review of the literature. *Journal of Informetrics, 5*(1), 14–26. https://doi-org.vu-nl.idm.oclc.org/10.1016/j.joi.2010.06.004

Zweekhorst, M. B. M., Broerse, J. E. W., & Bunders, J. F. G. (2001). Innovations for sustainable development: The need for transdisciplinary knowledge integration. *Interdisciplinary Environmental Review, 3*(1), 75–94. https://doi-org.vu-nl.idm.oclc.org/10.1504/IER.2001.053868

CHAPTER 6

Accompanying Transformation-Oriented Research: Contributions, Relations and Methods

Niko Schäpke

6.1 Introduction: Transformation-oriented and Accompanying Research

There has been far too little progress in overcoming socio-ecological problems and shaping a sustainable future in terms of scale and speed. Societal transformations, as fundamental changes in socio-ecological and socio-technical systems towards more sustainability and justice, are thus high on scientific and political agendas (e.g. Patterson et al., 2017; https://www.ipbes.net/transformative-change; Scoones et al., 2020; United Nations General Assembly, 2015). Transformational sustainability research is concerned with understanding and developing solutions to persistent socio-ecological problems, with a focus on intentionally shaping

N. Schäpke (✉)
Chair of Environmental Governance,
University of Freiburg, Freiburg, Germany
e-mail: niko.schaepke@ifp.uni-freiburg.de

RIFS – Research Institute for Sustainability, Helmholtz Centre Potsdam,
Potsdam, Germany

B. J. Regeer et al. (eds.), *Transdisciplinarity for Transformation*,
https://doi.org/10.1007/978-3-031-60974-9_6

societal change (Fazey et al., 2018; Wiek & Lang, 2016). There are various approaches to transformation-oriented research, including among others real-world, urban living or transformation labs, transition management, adaptive socio-ecological systems management, backcasting and transition experiments (McCrory et al., 2020; Wiek & Lang, 2016). Jointly, approaches aspire to make tangible contributions to benefit societal transformations towards sustainability (Schäpke et al., 2018a, b). They address problems in complex systems in future-oriented and participatory ways, aiming to catalyse learning and innovation via the integration of various kinds of knowledge (Caniglia et al., 2020).

Transformation-oriented sustainability research has roots in pragmatism (Caniglia et al., 2020). Here, the generation of knowledge is action-oriented and motivated to enable insights on how to tackle real-world problems. Knowledge is created in a transdisciplinary process that links action, reflection and capacity building (Caniglia et al., 2020). It is most effectively co-produced in context-based, pluralist and goal-oriented processes (Norström et al., 2020), hosted in spaces for social learning and interaction (Wittmayer & Schäpke, 2014). Accordingly, the design and facilitation of processes of knowledge co-production and integration become core activities in transformation-oriented sustainability research (Miller, 2013; Wittmayer & Schäpke, 2014). Processes are marked by constant iteration and adaptation, taking up new insights as well as unfolding needs and demands of participants.

Transformation-oriented research faces several challenges. This includes to design research processes oriented towards continuos emergence and iteration (see Regeer et al., this volume, Chapter 3). Decisions on adaptation should be well informed and aware of their implications for research and practice, which can be hard to foresee in daily research practice. Furthermore, developing overarching insights from transdisciplinary research, including on its processes and impacts, is inherently difficult, not least as engaged researchers have to balance the rigorous generation of knowledge and demands to appropriatly facilitate the process (Belcher & Halliwell, 2021; Schäpke et al., 2024). Moreover, transformation-oriented research such as in the form of real-world laboratories (RwLs) is still new to many researchers and a highly complex task, which in parts is at odds with disciplinary traditions and their good practice standards (Defila & Di Giulio, 2018a; Verwoerd et al., 2020). Challenges call for support of researchers engaged in transformation-oriented research, including in knowledge development, process facilitation and decision-making.

Here, I present accompanying research—namely, research that accompanies and researches others' research—as a possibility for complementing and supporting transformation-oriented research in transdisciplinary knowledge co-production and process reflexivity. Accompanying research can generally be applied to support science and research of various kinds. In my view, the specifics and challenges of transformation-oriented research make the complementary use of accompanying research particularly beneficial. To make my contribution more specific and tangible, I focus on accompanying research for RwLs. RwLs are characterized by transdisciplinary experimentation, aiming to research and contribute to the transformation of society. Thus, they can function as an ideal-typical example of transformation-oriented sustainability research (Schäpke et al., 2018b) and to discuss related accompanying research.

So, what is accompanying research and how can it benefit transformation-oriented sustainability science? What methods does accompanying research use and how does it relate to those being accompanied?

Following this introduction the chapter *first* outlines the objectives and contributions of accompanying research, and *second* goes on to elaborate on how to conceive and structure the relationships between accompanying research and relevant stakeholders, including necessary balancing acts. Together, objectives and relationships make it possible to describe and design the functioning of accompanying research (Defila & Di Giulio, 2018a). Conceptual reasoning is further complemented with suitable methods to deploy accompanying research, aiming to provide hands-on guidance. In the *third* section, and to deepen considerations, I present and discuss a case study of accompanying research to the Co-Creative Reflection and Dialogue Space (CCRDS) from UNFCCC climate conferences. The chapter ends with a concluding discussion.

Concerning methods and data, elaborations on objectives and relationships build on a synthesis of insights published by experts in accompanying research (including Defila & Di Giulio, 2018a; Freeth & Vilsmaier, 2019; Weith et al., 2019 and others) and on the wider literature on transformation-oriented research. I also include reflections based on my experience in the accompanying research of RwLs, including 14 so-called BaWü Labs from Baden-Württemberg in Germany, taking place from 2014 to 2018. The CCRDS case study combines self-reflection, reflections in the CCRDS team, and considerations of academic studies.

6.2 Understanding Accompanying Research and Its Contributions

Accompanying research accompanies and researches the research activities of third parties and is therefore characterized by the relationship to other people's research (Fiedeler et al., 2010). These may be the research activities of a single research project, for example, the various real-world experiments of a RwL, or different research projects within a research programme (Defila & Di Giulio, 2018a; Weith et al., 2019). Accompanying research is usually conducted by individuals who are not directly involved in the research activities being accompanied, but who are in engagement with them, observing, documenting, and thus being able to shape an ongoing process of reflection (von Wehrden et al., 2019). In this sense, accompanying research is independent research despite being related to others' research (Defila & Di Giulio, 2018a).

Accompanying research has different objectives, depending on the subject in question and the sponsor, be it a single research project, or the funding body of for instance an entire funding line. It can provide the funding body with further insights into the funded programme—or help those involved in a transformation-oriented research project to better understand it and its impact. This could include the 'reflexive generation of new knowledge' (Weith et al., 2019, p. 294), knowledge management, or communication and networking for the accompanied research.

For Freeth and Vilsmaier (2019) the promotion of learning processes of accompanied research teams lies at the centre of their approach to formative accompanying research. They distinguish learning about, with and for the team: *learning about* refers to the generation of transferable insights about the accompanied research; *learning with* occurs in the role of a team member of the accompanied research; and *learning for* aims at strengthening the research project and its results through interventions in the research process. In sum, accompanied research can inform, shape, and improve the research process[1] (Freeth & Vilsmaier, 2019).

[1] At times, accompanying research is called formative accompanying research, as it supports the improvement of the accompanied research in some form. I avoid introducing a formal distinction between formative and 'regular' accompanying research and treat both

How to obtain a structured overview of the various possible contributions and their implications for designing accompanying research? Based on many years of experience, Defila and Di Giulio (2018a) propose two main contributions of accompanying research: ***research and knowledge generation*** as well as ***procedural support***. These functions can be designed differently, depending on the specific objectives, competencies and funding of the accompanying research (ibid.). I next discuss the two generic types of contributions and their implications for design and suitable methods.

6.2.1 Research and Knowledge Generation

> This concerns the question: What scientific and practical knowledge is/shall be generated by the accompanying research (Defila & Di Giulio, 2018a)?

Three types of new knowledge that accompanying research generates are distinguished (ibid.).

Knowledge about fundamental topics of the accompanied RwL or RwL funding programme (K1): Here, the accompanying research generates complementary knowledge and fills possible thematic knowledge gaps (Defila & Di Giulio, 2018a). Accordingly, RwL accompanying research could be commissioned by the project leadership to contribute knowledge to a thematic area that is identified as particularly relevant in the course of the project, but which is not adequately covered by the existing sub-projects. A hypothetical example might include a RwL experimenting with autonomous driving for mobility transitions that has a rapid need for new knowledge on upcoming legal issues restricting autonomous driving, but lacks the resources to develop this knowledge. Another possibility is, that the funding agency decides that an additional topical area not covered by the originally funded RwLs is in need of research (Deflia & Di Giulio 2018a). The contribution of such fundamental knowledge is conceivable if the accompanying research starts the project with open capacities and has sufficient knowledge and competencies in the new topic area. This is unlikely to be a common situation. The accompanying research will itself often have a set research agenda, and lack in-depth

as broadly synonymous, generally assuming some form of support from accompanying to accompanied research.

expertise on specific topical areas of RwLs, but rather have capacities on process facilitation, reflexivity and knowledge integration per se. For this reason, this contribution will not be discussed in further depth. Suitable *methods* would be diverse, depending on the topic in question, and not specific to doing accompanying research.

Knowledge of processes within a RwL or RwL funding programme (K2): This knowledge is acquired from a (partly) independent perspective (Defila & Di Giulio, 2018a). The accompanying research takes on the role of critical, analytical research that aims to gain insights into activities of the RwL or programme. For example, it collects and structures methods for conducting experiments within a RwL or synthesises related success factors (Bergmann et al., 2021). Insights gained can support the design, orientation and optimization of an (upcoming) RwL and its processes (Freeth & Vilsmaier, 2019). As Defila and Di Giulio (2018a) point out, to develop K2 knowledge it is helpful if the accompanying research has specific expertise on observed scientific processes and aspects—for example, on methods of transdisciplinary research as used by RwLs (Defila & Di Giulio, 2018a, see also Bergmann et al., 2021).

Suitable *methods* are required to collect and compare data from different parts of the project, such as different experiments, or various RwLs of a funding line. In my experience, this includes literature reviews, surveys, different types of interviews, group discussions as well as the analysis of metadata, e.g. from project proposals, result documents of sub-projects or experiments or similar 'grey' literature. This may include socio-scientific data and their analysis, as well as technical-scientific data, for example on resource consumption or ecological indicators related to the project. An exchange with the members of the RwL team about (interim) results—such as in workshops—enables the research results to be compared with practical experience and to learn together (Bergmann et al., 2021). To facilitate and structure the comparison and synthesis of findings from different projects or experiments, an overarching analytical framework may be used (e.g. Luederitz et al., 2017; von Wirth et al., 2019; Williams & Robinson, 2020). Borrowing from reflexive monitoring in action, process tracing, reflexive process description or timeline and eye-opener workshops are some of the methods used to go more deeply into reflecting on the processes in a transdisciplinary project (van Mierlo et al., 2010). Given their co-creative character, these methods are suitable for the following knowledge contribution as well.

Integrated knowledge on topical and/or procedural aspects (K3) is generated in collaboration with and between different RwL stakeholders. Here, the accompanying research acts as a designer and facilitator of a 'continuous and systematic process of integration' (Defila & Di Giulio, 2018a, p. 99). The objective is to create synergies by linking knowledge bases from different activities and sub-projects (Defila & Di Giulio, 2018a) and to enable a joint learning process. Here, accompanying research can include meta- and comparative studies and generate cross-case knowledge (Weith et al., 2019). The corresponding competences for the organization of an inter- or transdisciplinary knowledge integration process should be available in the accompanying research team (ibid.).

In my experience, appropriate *processes and methods* often include the organization and structuring of events or series of events (e.g. workshops, seminars, conferences, regular project meetings) for exchange, joint learning, reflection and knowledge integration. The events serve the goal-oriented exchange and collective exploration of the (jointly defined) topics and issues. The starting point for generating cross-cutting insights should be the clarification of issues and questions of common interest, as well as the thematization and documentation of mutual expectations (Defila & Di Giulio, 2018a).

Depending on the objectives and the state of knowledge, the exchange may be more open and explorative and/or more goal- and result-oriented. A combination of open and result-oriented phases can be a good way to combine learning, new insights and securing the results. Elements for presenting existing knowledge (e.g. the status of work) are also part of this. The exchange could be organized internally or involve external actors (e.g. final conferences, thematic workshops), depending on the need for (additional) knowledge, confidentiality or publicity, as well as existing organizational resources. Examples from the BaWü Labs include a combination of events for broad and for focused exchange. We combined larger conferences, oriented towards broad public participation and aiming to raise interest in the research programme and RwLs as a new approach, with more focused symposia that connected actors from RwLs and experts on transformative research to discuss specific topics (for examples of events see Schäpke et al., 2017; Wagner et al., 2016). Events can additionally serve to create publicity and communication opportunities (see next section).

If the accompanying research is part of the RwL team and its regular working meetings, spontaneous interventions such as by mirroring back observations or information from surveys to team members are also possible (Freeth & Vilsmaier, 2019). This can advance learning and co-operation processes in the RwL (Freeth & Vilsmaier, 2019, see section on dynamic balance of accompanying research). Field notes, participant observations or simple daily reflection emails to a colleague are methods to track observations and to bring them to joint consideration in appropriate situations (Klassen et al., 2021; Wittmayer et al., 2014).

An important step for the generation of integrated knowledge is the creation of joint products that concretize and document the process of generating knowledge. Depending on the objective and time horizon, the documentation might include position or discussion papers, books or thematic booklets as well as result reports or press releases, blogs, podcasts or an internet presence. Accompanying research can either exclusively organize the process without taking on a content-related role (Defila & Di Giulio, 2018a) or contribute content, for example, in the form of own chapters or forewords to publications (Defila & Di Giulio, 2018a; for publication examples see Defila & Di Giulio, 2018b, 2019, Schäpke et al., 2018a). Weith and colleagues (2019) also cite joint authorship of accompanying research with other project stakeholders as an effective way to integrate knowledge. Co-authorship with multiple authors, including practitioners, is often said to be challenging with regard to coordination and motivation, for instance. In my experience, this can be alleviated if the publication process finds ways to acknowledge verbal and in-workshop contributions, multiple contributions are managed by drawing on digital knowledge management tools and there is a stringent, well-coordinated writing process (for a process example with more than 170 authors see Fazey et al. [2020]).

In addition to scientific publications, Weith and colleagues (2019) emphasize the role of practice- and policy-oriented publications, such as policy recommendations. These may be developed with less time investment or at least published without the often long publication processes of academic journals. In addition, they are highly relevant in terms of the societal impact of RwLs and similar formats. Scientific and practice-oriented publications can be built up synergistically (Weith et al., 2019). If accompanying research is involved in terms of content, a clearly delineated role in the publication is advisable in order not to jeopardize the critically independent role in the project (*see section on a dynamic-balanced accompanying research*).

6.2.2 Process-Related Contributions of the Accompanying Research

> This concerns the question: Which further activities are/ shall be implemented to support the research process of the accompanied research (Defila & Di Giulio, 2018a)?

In addition to knowledge generation, accompanying research can also bring further additional benefits to an accompanied project, such as a RwL. This might be, for example, communication, public relations and networking, including the dissemination of knowledge about the RwL and its results (Defila & Di Giulio, 2018a). For this, there might be different publications, such as editing thematic booklets, anthologies or websites and blog entries. Other possibilities are the organization of events involving an (interested) public and/or relevant experts or joint appearances at conferences (Defila & Di Giulio, 2018a; Schäpke et al., 2017). Accompanying research can also initiate the formulation of core results or policy recommendations and communicate them (Weith et al., 2019, see section on **K2** and **K3** above for relevant methods). In addition to this external communication, Weith and colleagues (2019) cite internal communication as a possible contribution of accompanying research, for example, in the context of (co-)designing RwL internal events.

Coaching and consulting of the RwL participants are another contribution of the accompanying research, such as on the possibilities of meeting emerging challenges (Defila & Di Giulio, 2018a). This is also a central aspect of the formative accompanying research approach developed by Freeth and Vilsmaier (2019). It can function to enhance reflexivity and informed decision-making in the accompanied project (Klaassen et al., 2021; van Mierlo et al., 2010). As RwLs are (still) novel for many researchers and practitioners, and working in them is highly complex, such support should be planned for and requested at the outset of a project to secure possibilities (Defila & Di Giulio, 2018a). Coaching and consultation can take place on an ad hoc basis at the request of project participants, as well as through regular dialogue and reflection workshops (see Defila & Di Giulio, 2018a for an exemplary format). While accompanying research can and should openly communicate the offer of coaching and consulting, it should then be based on the needs and demands of the project participants and based on the agreement of a clear framework.

A third area cited by Weith and colleagues (2019) is knowledge management. Here, the accompanying research can make contributions to the synthesis of the findings of the RwL and offers them to third parties for use in a structured manner via digital, openly available platforms, for example. For this, permanent financing of the underlying infrastructure is crucial (Schneidewind et al., 2018). Within RwL projects, it is conceivable that the accompanying research gets involved in the collection and structured storage of results from sub-projects and experts via databases internal to the project (see Table 6.1 for an overview on contributions).

Table 6.1 Contributions of accompanying research to transformation-oriented sustainability research, related methods and formats

Type of contribution	*Exemplary methods and formats*
Critical-analytical knowledge on processes of one or several RwL(s) (K2) Objectives: • Overarching insights into various activities of the RwL (and its' experiments) • Enable orientation, design and optimization of the processes of RwLs	Collection, critical analysis and comparison of data through: • Literature analysis, surveys, interviews, group discussions, (literature) analysis of metadata and 'grey' literature • Analysis of socio-economic, ecologic or technical data • Discussion and comparison of the results with the RwL team in workshops, etc. • Process tracing, reflexive process description or timeline and eye-opener workshops

(continued)

Table 6.1 (continued)

Type of contribution	*Exemplary methods and formats*
Integrated knowledge from the collaboration with different RwL actors (K3) Objectives: • Generate synergies by connecting knowledge assets from different actors and activities and enable a common learning process • Integration of various knowledges to co-produce shared understanding of RwL content and processes	• Organisation and structuring of internal, external or public events, e.g. workshops, seminars, conferences, project meetings • Mirroring back observations to team, based on field notes, participant observation, surveys, daily reflection emails • Organisation/ Co-creation of joint products with a scientific and/or socio-political target group: e.g. position or discussion papers, books, result reports, press releases, blogs, podcasts, websites, strategy papers, policy recommendations
Additional contributions of the accompanying research	
Communication, public relations and networking	• Publications (see K2) • Organization of external and internal events (see K3)
Coaching and consulting	• Regular dialogue and reflection workshops • Ad hoc consulting • Participation in internal project events as a critical observer or friend
Knowledge management	• Synthesis of the findings of the RwL (see K2) • Digital platforms • Project internal databases

Based on Defila and Di Giulio (2018a) in strongly adapted form, complemented by Freeth and Vilsmaier (2019), and Weith et al. (2019), methods based on multiple sources and the author's experiences. See main text for detailed references

6.3 Relationship with the Actors

This includes the question: What is/ shall be the nature of the relationship with the researchers/ stakeholders/ other projects and/or the funding body (Defila & Di Giulio, 2018a)?

Accompanying research has a special role within a research programme or project, as its objective, function and mode of operation are defined in relation to the activities being researched and accompanied (Fiedeler et al., 2010, cited in Defila & Di Giulio, 2018a). Defila and Di Giulio (2018a) discuss various types of relationship, three of which are presented here as being of particular relevance to RwL accompanying research.[2] These are:

- **(Sub-)Projects as research object**: The accompanying research has the processes and results operated by other actors of the RwL as the object of consideration in order to generate new knowledge (K2). In doing so, the accompanying research depends on the cooperation of those who are responsible for the RwL project research. Simultaneously, to a certain extent the latter also become the research object (Defila & Di Giulio, 2018a). At the same time, the accompanying research learns a lot of what happens in the RwL, which corresponds to a '*learning about*' the RwL in Freeth and Vilsmaier (2019). In order not to jeopardize the collaboration, Defila and Di Giulio (2018a) strongly advise to observe the principle of informed consent and to ensure anonymity and confidentiality.
- **Create the framework for the cooperation** of the other members of the RwL: Especially for the generation of integrated knowledge (K3), the accompanying research can invite the other team members to cooperate, even without making substantial content contributions (Defila & Di Giulio, 2018a). This corresponds to a '*learning for*' the RwL according to Freeth and Vilsmaier (2019). Here, the accompanying research has only an indirect influence on the quality of the results. This requires a solid process design to facilitate high-quality results as well as to adapt the objective of the cooperation to the capacities and interests of the RwL team (Defila & Di Giulio, 2018a).
- **Content-related collaboration** with the members of the RwL team: Likewise for the generation of integrated knowledge (K3), the

[2] The two types of relationships not discussed in depth are: (a) no specific relationship to the projects, with the accompanying research barely running in parallel to other projects in the same funding line; and (b) RwL as data sources for the accompanying project, without further interaction or collaboration (see Defila & Di Giulio 2018a).

accompanying research can also make its own content-related contributions, building on the design of the collaboration of the RwL team (Defila & Di Giulio, 2018a). This depends on having the appropriate competencies and corresponds to a '*learning with*' and '*learning for*' the RwL according to Freeth and Vilsmaier (2019). Therefore, a solid process design is important as well as assuring joint interests in and mutual benefits from the collaboration (Defila & Di Giulio, 2018a). Negotiating explicit and shared goals for the cooperation helps to secure this.

6.3.1 A Dynamically Balanced, Appropriately Related and Reflexive Design of Accompanying Research

The relationship between the accompanying researcher and the other members of the research team will be marked by three balancing acts, and should be designed according to three core practices (Freeth & Vilsmaier, 2019):

> First, this includes a balance between (more distanced) *observation* and (more involved) *participation* (Freeth & Vilsmaier, 2019). Oriented towards the core practice of "**dynamic proximity**", the accompanying research will continuously seek to find the 'right' distance to the accompanied project (Freeth & Vilsmaier, 2019, p. 62). For example, the accompanying research should be close enough to the research process to see the details and distant enough to see the bigger picture. Similarly, it should be close enough to identify opportunities "for team reflection, but not too close that this happens solely" at the suggestion of the accompanying researcher (Freeth & Vilsmaier, 2019, p. 62).
>
> Second, Freeth and Vilsmaier (2019) suggest a balance of *scientific curiosity* and *feeling responsible* and care for the success of the research process. The core practice here is ***critical (self-)reflexivity*** towards one's own roles, interests and influences (ibid.). In this balance, the accompanying research remains attentive in 'research mode' on the one hand, but waits for the right time to investigate so as not to disturb the team's research processes. They feel care for the research project and its members and the effect of the accompanying research on it, but without becoming overly engaged and

"overinvested in the team's research success" (Freeth & Vilsmaier, 2019, p. 63).

Third, the authors strike a balance between "*impartiality* and *investment*" (Freeth and Vilsmaier, 2019, p. 64). Impartiality means trying to be aware of one's own interests, but not taking sides, for example, in decisions about research project activities (ibid.). Involvement here describes openly taking a position when the accompanying researcher's own interests are directly affected. Here, accompanying research can be based on the core practice of ***appropriate relatedness***: it on occasion explicitly contributes as a quasi-independent observer, rarely introduces perspectives based on its insider experience of the project and often assumes a mediating position that supports an open exchange of different views in the project (Freeth & Vilsmaier, 2019).

Freeth and Vilsmaier (2019) emphasize that the role of the accompanying research, and also the way it shapes the three balancing acts, should be dynamically changeable in the course of the research project. While in some situations an observing, impartial role characterized by scientific curiosity may better support the participants' cooperation and learning, in other situations taking a stronger position and actively assuming responsibility may be more appropriate (ibid.).

In my experience, accompanying research to the BaWü Labs was marked by seeking to maintain the right balance and to enable dynamic proximity. At a higher level, accompanying research was offered by two different teams that had different focuses and operated in slightly different ways. This included a team from Basel University, who had a stronger focus on facilitating knowledge integration and systematic dialogues between the labs, offering coaching and consultancy (Defila & Di Giulio, 2018a). A second team, of which I was part of, comprised Leuphana University, the Institute for Social-Ecological Research (ISOE) and the Wuppertal Institute. It focused more strongly on developing overarching insights based on observations and surveys, systematic literature work and catalysing exchange between the labs with wider expert circles (Bergmann et al., 2021). Accordingly, both teams found different answers to the three balancing acts, for instance of being scientifically curious but not feeling (too) responsible, of observing and/or participating, and between investment and impartiality. As outlined by Defila and Di Giulio (2018a), these answers and related positioning corresponded to the differences in contributions sought. From a wider perspective, I assume that setting up an accompanying research in the form of different teams is a good way to clarify expectations and to offer accompanied projects and funders a range of contributions and relationship options.

6.3.2 Accompanying Research Is Not Evaluation

Accompanying research differs from classical evaluation or assessment of a research project. While the accompanying research does aim to gain overarching insights, including those based on comparative analysis, it does not take on an evaluative role on the basis of its own findings (Bergmann et al., 2021; Defila & Di Giulio, 2018a; Weith et al., 2019). Providing an evaluation of, for instance, achievements and failures of the accompanied project could fundamentally jeopardize the trusting cooperation with the accompanied research project. This is particularly true when insights are made available to others, such as the funder. Related conflicts of interest, including relations of responsibility and loyalty to actors with different interests, are difficult to resolve. As a rule of thumb, accompanying research should therefore maintain its independence, especially in relation to actors outside 'their' real-world labs (e.g. the funding agency) (Defila & Di Giulio, 2018a).

At the same time, it is conceivable that the accompanying research supports the process of reflexivity and self-assessment of those being researched (Freeth & Vilsmaier, 2019; see also van Mierlo et al., 2010). Accordingly, accompanying research provides the accompanied project with information on processes and performance, possibly acting as a 'critical friend'. In this function, accompanying research overlaps with a project's internal reflexive monitoring and evaluation (Klaassen et al., 2021; Verwoerd et al., 2020). It aims to be sufficiently close to fully understand the issues encountered, but with sufficient distance to legitimately and critically reflect on the process (Klaassen et al., 2021, p. 233). Overall, the accompanying researcher should bring in the insights of their own assessments, for instance on difficulties of collaboration in the observed research project, in balanced ways and based on prior agreement with the other team members (Freeth & Vilsmaier, 2019, see previous section, cf. also Defila & Di Giulio, 2018a).

In my experience with the BaWü Labs, the novelty of the RwL approach, including a high level of political and scholarly interest, may have led to a very clear need to distinguish between accompanying research and evaluation. This corresponds to the idea of establishing a safe niche space, to try out and test innovations as well as enable learning and improvement, while avoiding immediate 'market' selection pressure (Smith & Raven, 2012). In other situations, with less external interest and political sensitivity, boundaries might be drawn differently or less strictly (see examples of reflexive evaluation above). Transparency regarding the aims and relationships as well as continuous expectation management are advisable in any case (see Table 6.2 for an overview on relationships).

Table 6.2 Overview on types of relationships between accompanying and accompanied research actors

Type of relationship	*Practices and aspects to consider*
RwL process and results as object of observation in order to generate knowledge about processes of the real-world lab (K2)	• Principle of informed consent • Preservation of anonymity and confidentiality
Provide the framework for cooperation among members of the RwL to generate integrated knowledge (K3)	• Solid process design • Adaptation to the capacities and interests of the real-world lab team • Inter- and transdisciplinary process competence of the accompanying researchers
Content collaboration with RwL teams to generate integrated knowledge (K3)	• Solid process design • Adaptation to the capacities and interests of the real-world lab team • Inter- and transdisciplinary process competence of the accompanying researchers • Content competence of the accompanying researchers
Balancing acts, between: a. Observation and participation b. Scientific curiosity and feeling responsible c. Impartiality and Investment	Core practices guiding balancing acts: a. Dynamic proximity b. Critical (self-)reflexivity c. Appropriate relatedness

Based on Defila and Di Giulio (2018a), Freeth and Vilsmaier (2019), and Weith et al. (2019); complemented by Bergmann et al. (2021), Schäpke et al. (2018b), and Wamsler et al. (2020). See main text for detailed references

6.4 Accompanying the 'Co-creative Reflection and Dialogue Space'

6.4.1 Background of the Project

From 2019 to 2023 I was part of an inter- and transdisciplinary research and practice team that repeatedly offered an experimental and co-creative reflection and dialogue space (the 'Co-Creative Reflection and Dialogue Space', or CCRDS) at the UN Framework Convention on Climate Change (UNFCCC) conferences of the parties (COPs), specifically COP 25–28 (Mar et al., 2023; Wamsler et al., 2020, Bruhn et al., this volume, Chapter 7). The CCRDS was led by the Research Institute for Sustainability, Helmholtz Centre Potsdam (RIFS), and realized in cooperation with partners including Lund University, the University of East Anglia, Chalmers University of Technology and Freiburg University. Further partners came from various civil society organisations, international organisations and research bodies. Over the course of the two weeks of each COP, the CCRDS was located in the official 'blue zone' and was part of the side-event programme paralleling the negotiations.

As a co-creative effort, the CCRDS and its partners offered a total of more than 100 workshops to experiment with multiple forms and formats of communication. The focus was on enabling reflection, co-creation and dialogue (see Bruhn et al., this volume, Chapter 7). The overall aim of the CCRDS was to support a more relational and transformational culture of communication and collaboration at the COPs (Schäpke et al., 2023). Participants included negotiators as well as observers coming from many countries and geographical regions. Overall, the CCRDS efforts received very positive feedback from participants, highlighting its capacity to establish a safe, relational and reflexive space (ibid.).

As transdisciplinary and transformation-oriented efforts, the workshops as well as the CCRDS engagement more at large were objects of research. I was part of a small, varying group running the research. During COP 25 and 26, this group included researchers from partner universities of the RIFS, while during COP 27 and 28, it comprised both a RIFS researcher and partners from universities. As researchers, we sought to explore the current culture of collaboration at the conferences, to identify desirable changes from the point of participants, not; and most importantly, to understand and assess ways to change the communication culture. The latter primarily related to assessing the CCRDS workshops and the CCRDS more broadly. We designed and applied a mixed-methods approach. With some variation, the research design combined surveys, semi-structured expert interviews, participatory observation and reflective sessions in the team (Wamsler et al., 2020).

6.4.2 Contributions and Relations

In retrospect, my role in the team can be described as accompanying research, in that I was generally independent from the RIFS team facilitating the CCRDS and the workshops. The objective of the accompanying research was to support the generation of ***knowledge about processes in the CCRDS (K2)*** (Table 6.3). Related research questions included: How do the participants perceive the workshops and the applied methods? What possibilities for improvement do they see? To what extent is the CCRDS a good example of a desirable new culture of collaboration? We drew on participant observation as well as expert interviews with workshops hosts and COP decision-makers, complemented with participant surveys. Research had a longitudinal character, covering various COPs.

In addition, the accompanying research retrospectively aimed at generating **integrated** ***knowledge about the object of observation of the CCRDS (K3)***, i.e. the current and the desirable future culture of COP collaboration and communication. This included research questions such as: How can the current culture of communication and collaboration be characterized? Which role do relations of (dis)trust play to enable or restrict climate action? What underlying mindsets are prevalent in the present culture and how should this be different? Here, data contributions included surveys from participants and expert interviews with decision-makers from various organizations at the COPs. Furthermore, it included setting up joint reflection meetings as well as collaborative publications to integrate insights from different actors engaged in the CCRDS. Again, research had a longitudinal character, covering various COPs.

The CCRDS and related accompanying research had a strongly collaborative approach, meaning that there was a distribution of tasks, for example, between workshop facilitation and related research. Yet, during the process and when gathering results, everybody worked closely together. Roles and relationships as well dynamically evolved.

The forms of relationship between me as accompanying researcher and the other members of the team were diverse (see Table 6.3 for details).

Table 6.3 Accompanying research to the Co-Creative Reflection and Dialogue Space

Dimension to structure and design the accompanying research	
Aimed for contribution	*Methods and tools*
Knowledge about the process (K2): Understand and assess the CCRDS and its workshops	• Shared reflection sessions • Participant observations • Expert interviews with workshop hosts • Surveys from CCRDS participants
Integrated Knowledge about the topic of the CCRDS (K3): Understand and assess need for and ways towards a transformative communication culture at COPs	• Expert interviews with COP decision-makers • Surveys from CCRDS participants • Collaborative scientific publications • Collaborative policy/practice-oriented publications
Relationships to actors	**Practices and principles**
CCRDS as object of observation in order to generate knowledge about its processes and workshops (K2)	• Principle of informed consent from all partners • Preservation of anonymity and confidentiality of workshop participants and external workshop hosts • Co-development of publications with CCRDS team • Framework setting standards for joint publications
Provide the framework for cooperation among members of the CCRDS to generate integrated knowledge (K3)	• Develop overarching research design, including definition of conceptual frameworks and units of analysis, adjusted to practice needs/ understandings • Assure solid overall process design • Framework setting standards for joint publications
Content collaboration with CCRDS teams to generate integrated knowledge (K3)	• Develop concrete tools, including guidelines for integrative reflection, data collection • Co-create solid overall and in-depth process design and quality assurance

In part, they were oriented towards seeing the ***CCRDS as an object of research***, when aiming to be rigorous in the research design, or attending workshops as participating researcher. In part, it leant more towards ***setting up a framework or workshops for collaboration*** among CCRDS team members on specific areas of mutual interest. Lastly, it included guiding ***content-oriented collaboration*** on shared areas of interest, by providing research tools and instructions and co-leading shared publication processes (results include Fraude et al., 2021; Mar et al., 2023; Wamsler et al., 2020). Beyond this, the relationship between accompanying and accompanied research sometimes became blurred, such as when I provided feedback on workshop design, or was ad hoc supporting the realization of the CCRDS and specific workshops.

These plural relationships had ups and downs. They were benefical, for instance in allowing for immersive understandings of ongoing processes. Yet, they were also challenging due to the blurring of roles and the recurrent need to clarify the allocation of tasks and expectations. This also meant that communication was multifaceted. On the one hand, it often took place in the form of a very open and trusting exchange that as well allowed to address personal needs and challenges, building on joint reflection sessions and ad hoc meetings. On the other hand, it was content-oriented and task-driven, such as in the development of research tools or joint publications. These varying relationship and related tasks were partly clarified in advance, and partly emerged during the process.

6.4.3 Reflections on the Balancing Acts

Reflecting the CCRDS accompanying research experience through the lenses of the balancing acts and related core practices of dynamic proximity, critical (self-) reflection and appropriate relatedness (see Freeth & Vilsmaier, 2019) is instructive.

Dynamic proximity relates to the need of balancing participation and observation (ibid). Therefore, we as the wider CCRDS team at a higher level agreed on main areas of responsibility (research or facilitation) and related decision-making capacities. Furthermore, and given that the CCRDS team established an open and co-creative working culture, the overall relation in the team can be described as rather close and fluid (i.e. proximate). I was able to participate in workshops, which was one of the highlights of working with the CCRDS team, allowing me to gain in-depth, first-hand insights. To not unsettle the workshop aims and facilitation, participation should, however, be based on prior agreement with the workshop hosts and be made transparent to the other participants (e.g. active participation vs. participation as observer).

Critical (self-)reflexivity of one's own roles, interests and influences when balancing scientific curiosity and 'feeling responsible' (see Freeth &

Vilsmaier, 2019) proved to be rather challenging, particularly in the heat of the moment at climate conferences. Generally, a practice of team reflection and individual self-inquiry supported reflexivity and the continous evolution of the CCRDS design, its aims as well as given roles and responsibilities. A repeated challenge related to finding appropriate ways to assure that research and data collection actually took place, without disturbing the wider CCRDS process. This included finding good moments to ask participants to complete surveys or give interviews, fitting the flow of the workshops. Furthermore, it included asking colleagues for help to, for instance, collect data in a packed workshop programme and a very hectic overall conference, which makes self-organisation and coordination central.

From experience, allowing procedures to evolve gradually, can reduce unnecessary stress, as can institutionalizing continuous points of exchange with further team members to make sure that learning and iteration (can) actually happen. The practical aspects of conducting research proved very important (e.g. placing surveys visibly on site), to make research happen (easily) and free up capacities for participation. Developing an appropriate level of aspiration on what can be achieved, including a 'plan B', and embracing the idea of skilful improvisation, is helpful in ensuring appropriate accompanying research work under emergent and 'imperfect' conditions.

Appropriate relatedness, balancing impartiality and 'putting oneself (and one's research demands) first' (see Freeth & Vilsmaier, 2019), was a daily practice (and struggle). The CCRDS team established flat hierarchies and decision-making procedures, including to openly discuss accompanying research's and CCRDS's processes. It was at times difficult for me to bring in my perspective on various aspects of the CCRDS process, while abstaining from getting (too) engaged and taking positions in decisions beyond the domain of the accompanying research. This specifically occurred when a decision seemed particularly important for the course of the overall project, and/or I had strong opinions regarding what I deemed a good or bad idea. The established open communication culture made it possible to (often) bring up possible tensions and difficulties and resolve them. Relations with the CCRDS RIFS team thereby were constantly being re-negotiated. While generally beneficial, from my perspective the varying roles also created difficulties, leading to insecurity about appropriate action, misunderstandings and lack of planning security. A practice of constant (self-)reflection, negotiation and adaptation of tasks and roles is influenced by personal strengths and preferences and requires supportive conditions that give room for reflection and understanding. In my view, a middle ground between flexibility and adaptation and continuity is advisable.

Lastly, there is an inherent tension in working in a co-creative and dynamic process and environment, and the need for the prior planning and arrangement of research, including in relation to data-collection. COPs are likely to be chaotic and hectic, and we witnessed not only strict and ever-changing COVID-19-related regulations but also last-minute information policies on the part of the organizers, or challenging event organization.[3] Accordingly, as an accompanying researcher you and your work not only depend on the decisions of the project being researched (here the CCRDS), but you are also affected by wider developments influencing the accompanied project. In part, the dependence of the accompanying research can be mediated by adapting its aims, but it might as well lead to lasting research difficulties or even the (partial) failure of the accompanying project. Implications of dependencies and possible remendiations and changes should be made transparent and (where suitable and possible) agreed upon with the accompanied project team.

6.5 Concluding Remarks

Accompanying research is a promising complement to transformation-oriented research and its innovative and challenging research formats, offering a wide range of options for additional benefits. It can be used to support the generation of knowledge from transformation-oriented research both with regard to its procedural dimension (e.g. how to successfully do transformative research) and in relation to underlying topical aspects (e.g. how do social systems transform). Beyond knowledge generation, accompanying research has the capacity to enhance the actual performance of transformation-oriented research, including by providing opportunities and inputs to increase reflexivity, iteration and adaption of the accompanied research.

[3] As an example: although we had arranged and substantially paid for a professional pavilion set up during one of the COPs, on arrival on site, our pavilion was basically non-existent and the local organizers had very limited capacity to set it up any further, requiring our constant improvisation and new skills.

In this chapter, I drew on existing scholarship to present a differentiation of contributions accompanying research can make and showed a variety of methods to put this into practice. This includes different contributions to research and knowledge generation, as well as procedural support. Building on existing scholarship, I also presented possibilities for shaping the relationship between accompanied and accompanying research and discussed various balancing acts required to shape transparent and effective relationships that are accepted by the various actors. The different contributions and the forms of relationships can be used to reflect and design accompanying research work. I used them to both reflect experiences from accompanying the BaWü Labs (Baden-Württemberg, Germany), as well as the Co-Creative Reflection and Dialogue Space manifested at UNFCCC COPs. The differentiation of contributions of accompanying research as well as of the relationships with stakeholders, underlying balancing acts and core practices, proved to be highly useful dimensions for the reflection of experiences.

Looking ahead, these dimensions are promising in terms of orienting the design of accompanying research for future transformation-oriented research projects. This could include considering the knowledge contributions to be achieved, and related methods and processes. The reflection of possible relationships to the various stakeholders including the consideration of necessary balancing acts can inform decisions to shape relations, set expectations and foresee possible tensions and challenges. This process should include a reflection and exchange between accompanying and accompanied research on the contextual conditions that enable or constrain relationships of dynamic proximity and balancing observation and participation. Successful accompanying research—as does research more generally—depends on appropriate framing conditions, including funding. Given the inherent challenges of transformation-oriented research, and the urgency of understanding and working towards transformation, accompanying research in its various forms holds strong potential to be both an effective and meaningful contribution.

Acknowledgements A heartfelt thank you to my colleagues from the CCRDS team at RIFS and beyond, for allowing me to work with you and learn from you over the years. A similar cordial thanks to my colleagues from the accompanying project ForReal located at Leuphana University, Institute for Socio-Ecological Research and Wuppertal Institute, as well as the accompanying research team of University of Basel, for years of collaboration, advice and joint learning possibilities.

This is a strongly modified and translated version of: Schäpke, N. (2023). Begleitforschung in Reallaboren. In S. L'Orange Seigo, M. Probst, M.l Stauffacher, E. Lobsiger, & Y. Blumer (Eds.), *Interventionen in Reallaboren: Ein Handbuch für die Praxis* (pp. 264–281). ETH Zurich. https://doi-org.vu-nl.idm.oclc.org/10.3929/ethz-b-000608410

An initial translation of Schäpke, N. (2023), which was strongly modified during the publication process, had been created in 2022 using support of DeepL Pro.

References

Belcher, B., & Halliwell, J. (2021). Conceptualizing the elements of research impact: Towards semantic standards. *Humanities and Social Sciences Communications, 8*(1), 183. https://doi-org.vu-nl.idm.oclc.org/10.1057/s41599-021-00854-2

Bergmann, M., Schäpke, N., Marg, O., Stelzer, F., Lang, Sußmann, N., et al. (2021). Transdisciplinary sustainability research in real-world labs: Success factors and methods for change. *Sustainability Science*, *16*(2), 541–564. https://doi-org.vu-nl.idm.oclc.org/10.1007/s11625-020-00886-8

Caniglia, G., Luederitz, C., von Wirth, T., Fazey, I., Martín-López, Lang, D. J., et al. (2020). A pluralistic and integrated approach to action-oriented knowledge for sustainability. *Nature Sustainability*. https://doi-org.vu-nl.idm.oclc.org/10.1038/s41893-020-00616-z

Defila, R., & Di Giulio, A. (2018a). What is it good for? Reflecting and systematizing accompanying research to research programs. *GAIA—Ecological Perspectives for Science and Society*, *27*(1), 97–104. https://doi-org.vu-nl.idm.oclc.org/10.14512/gaia.27.S1.17

Defila, R., & Di Giulio, A. (2018b). *Transdisziplinär und transformativ forschen*. Springer. https://doi-org.vu-nl.idm.oclc.org/10.1007/978-3-658-21530-9

Defila, R., & Di Giulio, A. (2019). *Transdisziplinär und transformativ forschen. Eine Methodensammlung*. Band 2. Springer. https://doi-org.vu-nl.idm.oclc.org/10.1007/978-3-658-27135-0

Fazey, I., Schäpke, N., Caniglia, G., Patterson, J., Hultman, J., Wyborn, C., et al. (2018). Ten essentials for action-oriented and second order energy transitions, transformations and climate change research. *Energy Research and Social Science*, *40*. https://doi-org.vu-nl.idm.oclc.org/10.1016/j.erss.2017.11.026

Fazey, I., Schäpke, N., Caniglia, G., Hodgson, A., Kendrick, I., Young, H. R., et al. (2020). Transforming knowledge systems for life on Earth: Visions of future systems and how to get there. *Energy Research and Social Science, 70*. https://doi-org.vu-nl.idm.oclc.org/10.1016/j.erss.2020.101724

Fiedeler, U., Nentwich, M., Simko, M., & Gazso, A. (2010). What is accompanying research on nanotechnology? *NanoTrust Dossiers*, *11*(July), 1–5. epub.oeaw.ac.at/ita/nanotrust-dossiers

Fraude, C., Bruhn, T., Stasiak, D., Wamsler, C., Mar, K., Lawrence, M., et al. (2021). Creating space for reflection and dialogue: Examples of new modes of communication for empowering climate action. *GAIA-Ecological Perspectives for Science and Society, 30*(3), 174–180. https://doi-org.vu-nl.idm.oclc.org/10.14512/gaia.30.3.9

Freeth, R., & Vilsmaier, U. (2019). Researching collaborative interdisciplinary teams. *Science & Technology Studies*, *33*(3), 57–72. https://doi-org.vu-nl.idm.oclc.org/10.23987/sts.73060

Klaassen, P., Verwoerd, L., Kupper, F., & Regeer, B. (2021). Assessment of responsible innovation. In E. Yaghmaei & I. Van De Poel (Eds.), *Assessment of responsible research and innovation. Methods and practices*. Routledge.

Luederitz, C., Schäpke, N., Wiek, A., Lang, D. J., Bergmann, M., Westley, F. R., et al. (2017). Learning through evaluation—A tentative evaluative scheme for sustainability transition experiments. *Journal of Cleaner Production, 169*. https://doi-org.vu-nl.idm.oclc.org/10.1016/j.jclepro.2016.09.005

Mar, K. A., Schäpke, N., Fraude, C., Bruhn, T., Wamsler, C., Lawrence, M. G., et al. (2023). *Learning and community building in support of collective action: Toward a new climate of communication at the COP. July 2022*, 1–12. https://doi-org.vu-nl.idm.oclc.org/10.1002/wcc.832

McCrory, G., Schäpke, N., Holmén, J., & Holmberg, J. (2020). Sustainability-oriented labs in real-world contexts: An exploratory review. *Journal of Cleaner Production*, 277. https://doi-org.vu-nl.idm.oclc.org/10.1016/j.jclepro.2020.123202

Miller, T. R. (2013). Constructing sustainability science: Emerging perspectives and research trajectories. *Sustainability Science, 8*(2), 279–293. https://doi-org.vu-nl.idm.oclc.org/10.1007/s11625-012-0180-6

Norström, A. V., Cvitanovic, C., Löf, M. F., West, S., Wyborn, C., Österblom, H., et al. (2020). Principles for knowledge coproduction in sustainability research. *Nature Sustainability, 3*(3), 182–190. https://doi-org.vu-nl.idm.oclc.org/10.1038/s41893-019-0448-2

Patterson, J., Schulz, K., Vervoort, J., van der Hel, S., Widerberg, O., Barau, A., et al. (2017). Exploring the governance and politics of transformations towards sustainability. *Environmental Innovation and Societal Transitions, 24*, 1–16. https://doi-org.vu-nl.idm.oclc.org/10.1016/j.eist.2016.09.001

Schäpke, N., Bergmann, M., Stelzer, F., & Lang, D. J. (2018a). Labs in the real world: Advancing transdisciplinary research and sustainability transformation: Mapping the field and emerging lines of inquiry. *GAIA, 27*(1), 8–11. https://doi-org.vu-nl.idm.oclc.org/10.14512/gaia.27.S1.4

Schäpke, N., Stelzer, F., Caniglia, G., Bergmann, M., Wanner, M., Lang, D. J., et al. (2018b). Jointly experimenting for transformation? Shaping real-world laboratories by comparing them. *GAIA*, *27*. https://doi-org.vu-nl.idm.oclc.org/10.14512/gaia.27.S1.16

Schäpke, N., Stelzer, F., Marg, O., Bergmann, M., Miller, E., Lang, D. J., et al. (2017). Urban BaWü-Labs: Challenges and solutions when expanding the real-world lab infrastructure. *GAIA*, *26*(4). https://doi-org.vu-nl.idm.oclc.org/10.14512/gaia.26.4.19

Schäpke, N., Beyers, F., Fraude, C., Mar, K., Schroeder, H., Lawrence, M. G., et al. (2023). *Research and practice to scale co-creation and reflection* (Issue November). https://publications.rifs-potsdam.de/rest/items/item_6003201_2/component/file_6003202/content

Schäpke, N., Wagner, F., Beecroft, R., Rhodius, R., Wanner, M., Parodi, O., et al. (2024). Impacts of real-world labs in sustainability transformations—Forms of impacts and ways to create it, associated challenges and methodological advances. *GAIA—Ecological Perspectives for Science and Society, SI*.

Schneidewind, U., Augenstein, K., Stelzer, F., & Wanner, M. (2018). Structure matters: Real-world laboratories as a new type of large-scale research infrastructure: A framework inspired by Giddens' structuration theory. *GAIA—Ecological Perspectives for Science and Society*, *27*(1), 12–17. https://doi-org.vu-nl.idm.oclc.org/10.14512/gaia.27.S1.5

Scoones, I., Stirling, A., Abrol, D., Atela, J., Charli-Joseph, L., Yang, L., et al. (2020). Transformations to sustainability: Combining structural, systemic and enabling approaches. *Current Opinion in Environmental Sustainability, 42*, 65–75. https://doi-org.vu-nl.idm.oclc.org/10.1016/j.cosust.2019.12.004

Smith, A., & Raven, R. (2012). What is protective space? Reconsidering niches in transitions to sustainability. *Research Policy, 41*(6), 1025–1036. https://doi-org.vu-nl.idm.oclc.org/10.1016/j.respol.2011.12.012

United Nations General Assembly. (2015). *Transforming our world: The 2030 agenda for sustainable development*.

van Mierlo, B., Regeer, B., van Amstel, M., Arkestejin, M., Beekman, V., Leeuwis, C., et al. (2010). *Reflexive monitoring in action. A guide for monitoring system innovation projects*. Wageningen University and Vrije Universiteit Amsterdam

Verwoerd, L., Klaassen, P., Veen, S. C. Van, & Regeer, B. J. (2020). Combining the roles of evaluator and facilitator: Assessing societal impacts of transdisciplinary research while building capacities to improve its quality. *Environmental*

Science and Policy, *103*(November 2019), 32–40. https://doi-org.vu-nl.idm.oclc.org/10.1016/j.envsci.2019.10.011

von Wehrden, H., Guimarães, M. H., Bina, O., Varanda, M., Lang, D. J., Lawrence, R. J., et al. (2019). Interdisciplinary and transdisciplinary research: Finding the common ground of multi-faceted concepts. *Sustainability Science*, *14*(3), 875–888. https://doi-org.vu-nl.idm.oclc.org/10.1007/s11625-018-0594-x

von Wirth, T., Fuenfschilling, L., Frantzeskaki, N., & Coenen, L. (2019). Impacts of urban living labs on sustainability transitions: Mechanisms and strategies for systemic change through experimentation. *European Planning Studies*, *27*(2), 229–257. https://doi-org.vu-nl.idm.oclc.org/10.1080/09654313.2018.1504895

Wagner, F., Schäpke, N., Stelzer, F., Bergmann, M., & Lang, D. J. (2016). BaWü-labs on their way: Progress of real-world laboratories in Baden-Württemberg. *GAIA*, *25*(3). https://doi-org.vu-nl.idm.oclc.org/10.14512/gaia.25.3.21

Wamsler, C., Schäpke, N., Fraude, C., Stasiak, D., Bruhn, T., Mundaca, L., et al. (2020). Enabling new mindsets and transformative skills for negotiating and activating climate action: Lessons from UNFCCC conferences of the parties. *Environmental Science and Policy*, *112*, 227–235. https://doi-org.vu-nl.idm.oclc.org/10.1016/j.envsci.2020.06.005

Weith, T., Rogga, S., Zscheischler, J., & Gaasch, N. (2019). Beyond projects: Benefits of research accompanying research: Reflections from the research programme Sustainable Land Management. *Gaia*, *28*(3), 294–304. https://doi-org.vu-nl.idm.oclc.org/10.14512/gaia.28.3.10

Wiek, A., & Lang, D. J. (2016). Transformational sustainability research methodology. In H. Heinrichs (Ed.), *Sustainability science* (pp. 31–41). Springer. https://doi-org.vu-nl.idm.oclc.org/10.1007/978-94-017-7242-6_3

Williams, S., & Robinson, J. (2020). Measuring sustainability: An evaluation framework for sustainability transition experiments. *Environmental Science & Policy*, *103*(October 2019), 58–66. https://doi-org.vu-nl.idm.oclc.org/10.1016/j.envsci.2019.10.012

Wittmayer, J. M., & Schäpke, N. (2014). Action, research and participation: Roles of researchers in sustainability transitions. *Sustainability Science*, *9*(4). https://doi-org.vu-nl.idm.oclc.org/10.1007/s11625-014-0258-4

Wittmayer, J. M., Schäpke, N., van Steenbergen, F., & Omann, I. (2014). Making sense of sustainability transitions locally: How action research contributes to addressing societal challenges. *Critical Policy Studies*, *8*(4). https://doi-org.vu-nl.idm.oclc.org/10.1080/19460171.2014.957336

CHAPTER 7

Safe Spaces in Unsafe Environments—Experiences from COP26 About Hosting Inclusive Spaces for Deep Encounters and Reflection

Thomas Bruhn, Niko Schäpke, Carolin Fraude, Lena Bendlin, Azza Dirar, Man Fang, Karen Finn, Maureen Goodman, Laureline Simon, Stefanie Kunkel, Shantanu Mandal, Natalie Mangondo, Kathleen A. Mar, Christine Peringer, Heike Schroeder, Joseph A. Siegel, Juliette Tronchon, Calder Tsuyuki-Tomlinson, Malika Virah-Sawmy, Christine Wamsler, and Mark G. Lawrence

7.1 Introduction—The Need for 'Safe-Enough' Communication Spaces in Transdisciplinary Research

There has been steadily growing attention, both academic and in the broader society, to sustainability and socio-ecological transformation (Clark et al., 2005; Kates et al., 2001). Increasingly, researchers believe

T. Bruhn (✉) · N. Schäpke · C. Fraude · M. Fang · S. Kunkel · K. A. Mar · M. Virah-Sawmy · M. G. Lawrence

B. J. Regeer et al. (eds.), *Transdisciplinarity for Transformation*,
https://doi.org/10.1007/978-3-031-60974-9_7

that the current challenges of sustainability and global warming are best described and addressed from an understanding of complex adaptive

Research Institute for Sustainability (RIFS)—Helmholtz Centre Potsdam, Postdam, Germany
e-mail: thomas.bruhn@rifs-potsdam.de

N. Schäpke
e-mail: niko.schaepke@ifp.uni-freiburg.de

C. Fraude
e-mail: carolin.fraude@rifs-potsdam.de

S. Kunkel
e-mail: stefanie.kunkel@rifs-potsdam.de

M. Virah-Sawmy
e-mail: malikavs@gmail.com

M. G. Lawrence
e-mail: mark.lawrence@rifs-potsdam.de

N. Schäpke
Freiburg University, Freiburg Im Breisgau, Germany

C. Fraude · M. Goodman · S. Mandal
Brahma Kumaris, London, UK
e-mail: maureen.goodman@uk.brahmakumaris.org

L. Bendlin
Process Management Office of Berlin's Districts, Berlin, Germany

A. Dirar · H. Schroeder
University of East Anglia, Norwich, England

M. Fang
School of Public Policy and Management, Tsinghua University, Beijing, China

K. Finn
University of Bedfordshire, Luton, England

L. Simon
One Resilient Earth, Berlin, Germany
e-mail: laureline@oneresilientearth.org

N. Mangondo
Ministry of Environment, Climate, and Wildlife, Harare, Zimbabwe

systems (Clark & Harley, 2020; Espinosa & Porter, 2011; Kay et al., 1999; Liu et al., 2007; Ravetz, 2006; Steffen et al., 2011; Waltner-Toews et al., 2008). In academic institutions this has led to a growing call for transdisciplinary research (TDR) as a mode of addressing complex social challenges in a more encompassing way (Bergmann et al., 2012; Felt, 2010; Hirsch Hadorn et al., 2006; Jahn, Lang et al., 2012; Thompson Klein, 2004). The intention of transdisciplinary approaches is to generate results that draw on a more inclusive knowledge base and are hence better able to provide viable responses to complex, 'wicked' (Rittel & Webber, 1973) and ill-defined problems (Lawrence et al., 2022). Involving non-academic perspectives from the outset and throughout all stages of the research process is deemed an appropriate way to generate knowledge that adequately meets sustainability questions, which involve multiple different stakeholders and epistemologies (Newig et al., 2019). Since TDR researchers have to engage with non-academic perspectives throughout the research process, this chapter focuses on the kind of communication practices that are appropriate and effective for engaging with people from potentially very different backgrounds.

The diversity of perspectives is considered essential to ensure a just and fair research process and is valued as a necessary resource and contribution to develop a comprehensive understanding of an issue of shared interest among everyone included in the research process. Ensuring that

C. Peringer · J. A. Siegel
Mediators Beyond Borders International, Washington, D.C, USA
e-mail: christine@peringer.ca

J. A. Siegel
e-mail: jsiegel@law.pace.edu

J. Tronchon
ProVeg International, Berlin, Germany
e-mail: juliette.tronchon@proveg.com

C. Tsuyuki-Tomlinson
Blavatnik School of Government, University of Oxford, Oxford, UK

C. Wamsler
Lund University Center for Sustainability Studies (LUCSUS), Lund, Sweden
e-mail: christine.wamsler@lucsus.lu.se

participants can engage meaningfully with the different perspectives does, however, present several practical challenges, including how to design and host the communication process. Recommendations for ideal-type TDR process suggest a careful process design, considering whom to include at which points and how in order to achieve effective and fluid interactions. Real-life transdisciplinary processes, however, often take place under non-ideal circumstances. For effective transdisciplinary processes it is important that those involved establish a communication and relationship culture in which their different perspectives can connect and interact meaningfully. This in turn requires everyone involved to engage in dialogue about controversies and differences in a constructive way. While it might not be necessary to achieve a full consensus, it is crucial that potentially different or even mutually exclusive perspectives can be voiced and managed in a way that is respectful to everyone involved. Consequently, research processes need to be designed in such a way that participants can establish mutual trust as the basis of a good and effective relationship. We could refer to such spaces as 'safe enough', meaning that all participants feel comfortable to engage in differences and vulnerabilities or uncertainties, etc., with no fear of getting hurt but being able to stay in constructive relationship with the other participants.

7.1.1 Specific Challenges in Transdisciplinary Communication and Interaction

Against this background, scientists have paid increasing attention to modes of communication and interactions among scientists, policymakers and civil society with respect to their co-creative potential (Bruhn et al., 2019; Lawrence et al., 2022; Nanz et al., 2017). Participants with different backgrounds and perspectives (such as their academic discipline, culture, ontological and cosmological differences, etc.) may construct knowledge and meaning differently and hence run into misunderstandings and related conflicts. All participants in a transdisciplinary process are challenged to—at least temporarily—leave behind their seemingly clear and well-founded understandings and assumptions about their own knowledge. Through the lenses of different perspectives, the same phenomena might be interpreted differently. This may create situations outside the participants' ´comfort zone´. For the researchers it means that in these communication processes they need to acknowledge the non-objective nature of their research and include reflexive practices in the research process (Fazey et al., 2018; Lang, Wiek et al., 2017; Popa et al., 2015). Engaging in communication with people with diverse perspectives brings various challenges to a research process with which academics from conventional disciplinary settings are usually unfamiliar. This may also be influenced by the level of experience of those involved, for instance if for some researchers a certain disciplinary rigour is critical to establishing their identity or career path.

Besides an interest in procedural and institutional aspects, there is also a growing interest in the role of mindsets and mental models for effective transdisciplinary interaction. This includes both the relevance of, and possible engagement with the participants' inner lives—notably beliefs, values, worldviews, emotions and motivations—in transdisciplinary communication (Brink et al., 2019; Creutzig & Kameier, 2020; Grothmann, 2018). It is claimed that these often-disregarded dimensions of transformation have strong leverage in driving change towards sustainability—not only in relation to individual agency, but also to groups in all sectors, including business, government and education (Wamsler, 2020; Woiwode et al., 2021). At the same time, the term 'inner' might be misleading as it suggests a dichotomy between 'inner' and 'outer' dimensions of change. Many writers therefore tend to integrate the more subjective 'inner' aspects into a 'relational' understanding of transformation (Walsh et al., 2020; West et al., 2020).

This raises questions about how to host communication formats or 'spaces' in which people with different perspectives and viewpoints can interact in such a way that they can not only talk *at* each other but engage in genuine mutual learning *with* each other. Co-creation of knowledge may require an atmosphere in which participants do not feel pushed to defend their positions but feel safe enough to openly explore the methodological and normative assumptions underlying the various positions on a given topic. Spaces and formats that emphasize honesty, openness and trust as foundations of their communication culture can strengthen connectedness to oneself and others (Wamsler et al., 2020). When these spaces, which are usually based on experiential reflection and communication, are safe enough for disagreements or mistrust to surface and be addressed, they can initiate profound change in a person's life and, consequently, support cultural transformation (Pereira et al., 2020). We further suggest that such formats can enhance the perception and understanding of deeper, common concerns that underlie what appear to be conflicting interests and can help overcome polarization and opposition (Mar et al.,

2021, 2023; Wamsler et al., 2020). In times of increasing social division, it seems especially important to foster a mode of communication that works constructively with potential differences and generates mutual and respectful understanding. Within this context, the former Institute for Advanced Sustainability Studies (IASS) (from 2023 the Research Institute for Sustainability—Helmholtz Center Potsdam (RIFS) has made efforts to experiment with spaces and formats for communication that allow for self-reflection and reciprocal dialogue among stakeholders (Fraude et al., 2021).

It is key for the effectiveness of such spaces that participants, including the researchers, can question definitions and explore potential inconsistencies and vulnerabilities in their interactions. All of this requires a degree of trust. Participants in such a process need to be able to relate to each other on a basis of mutual respect and listening. In practice, however, the presence of diverging or conflicting positions can trigger defence mechanisms in people and groups, particularly when the issue is perceived as a risk or even threat to some participants. This calls for careful efforts to design safe enough conditions for effective communication when researchers try to engage in critical fields such as sustainability. Learning how to design and host such safe spaces or transformative spaces (Pereira et al., 2020) is an ongoing challenge for researchers in transdisciplinary processes and needs to be highly adaptive to each specific research context.

7.1.2 *The UNFCCC COPs as an Exemplary Context for Research on Transdisciplinary Communication*

One of the most pressing issues in relation to sustainability is global warming. Every year, the secretariat of the United Nations Framework Convention on Climate Change (UNFCCC) hosts a major two-week Conference of the Parties (COP). These COPs usually gather some 15,000 to 20,000 representatives from the 193 UN member states and accredited local, national and international organizations. The COPs involve official political negotiations as well as a broad range of side events. Here, a wide spectrum of stakeholders including, among others, academics, civil society organizations (CSOs), Indigenous groups, businesses and youth groups present their knowledge and positions to anyone attending the COP. These side events are dominated by conventional formats, usually in slide-based presentations, or conventional panel discussions, with limited time for questions or audience interaction (Mar, Fraude et al., 2021). More specifically, research on the communication culture at the COPs found that the current culture of communication and negotiation fosters deep-rooted distrust between different stakeholder groups (Wamsler et al., 2020). At the same time, the COPs represent a unique setting for researchers to engage with an exceptionally large spectrum of perspectives and expertise. So, they present an interesting context for researchers to experiment with formats for transdisciplinary communication.

This chapter reflects on our experiences of hosting a space for co-creative reflection and dialogue at COP26 in Glasgow in 2021. Tens of thousands of people from all over the world and all kinds of backgrounds come to the COPs to address climate change. Some part of the COP is dedicated to high-level political negotiations while the COP is also a major venue for delegations of organizations with observer status. The conference takes place in an overall atmosphere of growing urgency and threat of ecological collapse, conflicting (political) interests, uneven historic responsibilities and unfairly distributed impacts of climate change. This tense environment makes it difficult for participants to engage openly with each other, and most sessions are designed around the one-way dissemination of knowledge to participants. Consequently, there is a need for more dialogical exchange and trust-building among stakeholders. Responding to this need, the Co-Creative Reflection and Dialogue Space (CCRDS) offered a series of sessions that invited all COP26 participants to (spontaneously) engage in joint reflection and mutual learning on a range of topics.

The CCRDS represents a transdisciplinary experiment in various ways. First, the individual sessions hosted at the COP are spaces for transdisciplinary dialogue, meaning that they are designed to facilitate conversation and meaning making for diverse participants, including academics. This enables us to observe these sessions and learn something about principles of communication in transdisciplinary settings. Second, the observation and evaluation follow transdisciplinary principles. The findings from methods drawn from the social sciences (participant observation, qualitative surveys) are integrated with the reflections and observations of those involved in hosting sessions, such as the co-authors of this chapter. And third, the entire project of the CCRDS is set up as a TDR project. It began in 2018 based on consultations among scientists and non-scientific stakeholders such as the UNFCCC and several CSOs. Under the formal leadership of the IASS Potsdam (now RIFS) a transdisciplinary group designed the overall process and every year a transdisciplinary group forms around the preparation of the new iteration of the experiment.

We will first describe the COP as a context for transdisciplinary work and the CCRDS as an experimental intervention of transdisciplinary communication. In outlining our findings, we explore the challenges facilitators faced in hosting inclusive spaces in the COP context. For example, aspects such as the noise level and the physical venue affect how far it is possible to offer spaces that aim to be sufficiently safe and inclusive for relationship building as part of a TDR process and knowledge integration.

We also outline some of the specific challenges that hosts encountered when trying to provide and maintain a 'safe enough' atmosphere that would allow participants to engage in conversation outside their usual 'comfort zone' and disclose potential learning edges or vulnerabilities. We describe practical approaches, in terms of the mindset, skillset and toolset that hosts used to tackle these challenges. Based on these experiences and reflections, we offer recommendations on how to design and host inclusive, safe enough spaces in politically contested context such as the COPs.

Our reflections are aimed at researchers who want to experiment with more interactive, dialogical and reflective communication formats, and to facilitators who are not grounded in research but are eager to contribute their expertise in settings such as the COPs. We also hope our findings will be valuable for researchers who are rather new to TDR practices and motivated to contextualize their work in the more 'messy' conditions outside conventional academic research, particularly in a project funded by a third party. We would also hope the chapter will help interested readers to avoid some of the pitfalls when engaging in communication with non-academic perspectives in non-ideal settings. Finally, we aim to motivate researchers by sharing the rewarding experiences of our experimental set-ups in such challenging conditions.

7.2 Investigating the CCRDS at COP26: Background and Methods

Responding to the opportunity to improve the communication culture at the COPs, the IASS Potsdam (now RIFS Potsdam), together with several partners, decided to experiment with and research alternative forms of

Fig. 7.1 Impression of the typical conversation atmosphere in the CCRDS at COP 26

communication at the COPs, where we have developed various iterations of communication formats.

7.2.1 *First Experiments at COP24 and COP25*

The initial experiments with communication formats led by the IASS Potsdam took place during the COP24. They ran in parallel with research activities to assess the demand for alternative forms of communication at the COPs to enhance mutual learning and consequently enable more effective climate action. The experiment started with only a few sessions at COP24 in which participants were invited into genuine encounters and dialogue about affective aspects and potential vulnerabilities regarding climate change, such as climate anxiety or climate grief. These first experimental sessions involved 40 diverse participants (e.g. youth, government representatives and non-government organizations (NGOs)) and the overall feedback was very positive. In particular, participants supported the IASS researchers' assumption regarding an (at least partly) dysfunctional communication culture at the COPs. Positive participant feedback encouraged further research and engagement activities in this direction (Fig. 7.1).

When assessing the options for the COP25 in Madrid, it became clear that the facilities that IASS researchers could rent for limited time-slots at the COP venue would be unsuitable for the intended interactive and dialogical sessions of the type in which the IASS was interested. The pavilions or side-event rooms were designed for conventional presentations or panels and it was difficult or forbidden to rearrange chairs for a conversational session. Therefore, the IASS decided to establish its own science communication space at COP25 in Madrid: The Co-Creative Reflection and Dialogue Space (CCRDS). We rented a 20m^2 room near to the office and pavilion spaces of the different delegations where we hosted 20 interactive workshops, including dialogical sessions, reflective practices (such as silent journaling, guided meditation, associative drawing and other formats) and action-oriented workshops (Fraude et al., 2021). We used a comprehensive research approach, triangulating surveys, participatory observation and expert interviews. Our results confirmed our initial assumptions: COP participants who connected with us in the context of the CCRDS saw a clear need for different forms of communication. They also expressed a wish for a changed communication culture, particularly including dialogical and reflexive processes to foster an atmosphere of deeper trust and collaboration (Wamsler et al., 2020). Once again, CCRDS practices were found to be valuable.

7.2.2 *The CCRDS at COP26*

Consequently, at COP26 in Glasgow, we offered a new iteration of the CCRDS, building on the experiences at previous COPs. This time, the IASS rented a 25m^2 space in the pavilion area in which countries or organizations hold presentations in booths of various shapes and sizes. Two sides of the space were open to the hallways and the other two closed by the neighbouring pavilions. One reason for this set-up was to respect the requirements of social distancing due to COVID-19 and ensure sufficient air flow. The standard arrangement in this space was a circle of 14 movable chairs (see Fig. 7.2).

In this pavilion, a total of 40 sessions were hosted, 15 of which were hosted by IASS staff. The other workshops ('guest sessions') were hosted by people and organizations selected by the IASS before and during the COP. In the lead-up to the conference, the IASS invited organizations/partners to submit proposals for sessions to be hosted in the CCRDS. Altogether, the sessions were attended by ~ 200–250 participants. Further

Fig. 7.2 Set-up of the CCRDS at COP 26, located rather exposed at the crossroads of two highly frequented hallways

details about the workshops can be found elsewhere (Schäpke et al., 2023) and will be published separately. The participants' backgrounds were roughly comparable to the data published about the CCRDS at COP25 in Madrid (Wamsler et al., 2020).

The key idea behind these guest sessions was to enhance the diversity of approaches and the professional backgrounds of hosts and organizations involved in experimenting with communication formats within the framework for sessions within the CCRDS. The IASS also offered two preparation sessions in which those interested in hosting were told about the principles of dialogue and reflective interaction that were meant to be at the centre of the CCRDS. The IASS did not influence which *topics* the applicants chose to address, but gave the applicants instructions about *what kind of values, rules and forms of communication* should be applied in the sessions. The ideas for this design and invitation drew on principles from selected facilitation approaches and concepts such as the *Art of Hosting*, the *Manifesto for Slow Thinking* and *Transformative Learning* (Habermann & Schmidt, 2018; Mezirow & Taylor, 2009; Pogatschnigg, 2021).

Box 1. Principles for good dialogue displayed in the CCRDS at COP 26

- We will treat personal stories and material confidentially.
- We listen to each other with compassion and curiosity.
- We speak with intention: noting what has relevance to the conversation in the moment.
- We stay aware of the impact of our contributions to the circle.
- Suspend judgements, assumptions, and certainties—It is not about knowing who is right or wrong. It is about exploring together and surfacing what we do not know or see yet.
- Connect your thoughts to what was said before.
- Listen together for insights and deeper patterns or questions. Maybe we discover meaning behind meaning.
- Listen together for insights and deeper patterns or questions. Maybe we discover meaning behind meaning.
- Accept that diverging opinions are OK—We do not always need to reach consensus. New ideas can come from putting different perspectives together.
- Focus on what really matters.
- Play, doodle & draw—It can be helpful to use a large sheet in the middle of the group as a space to capture the essence of our collective thinking and reflection.
- Contribute your full self with mind and heart. All of us are invited to be both a professional and a human being.
- Listen with attention.
- Have fun!

The call for guest session proposals attracted stakeholder groups who shared the general intention, hypothesis and interest of the IASS, but also brought their own interpretations on how to translate these into specific communication formats. Many of the guest hosts had limited or no prior experience with hosting dialogical and reflective formats at the COP. Those who were familiar with hosting safe enough, participatory and inclusive communication formats had done so in contexts and under conditions that they had designed themselves, often for their own events. In fact, they would usually consider it crucial to ensure certain ideal-type conditions to provide a certain safety within a space or process.

So, offering a format for deep encounter at a venue like the COPs was a new challenge for most of the guest session hosts. The specific challenges and context factors will be discussed in the findings section.

7.2.3 Workshops to Investigate Hosts' Experiences in the CCRDS at COP26

One of the key research interests for the IASS was to examine how facilitators (either of guest sessions or of sessions hosted by IASS staff) design and host communication spaces and formats that aimed at being inclusive and participatory as a means to enable deep encounters and trust-building among the participants. Therefore, after COP26 and outside the COP venue, the IASS hosted three online workshops in which everyone who had hosted at least one session in the CCRDS discussed the following:

1. How did you perceive COP as context for hosting workshops that aim to be safe, inclusive etc.? Which aspects were essential in influencing the form/quality etc. of providing space?
2. What challenges did you face with respect to providing and maintaining a 'safe-enough space' in this environment?
3. Which practical approaches did you use to tackle these challenges?
4. What specific recommendations would you give others aiming at facilitating 'safe enough' spaces at the COP or similar venues, i.e. where one is not used to this kind of communication, but where it would be very important in light of the challenges facing societies?

The participants discussed these questions both in small groups and in plenary during the workshop session. The responses were collected on in an online whiteboard and participants were invited to add further comments after the workshop.

When discussing their practical approaches, the hosts were invited to cluster their responses in relation to the three competencies *mindset, skillset and toolset* (Fraude, 2021) defined as follows:

> *Mindset* refers to the internal lens through which people see and navigate life, which influences perspectives and attitudes (Wamsler et al., 2020), and covers the ability to observe and understand one's own attitudes, thoughts, feelings, perceptions, and reaction patterns and consciously steer these. It includes the ability to adapt one's inner self, respond to others, and respect

any kind of developments during a session. It also encompasses how one personally embodies the principles and qualities in any given method.

Skillset refers to communication and social skills, such as the ability to design reflection and dialogue processes and related understanding. It also includes the ability to facilitate a diverse group, even if there is internal disagreement. Here, practical experience is crucial, for example, when handling culturally sensitive issues or people who are emotionally more sensitive or reactive than one would usually expect.

Toolset relates to one's repertoire of tools, methods, techniques, instruments, and technologies for reflection and communication in the context of climate- and sustainability-related conferences. For example, it also refers to a practical recognition of the right moment to use a certain method, tool, or technique, and its limitations.

The present chapter is part of the hosts' self-reflexive exercise, led by the IASS-based organizers of the CCRDS. The first author synthesized the workshop findings were synthesized as a draft chapter, on which everyone who had hosted sessions at COP 26 CCRDS were invited to comment, and if they so wished to further collaborate on the chapter as co-authors. This various iterations of the draft. In the following, we present the reflections of hosts on their experiences, practices and recommendations.

7.3 Findings from Experiences of Hosting Inclusive Spaces at COP26

7.3.1 COP as an Overall Context for Hosting

The hosts perceived the conditions at the COP26 venue as extraordinarily challenging in many ways (Fig. 7.3).

On arriving, they felt 'tired from the commute' and from the process of getting into the venue, including long queues and security checks. Hosts felt a 'sense of being uninvited' and it took a 'long time to acclimatize' to the venue. An impression was that the COP 'needs people to feel welcome' but the 'human part [is] put in the corner'.

The general atmosphere was characterized by a confusing sense of breathless busyness, with a 'very up-paced atmosphere' with a 'constant lack of time' and people 'always running around'. One host mentioned that 'the trade fair setup was perfect for retail, to walk around and see what was available *to buy*'. A particular aspect that several hosts emphasized was a sense of inauthenticity and ingenuity. The venue appeared

Fig. 7.3 Surrounding of the CCRDS at COP26. The CCRDS was set up in an open fair venue with high noise levels and the hallways directly at the CCRDS were highly populated

'polished' and 'perfectionist' with 'indifferent people' and 'lots of hollow words'. The impression was that 'people [were] wearing masks not just for Covid'. Despite the presence of people and stakeholder groups from various backgrounds, hosts perceived 'diversity as a pretence'. Others described the COP as 'pretending to be inclusive', but just 'to feel-good' and with 'no real interest'. Some expressed the feeling that the atmosphere was shaped by 'a pressure to be busy, to be important' and wondered whether this busyness was actually some kind of 'protection from actually thinking about the issues'.

A key aspect that many hosts mentioned was the impression of a disconnect between people inside the venue and those who could not get in. One pointed out that 'COP is an exclusive space'. Consequently, some hosts were 'sad about people who were not able to be there' and felt a 'pressure [due to the] privilege to be there'. But even inside the venue, hosts noticed 'many walls and barriers' that made them feel 'isolated'. In the usual communication formats, hosts experienced a 'wall built between speakers and audience'. One said that 'the rows of seats looking at the *expert space* at the front did not encourage listeners to take part—there was often no dialogue, no discussion—just several monologues'. The disconnectedness also related to a sense of 'much separation (between civil society and policy)' and that a large part of the conference was 'disconnected to negotiations'. One host perceived the presence of 'traumatic memories'.

Combined with the 'cold hallways' and the high noise level, hosts experienced the atmosphere as 'confusing' and 'triggering lots of uncertainty' were 'constantly figuring out *what is going on*?' wondering 'what are people's objectives'.

Summarizing their subjective and emotional experience of the venue, hosts described their experience at COP as 'painful', 'very uncomfortable' and 'extremely unsafe'. It was emphasized that 'safety requires the opportunity to be easily heard' but that 'the noise and the busyness prevented that'. At the same time, several hosts reflected that this sense of unsafeness was a subjective perception that was not necessarily shared by all COP participants. While some hosts emphasized that it was 'impossible to acknowledge the profound existential unsafety' of the COP setting, while others felt that 'many COP participants flourished in the noisy and busy atmosphere' with no impression of unsafety. So, in conclusion, it was emphasized that the subjective sense of safety or its absence for a certain kind of activities was very much an individual matter.

7.3.2 Typical Challenges When Hosting Inclusive Spaces at the COP

Hosting a session in the CCRDS with the aim of offering an inclusive and safe enough space for deep encounter presented many specific challenges to the hosts that required specific attention and care. These ranged from dealing with the hectic conditions of the environment described above to challenging dynamics in the sessions due to the open/free set-up of the space to the presence of participants with strongly conflicting perspectives or backgrounds.

Overall, the CCRDS hoped to facilitate 'deep encounters' where people could attend and relate not only as bearers of knowledge and power or as stakeholder representatives but rather as human beings with their own emotions, ambiguities and struggles. The two main challenges in trying to facilitate such encounters were the venue's noisiness and its open setting, particularly as it was not enclosed on two sides, giving it a 'crossroads feeling' because it was situated at the entrance of the hall.

Almost all hosts found the presence of 'lots of distracting noise' difficult to cope with, not only in creating an atmosphere of focus and mutual attention but also in terms of making sure that everyone could understand each other properly. There was even a sense that the sessions in the CCRDS were 'competing with noise all around'. Hosts were wondering 'how will we engage people and get them to speak in an open noisy space?'.

The physical openness of the space posed additional challenges. Because there were no doors, participants saw the 'constant movement of people passing by'. During the sessions 'people [were] walking in and out' or 'people came in during the session or left in-between'. So, while aspiring to be inclusive it was 'difficult for them to come into the meeting' in the middle of a session. It also entailed extra efforts to deal with people who joined spontaneously out of curiosity. Thus hosts wondered 'how to include those that do not come intentionally'. Another common issue was that 'people [were] taking pictures during [a] session'. Such ambiental disturbance as well as during the sessions made it 'difficult to create a sense of containment' (Fig. 7.4).

Various factors arising from the setting were also brought into the session through the participants. Hosts noticed that 'people arrived tired

Fig. 7.4 Perspective from inside the CCRDS showing COP participants passing by during a session

from context' and that participants with leadership roles or perceived authority had a particular impact on the dynamics in the sessions. It was mentioned that it was 'impossible to hold settings against disturbances [from leaders]'. For example, 'leaders brought some *busy* energy into the space' and hosts witnessed that the participants' 'attention [was] unconsciously drawn to men with perceived authority'.

The hosts also struggled with their own inner state. The circumstances were perceived as 'overwhelming' which made it 'hard to focus'. The busy and polished atmosphere also triggered 'self-doubt', inner 'self-talk' and lack of clarity about whom the hosts were actually serving. They found themselves 'comparing' and 'becoming judgmental' about themselves and their hosting rather than staying mindful and appreciating what was present.

A whole field of challenges related specifically to the sessions that attempted to combine online participation with hosting deep dialogues at the COP. Several hosts made considerable efforts to '[be] digitally inclusive' in the sense that they tried to allow people who could not participate in person to join online (video-call or Twitter spaces). Due to the various distractions, hosts experienced an 'attention split between digital and physical' spaces, both for the participants and for themselves.

In terms of participation, online sessions made it easier to join and it also became clear that 'nobody walks *randomly* into the digital space', unlike in physical sessions when people passing by often chose to stay spontaneously. The wish to create safe enough spaces and simultaneously be inclusive also created a tension between the intimacy of an open atmosphere that provided space for confidential conversation and the need to consider aspects of data protection and privacy. For example, people felt unclear about 'who was secretly listening in'. Another issue was the depth of engagement, particularly when trying to host a dialogical session in which in-person participants interacted with the participants of a Twitter space. While online participation was in principle easier, hosts noticed that 'Twitter has to be short [superficial]' which conflicted with the aim to facilitate deep and reflective conversations.

In summary, the hosts perceived many 'tensions and trade-offs' such as 'safety vs. openness' and 'inclusivity vs. commitment'. They noted the 'collision of two modes', i.e. the 'attention economy/commercial' mode of the COP environment and the 'search for depth' in the CCRDS.

7.3.3 *Responses to These Challenges with Respect to Mindset, Skillset and Toolset*

During their sessions, hosts found different ways of responding to the challenges described above. We clustered these around their responses on their mindset, skillset or toolset.

Mindset

With respect to responding through their **mindset**, hosts described various ways of staying consciously in touch with the various tensions and challenges rather than judging them and going into resistance.

Specifically, some hosts described 'dropping all expectations on a result' or went into a mindset of 'openness to whatever happens as what should happen here'. Rather than attaching to a specific goal they 'accepted that this was not planned' and chose to 'focus on how to make the most of it'. Also, for specific challenges like 'people leaving mid-session' they cultivated a 'deep acceptance' and 'advised participants to take care of themselves and leave when they needed to'.

They reminded themselves that they were 'responsible only for how I show up, so I'll embody what I want to communicate [deep listening, open sharing]'.

Some hosts prepared for these kinds of challenges by a 'private meditation before facilitation to prepare my practice'. For several hosts it was helpful to make themselves 'aware of the trade-off between safety and fluidity, flexibility and inclusivity in an unsafe and busy space like that' and accepted the imperfection of their sessions and processes by 'making [a] choice on this trade-off for each session'.

In order not to be drawn into certain dynamics of the environment or session some hosts practised 'identifying and labelling unconscious patterns being reinforced' such that they could consciously deal with these patterns while facilitating. Particularly in the face of a perceived lack of safe circumstances, hosts were 'opening [their] heart fully as a space holder to support the participants in this unsafe space'.

Acknowledging the perceived lack of safety of the environment they also recognized that it would require more 'ability to set rules/structure'. This referred, for example, to 'being better prepared facilitating decolonizing and diversity practice', including clearly facilitated exercises on how to practise communication patterns that did not reproduce patterns of established power-imbalances or transgressions.

Skillset

In terms of their **skillset** hosts practised various techniques, particularly to bring the challenges consciously into the space or into the conversation. Rather than just silently dealing with them many hosts chose to voice the aspects of distraction and lack of perceived safety.

For example, hosts mentioned that it was helpful to 'play with the fact that it is unsafe and distracting', for instance by inviting participants to 'wear[ing] a hat if you are distracted, wear[ing] a veil if you are unsafe'. They practised 'voicing how the space distracts and name the distractions' and described it as 'a relief for the whole group when people name their lack of focus'.

Generally, hosts consciously drew the participants' attention to the various aspects that made the space unsafe or uncomfortable. For example, hosts were 'asking questions on how unsafety feels inside and outside'. To make these aspects transparent to the participants also required 'more regular reflection on unconscious and unsafe patterns we are reinforcing as facilitators'. This was especially so for online sessions, where some hosts used their struggles with technology or the imperfect technical set-ups as a prompt to reflect with participants about the pressures for perfection as experienced in the surrounding venue.

Another host used 'reframing an interaction that is giving rise to negative feelings in the group into something positive and generative'. For example, the rather harsh and aggressive intervention of one activist on another participant was reframed 'as a sign of her deep love for the work she is engaged in, rather than as a disruption'. In the dynamics of the session, it helped 'providing a new perspective with which to view the situation and to empower a transformative energy'.

Compared to their usual facilitation, hosts emphasized that hosting a session in the CCRDS at COP required 'doing more emotional release work' such as 'co-regulating together' 'through embodied practice (e.g. breathing together)'. Hosts mentioned that they 'recognized emotions' and 'welcomed silence'. For their own support, several hosts also preferred 'having several space holders' by their side in the workshop, meaning trusted persons to support the facilitator, and were keeping 'eye contact with colleagues' or applied 'self-talk to overcome doubts about perfection'.

Toolset

With respect to their **toolset**, hosts mentioned several practices that allowed the group to cope better with the specific challenges they experienced at the COP.

Following the above-mentioned intention to consciously acknowledge the distractions and surrounding unsafety, hosts often started with a 'check-in' explicitly 'asking for people's feelings' or 'how do you feel/ experience at COP?'. A popular tool was to split up the participants into 'smaller groups' which made it 'closer to talk', particularly to cope with the 'sound level'. Given the multiplicity of languages these small groups could also function as 'translation groups'.

Several hosts described 'meditating with the noise'. One also 'started with a meditation on the noise reframing it as expression of intensity and number of people caring about climate change. So, the context became positive and showed that people cared'. This included 'asking the participants to listen to all sounds at once without focusing on particular voices and feel the positive energy of many conversations in the venue'. By naming the distraction of noise they helped participants with 'finding out how to remain focused when it's loud and people [are] moving in and out'.

Given the circumstantial influences, several hosts were 'announcing very clear communication rules', suggesting for instance 'five seconds between people speaking' or 'not speak for 10 sec. after someone spoke' as a way to create spaces of silence to allow participants to process what had been said before jumping to an immediate (and potentially less conscious) response.

7.3.4 Recommendations for Future Hosting

Based on their experiences with hosting one or several sessions in the CCRDS at COP26, hosts made a several recommendations.

Various recommendations related to the space itself such as 'include nature in the space (plants, sounds, smells)' or 'install a curtain to be flexible' with respect to moving between and open and closed space more easily. One recommendation was also to 'have a few comfortable chairs and a small table with information about the aims of the space on the periphery of the main discussion space to encourage people to stop and find out more'. This would make it possible to 'begin dialogue already outside or before the formal start of a session'.

The wish to better manage the boundary between the inside of the CCRDS and the external circumstances also has implications for the staffing of the CCRDS. Several hosts recommended including 'a supporting *bridge* role', namely someone in charge of receiving late arrivals who want to join the session. Another host suggested having '[an] *admin* person assigned to speak to any people who are passing by but stop to find out what's happening'. This recommendation picked up on observations made by some hosts that 'there was a significant flow of people past the space that resulted in increased attendance' due to the openness of the space with only two walls unlike the closed space at COP25. Related to the constant transition of people during a session, it was recommended that a visible 'small *manifesto* on rules, etc.,' would help give people an orientation regarding the desired communication culture before joining.

With respect to online sessions, the hosts recommended we 'explicitly contact people who have commented online on previous COP' or to 'inquire [about the] needs of those who will be absent' for example through a 'public call for the *excluded voices*'.

There were many recommendations on how to stay in the state of mind that allows effective facilitation under the conditions at the COP. Hosts recommended, for example, the need to 'free yourself from any expectation of the outcome' and 'stay centred in the energy you want to represent'. Hosts should 'admit explicitly that this [the COP venue] is an unsafe space' and 'guide attention to unsafety' or 'play with this unsafety' but 'don't make it invisible'. Generally, it was emphasized that the host's perceived safety and comfort would significantly influence the atmosphere at their session.

7.4 Discussion—Being Conscious About the Unsafety Makes Spaces Safer

7.4.1 Bringing Awareness to Experiences of 'Unsafety' and Challenge

The obvious special feature of the CCRDS at COP was its direct connection with the unsafe surroundings of the COP venue, both in terms of the physical set-up and the atmosphere and communication culture. The CCRDS was not an ideal-type remote place for people who intentionally wanted to enter a safe and inclusive space for deep encounters but was an open, almost 'unprotected' space in the middle of a venue that rarely fosters similar aims and values.

The environment of the COP was perceived as busy, noisy, painful, confusing and unsafe, dominated by disconnectedness and hollow encounters or words. In summary, despite the diversity of perspectives at the COPs, diversity and inclusion were perceived as a pretence. Usually, hosts who aim at providing inclusive spaces for deep encounters spend a lot of thought and care in designing 'ideal' conditions to minimize disturbances and provide participants with enough time to build a trust with each other. In the CCRDS the situation was dramatically different. The atmosphere and influence of the environment was a constant intrusion for the hosts and participants.

At the same time, this contrast between the atmosphere inside and outside the CCRDS was reflected on in a critical fashion. We assume that many COP participants appreciated the design and venue of the COP and may not have perceived it as unsafe. These participants might not be expected to join sessions in the CCRDS or in turn even perceive these as 'unsafe' if they did so. We assume that most people who join the CCRDS sessions are looking for 'something different' and have a strong longing for more authentic encounters—and potentially a disdain for the currently dominant communication culture. Consequently, it is fair to assume some self-selection bias among the participants of the sessions in the CCRDS.

The hosts' reflections clearly showed the struggles arising from the tensions between a turbulent, busy and unsafe environment and the aspiration to offer a safe and inclusive space for deep encounters and various disturbances into the space, ranging from the noise level or interruptions as people entered or left during a session, adding to the hosts' confusion and self-doubts caused by the environment or the participants who brought the busyness of the venue with them into the CCRDS sessions. It becomes clear that it requires specific considerations and skills to design and host safe enough spaces that relate to this tension constructively.

An important finding emphasized by several hosts was that consciously acknowledging the unsafety of the setting made it feel safer. If there were obvious factors such as noise, late arrivals, etc., it was helpful not to ignore these but make them and their influence on the process explicit. As mentioned above, some hosts even chose to include meditation explicitly on the noise as a way to transition people into their encounter with each other.

This acknowledgement of the unsafe conditions (instead of ignoring them) was also reflected in the practices hosts applied to their own mindset. It seemed that in the presence of the polished and perfectionist environment of the COP many hosts felt a certain pressure leading to self-doubt and self-judgement. They found it helpful to consciously accept the imperfection of the conditions and the process they were facilitating and go into a mindset of deep acceptance, not complaining about the non-ideal conditions or feeling pushed by specific expectations. Just as the hosts supported themselves in keeping their presence, they also took extra care, such as by establishing stricter communication rules, to help participants keep their focus who otherwise might be drawn into the kind of behaviour that dominated the rest of the COP setting (long monologues, competitive talk, etc.).

It was also considered important to mediate the transition of participants between the unsafe setting outside the CCRDS and the reflective and dialogical atmosphere inside it. It was found helpful to include an extra person—often called 'bridge'—who could welcome late arrivals and explain what was happening. This helps ensure that people would not just 'crash' into an ongoing deep dialogue but could enter somewhat prepared and integrate themselves more smoothly. Such a 'bridge' allowed the hosts to keep their focus on what was going on inside the space and feel less distracted about what was going on around it (see Figure 7.5).

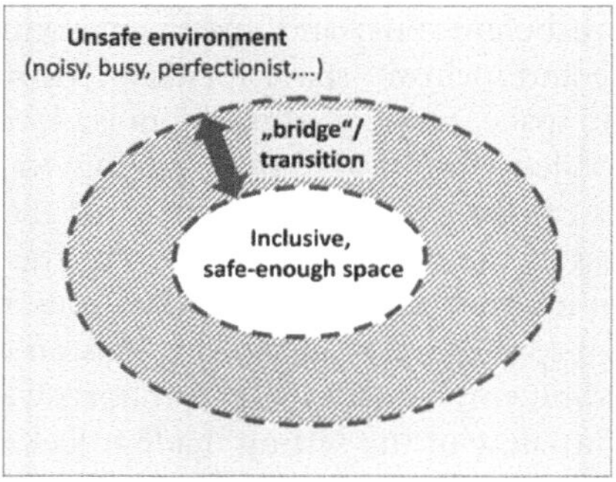

Fig. 7.5 The need for 'bridging' between the unsafe environment and the inclusive, safe-enough space

The transition between the COP environment and the CCRDS was, however, also a topic for the beginnings and endings of a session. All participants joined a session with the experience of the wider environment of the COP. It became obvious that it requires special care to help them transition into an atmosphere that is safe enough to facilitate deep and inclusive dialogue.

7.4.2 Reflection on Methods: Safe Enough for What?

A central question in these reflections concerns the meaning, impact and outcome of the CCRDS in this setting. Most of the challenges and efforts the facilitators/hosts endured are related to the fact that the CCRDS offers its sessions in a non-ideal surrounding to people who do not expect to come across such sessions at the COP. In fact, many participants had little prior experience with the kind of format in which they were involved. Obviously, the CCRDS does not follow the dominant culture of its environment. Already from the distance, attendants of the COP can see that there is something 'different' about the CCRDS, potentially only starting with its imperfect appearance or that some people are sitting in a circle or even on the floor. Yet, the strategic decision behind the experiment is to not displace the deep encounters and more inclusive conversations into sessions 'outside' the COP venue where the only people who would join are interested in these kinds of spaces anyway. Rather, the idea is to make its aspiration and culture visible to those passing by and being unaware that a more reflexive, safe and inclusive mode of interaction is possible. Merely by its presence the CCRDS reflects back to the surrounding environment that the dominant communication culture, of one-way presentations, panel discussions with passive audience, is not the only way to host encounters at the COP.

Importantly, the analysis is based on our hosts' perceptions of safety. While this may relate to actual threats in physical terms or regarding career potential or social acceptance, the reasons for the perceived (lack of) safety were less prominent. This understanding relates to the aim of the present communication approach to allow learning based on open reflection, including on values, norms and emotions, as well as on surfacing and addressing disagreement and conflict. In this regard, perceptions of (un) safety have strong implications for the possibility of opening up and engaging with one another in a more trusting way. It is in this sense that we investigated challenges and possibilities of establishing and maintaining a safe enough space for reflection and dialogue. On a critical note, the preoccupation with noise levels and a distracting atmosphere can be questioned in a situation where participants actually engage with issues that are critical for the physical, economic and social safety of millions of people, very few of whom attend the COPs. Here, being busy with one's own safety might even unconsciously work as a welcome distraction from confronting the severity of the actual situation. For this very reason, it remains critical to understand the establishment of safe enough spaces to enable deeper engagement with situations that would be even harder to bear and to relate to otherwise. Thus, building a safe enough, relational space needs to remain oriented towards supporting effective climate action.

As an action-oriented transdisciplinary experiment, the CCRDS combined practice and related research. The presented three iterations of the CCRDS generated insights on the actual demand for formats of communication and collaboration that enable more relational and collaborative encounters. On a general note, research supported the relevance of the practical approach of the CCRDS. More specifically, CCRDS at COP26 revealed important insights about how to host inclusive encounters in safe enough environments given the specific circumstances. The feedback from the participants continues to be very appreciative. Yet, it remains an open question how exactly the CCRDS aspiration to host safe and inclusive spaces at the COP is contributing towards more effective climate action. Future research will aim to understand more clearly what kind of experience participants have in the CCRDS and how these experiences can be a resource for their engagement towards climate action.

7.4.3 Towards Changing the Established Communication Culture

We hope the experiences from COP26 will be useful to all researchers or non-academic stakeholders who see the need for more inclusive engagement in contexts where such processes are seldom foreseen. We believe that our lessons may not only be supportive in the context of the COPs, but also illustrate a broader tension between an established mode of communication that is often a one-way, hierarchical and competitive discourse and an emerging trend towards more inclusive, reflexive and co-creative formats. In our experience, various stakeholders would like to work in more inclusive and co-creative ways, but their contexts essentially prevent these forms of communication. This includes early career researchers, who call for more interactive formats, transdisciplinary approaches and profound engagement with stakeholders in the different phases of their research process (Care et al., 2021; Schrot et al., 2020).

There is also a growing recognition that approaches for transformative action need to consider multiple forms of knowing as relevant aspects of learning and transformation processes (Wamsler et al., 2021), including, for example, emotions, affects or intuitive knowledge.

Confronted with such tensions, one option is to leave the established settings and launch new ones that may be more conducive to these kinds of engagement. For everyone seeing important potential in engaging in more conventional settings and aspiring to host spaces that emphasize co-creation, inclusiveness and safety, our findings may be of help. They may sensitize the organizers and hosts of such settings on how to handle the experience of unsafety of their environment, not only in order to accept an annoying aspect of the context, but rather as a manifestation of the communication culture that is subject to transformation. Seeing this as a focus for further research and engagement strategies could be a promising route for future iterations of the CCRDS. The immediate unsafety of the dominant communication culture could offer a powerful experience base for participants to reflect this atmosphere and its significance with respect to fundamental aspects of sustainability-related transformations. Specifically in relation to hosting practices, acknowledging and embracing the unsafety of the environment helps people connect around their shared experience of unsafety and related feelings of discomfort and fear. This may make the space to feel safe enough, welcoming deep encounter across apparent differences and vulnerabilities. In this sense, experiences of unsafety do not get in the way of more relational encounters and transformative learning, but they are actually the (only) way there.

In terms of TDR, these findings generate several interesting new perspectives. The task for the future may not necessarily be to design ideal processes for transdisciplinary communication. Rather, it seems promising to provide formats that help participants to acknowledge the unsafety and imperfection of any transdisciplinary communication process and integrate these reflections constructively. It may also be helpful to include a phase in TDR processes in which the participants can explore explicitly how safe they feel in a given context—which of their perspectives can comfortably be addressed, and which cannot. This might support processes of re-evaluating our own positions or convictions without the fear of losing face or inadequately exposing vulnerabilities to other participants. The entire phase of 'problem transformation' (Jahn et al., 2012) in a transdisciplinary research process might benefit substantially from this kind of deepening of the communication during in this phase.

7.5 Conclusion

In this chapter we have reflected on experiences of hosts of sessions in the CCRDS that aimed to be safe and inclusive under the prevailing circumstances at COP26. We have summarized perceptions of the COP as cold, noisy, perfectionist and unsafe and gave an overview on the challenges that this environment meant for hosting. We collected various lessons about the mindsets, skillsets and toolsets required, giving insights into how to deal with various kinds of disruption. A first key learning point was the relevance of being able to go into a deep acceptance of the imperfection of the situation and the process. Letting go of specific expectations helped hosts to remain present and hold the focus of the process. A second key insight was that it helped significantly to bring the disturbances and unsafety of the surrounding explicitly into the awareness of the participants and to make it part of the reflections during the session. Acknowledging and working consciously with the experience of unsafety made the atmosphere in the space feel safer. Finally, the reflections show the importance of including an extra person to function as a 'bridge' to receive and potentially transition people who approach the space and want to observe or join midway through a session.

The insights offer a valuable basis for researchers who want to experiment with inclusive and reflexive formats in settings that are not (yet) conducive to these kinds of formats. Our findings may be helpful in designing such spaces and better preparing to host them.

Acknowledgements The authors gratefully acknowledge the work of many people and administrative structures in various organizations that made it possible to set up and host the CCRDS at COP26. The work of the IASS is funded by the German Federal Ministry for Education and Research (BMBF) and the State of Brandenburg Ministry for Science, Research and Culture (MWFK).

References

Brink, E., & Wamsler, C. (2019). Citizen engagement in climate adaptation surveyed: The role of values, worldviews, gender and place. *Journal of Cleaner Production, 209*, 1342–1353. https://doi-org.vu-nl.idm.oclc.org/10.1016/j.jclepro.2018.10.164

Bruhn, T., Herberg, J., Molinengo, G., Oppold, D., Stasiak, D., & Nanz, P. (2019). Grounded action design—Transdisciplinary co-creation for better transformative processes. Frameworks for transdisciplinary research #9. *GAIA-Ecological Perspectives for Science and Society, 28*(4), 336–336. https://doi-org.vu-nl.idm.oclc.org/10.14512/gaia.28.4.3

Care, O., Bernstein, M. J., Chapman, M., Reviriego, I. D., Dressler, G., Felipe-Lucia, M. R., & Zähringer, J. G. (2021). Creating leadership collectives for sustainability transformations. *Sustainability Science, 16*(2), 703–708. https://doi-org.vu-nl.idm.oclc.org/10.1007/s11625-021-00909-y

Clark, W. C., Crutzen, P. J., & Schellnhuber, H. J. (2005). *Science for global sustainability: toward a new paradigm.* CID Working Paper 120. Harvard University. https://doi-org.vu-nl.idm.oclc.org/10.2139/ssrn.702501

Clark, W. C., & Harley, A. G. (2020). Sustainability science: Toward a synthesis. *Annual Review of Environment and Resources, 45*(1), 331–386. https://doi-org.vu-nl.idm.oclc.org/10.1146/annurev-environ-012420-043621

Creutzig, F., & Kapmeier, F. (2020). Engage, don't preach: Active learning triggers climate action. *Energy Research and Social Science, 70*, 101779. https://doi-org.vu-nl.idm.oclc.org/10.1016/j.erss.2020.101779

Espinosa, A., & Porter, T. (2011). Sustainability, complexity and learning: Insights from complex systems approaches. *Learning Organization, 18*(1), 54–72. https://doi-org.vu-nl.idm.oclc.org/10.1108/09696471111096000

Fazey, I., Schäpke, N., Caniglia, G., Patterson, J., Hultman, J., van Mierlo, B., & F. Säwe. (2018). Ten essentials for action-oriented and second order energy transitions, transformations and climate change research. *Energy Research & Social Science, 40*, 54–70. https://doi-org.vu-nl.idm.oclc.org/10.1016/j.erss.2017.11.026

Felt, U. (2010). Transdisciplinarity as culture and practice. *GAIA-Ecological Perspectives for Science & Society, 19*(1), 75–77.

Fraude, C., T. Bruhn, T., Stasiak,D., Wamsler, C., Mar, K., Schäpke, N., Schroeder, H., & M. Lawrence, M. (2021). Creating space for reflection and dialogue: Examples of new modes of communication for empowering climate action. *GAIA-Ecological Perspectives for Science & Society, 30*(3), 174–180. https://doi-org.vu-nl.idm.oclc.org/10.14512/gaia.30.3.9

Grothmann, T. (2018). *Wege für eine handlungsmotivierende Klimakommunikation—Ergebnisse psychologischer Forschung, 101*, 15–19.

Habermann, F., & Schmidt, K. (2018). *Over the fence: Rediscover the joy of projects, develop new ideas better, and have more fun working together.*

Hirsch Hadorn, G., Bradley, D., Pohl, C., Rist, S., Wiesmann, U., & U. (2006). Implications of transdisciplinarity for sustainability research. *Ecological Economics, 60*(1), 119–128. https://doi-org.vu-nl.idm.oclc.org/10.1016/j.ecolecon.2005.12.002

Jahn, T., Bergmann, M., & Keil, F. (2012). Transdisciplinarity: Between mainstreaming and marginalization. *Ecological Economics, 79*, 1–10. https://doi-org.vu-nl.idm.oclc.org/10.1016/j.ecolecon.2012.04.017

Kates, R. W., Clark, W. C. , Corell, R., Hall, J. M., Jaeger, C. C. , … & Svedin, U. (2001). Sustainability science. *Science, 292*(5517), 641–642. https://doi-org.vu-nl.idm.oclc.org/10.1126/science.1059386

Kay, J. J., Regier, H. A., Boyle, M., & Francis, G. (1999). An ecosystem approach for sustainability: Addressing the challenge of complexity. *Futures, 31*(7), 721–742. https://doi-org.vu-nl.idm.oclc.org/10.1016/S0016-3287(99)00029-4

Lang, D. J., Wiek, A., Bergmann, M., Stauffacher, M., Martens, P., Moll, P., Swilling, M., & Thomas, C. J. (2012). Transdisciplinary research in sustainability science: Practice, principles, and challenges. *Sustainability Science, 7*(SUPPL. 1), 25–43. https://doi-org.vu-nl.idm.oclc.org/10.1007/s11625-011-0149-x

Lang, D. J., Wiek, A., & von Wehrden, H. (2017). Bridging divides in sustainability science. *Sustainability Science, 12*(6), 875–879. https://doi-org.vu-nl.idm.oclc.org/10.1007/s11625-017-0497-2

Lawrence, M. G., Williams, S., Nanz, P., & Ren, O. (2022). Characteristics, potentials, and challenges of transdisciplinary research. *One Earth, 5*(1), 44–61. https://doi-org.vu-nl.idm.oclc.org/10.1016/j.oneear.2021.12.010

Liu, J., Dietz, T., Carpenter, S. R., Alberti, M., Folke, C., Moran, E., Pell, A. N., Deadman, P., Kratz, T., & Lubchenco, J. (2007). Complexity of coupled human and natural systems. *Science, 317*(5844), 1513–1516. https://doi-org.vu-nl.idm.oclc.org/10.1126/science.1144004

Mar, K. A., Fraude,C., Bruhn, T., Schäpke, N., Stasiak, D,. Schroeder, H., Wamsler, C., & Lawrence, M. G. (2021). Fostering reflection, dialogue and collaboration among actors at the UN Climate Change Conferences. IASS Policy Brief 5/2021. Institute for Advanced Sustainability Studies. https://doi-org.vu-nl.idm.oclc.org/10.48481/iass.2021.028

Mar, K. A., Schäpke, N., Fraude, C., Bruhn, T., Wamsler, C., Stasiak, D., Schroeder, H., & Lawrence, M. G. (2023). Learning and community building in support of collective action: Toward a new climate of communication at the COP. *Wires Climate Change, 14*(4), e832. https://doi-org.vu-nl.idm.oclc.org/10.1002/wcc.832

Mezirow, J., & Taylor, E. W. (Eds.). (2009). *Transformative learning in practice: Insights from community, workplace, and higher education*. John Wiley & Sons.

Nanz, P., Renn, O., & Lawrence, M. G. (2017). Der transdisziplinäre Ansatz des Institute for Advanced Sustainability Studies (IASS): Konzept und Umsetzung. *GAIA-Ecological Perspectives for Science and Society, 26*(3), 293–296. https://doi-org.vu-nl.idm.oclc.org/10.14512/gaia.26.3.19

Newig, J., Jahn, S., Lang, D. J., Kahle, J., & Bergmann, M. (2019). Linking modes of research to their scientific and societal outcomes: Evidence from 81 sustainability-oriented research projects. *Environmental Science & Policy, 101*, 147–155. https://doi-org.vu-nl.idm.oclc.org/10.1016/j.envsci.2019.08.008

Pereira, L., et al. (2020). Transformative spaces in the making: Key lessons from nine cases in the Global South. *Sustainability Science, 15*, 161–178. https://doi-org.vu-nl.idm.oclc.org/10.1007/s11625-019-00749-x

Pogatschnigg, I. M. (2021). *The art of hosting: wie gute Gespräche Führung und Zusammenarbeit verbessern*. Vahlen.

Popa, F., Guillermin, M., & Dedeurwaerdere, T. (2015). A pragmatist approach to transdisciplinarity in sustainability research: From complex systems theory to reflexive science. *Futures, 65*, 45–56.

Ravetz, J. R. (2006). Post-normal science and the complexity of transitions towards sustainability. *Ecological Complexity, 3*(4), 275–284. https://doi-org.vu-nl.idm.oclc.org/10.1016/j.ecocom.2007.02.001

Rittel, H., & Webber, M. M. (1973). 2.3 planning problems are wicked. *Polity, 4*, 155–169. http://www.jstor.org.vu-nl.idm.oclc.org/stable/4531523

Schäpke, N., Beyers, F., Fraude, C., Mar, K. A., Schroeder, H., Voggenreiter, V., Herzog, N., Wamsler, C., Bruhn, T., & Lawrence, M. G. (2023). *Research and practice to scale co-creation and reflection: how can we strengthen a relational culture of communication for climate action at COP28 in Dubai?* RIFS Discussion Paper, November. Research Institute for Sustainability–Potsdam. https://doi-org.vu-nl.idm.oclc.org/10.48481/rifs.2023.033

Schrot, O., Krimm, H., & T. Schinko, T. (2020). Enabling early career sustainability researchers to conduct transdisciplinary research: Insights from Austria. *Challenges in Sustainability, 8*(1). https://doi-org.vu-nl.idm.oclc.org/10.12924/cis2020.08010030.

Steffen, W., Grinevald, J., Crutzen, P., & McNeill, J. (2011). The anthropocene: Conceptual and historical perspectives. *Philosophical Transactions of the Royal Society a: Mathematical, Physical and Engineering Sciences, 369*(1938), 842–867. https://doi-org.vu-nl.idm.oclc.org/10.1098/rsta.2010.0327

Thompson Klein, J. (2004). Prospects for transdisciplinarity. *Futures, 36*(4), 515–526. https://doi-org.vu-nl.idm.oclc.org/10.1016/j.futures.2003.10.007

Waltner-Toews, D., Kay, J. J., & Lister, N.-M.E. (2008). *The ecosystem approach: Complexity, uncertainty, and managing for sustainability*. Columbia University Press.

Walsh, Z., Böhme, J., & Wamsler, C. (2020). Towards a relational paradigm in sustainability research, practice, and education. *Ambio, 1–11*. https://doi-org.vu-nl.idm.oclc.org/10.1007/s13280-020-01322-y

Wamsler, C., Osberg,G., Osika, W., Herndersson, H., & Mundaca, L. (2021). Linking internal and external transformation for sustainability and climate action: towards a new research and policy agenda. *Global Environmental Change, 71*, article 102373. https://doi-org.vu-nl.idm.oclc.org/10.1016/j.gloenvcha.2021.102373

Wamsler, C., Schäpke, N., Fraude, C., Stasiak, D., Bruhn, T., Lawrence, M., Schroeder, H., & Mundaca, L. (2020). Enabling new mindsets and transformative skills XE "Skills" for negotiating and activating climate action: Lessons from UNFCCC conferences of the parties. *Environmental Science & Policy, 112*, 227–235. https://doi-org.vu-nl.idm.oclc.org/10.1016/j.envsci.2020.06.005

West, S., Haider, L. J., Stålhammar, S., & Woroniecki, S. (2020). A relational turn for sustainability science? Relational thinking, leverage points and transformations. *Ecosystems and People, 16*(1), 304–325. https://doi-org.vu-nl.idm.oclc.org/10.1080/26395916.2020.1814417

Woiwode, C., et al. (2021). Inner transformation to sustainability as a deep leverage point: Fostering new avenues for change through dialogue and reflection. *Sustainability Science, 16*, 841–858. https://doi-org.vu-nl.idm.oclc.org/10.1007/s11625-020-00882-y

PART II

Diversities and Inclusion

CHAPTER 8

Challenges for Inclusion and Diversities: Opening up and Closing Down in Collaborative Research and Practice

Pim Klaassen and Anne Loeber

> Probably the most dangerous thing about an academic education [...] is that it enables my tendency to over-intellectualize stuff, to get lost in abstract argument inside my head, instead of simply paying attention to what is going on right in front of me, paying attention to what is going on inside me.
>
> Chimamanda Ngozi Adichie.[1]

[1] TED talk: *The danger of a single story*. Accessible through https://www.ted.com/talks/chimamanda_ngozi_adichie_the_danger_of_a_single_story. Last accessed 13/12/2023.

P. Klaassen (✉)
Stichting Stadslab, Amsterdam, The Netherlands
e-mail: pim@hetstadslab.nl

A. Loeber
Vrije Universiteit Amsterdam (VUA), Amsterdam, The Netherlands
e-mail: a.m.c.loeber@vu.nl

B. J. Regeer et al. (eds.), *Transdisciplinarity for Transformation*,
https://doi.org/10.1007/978-3-031-60974-9_8

In brief, the function of knowledge is to make one experience freely available to other experiences.

John Dewey.[2]

8.1 Introduction

Transdisciplinary approaches serve the dual aim of delivering new knowledge *about* a situation at hand, and actually interacting *with* and, hopefully, ameliorating that situation. This 'action-orientation' comes with many methodological and practical challenges as to the 'how' of transdisciplinarity, but also with a moral obligation to keenly look into the 'who'-question: whose knowledge counts and comes to bear on dealing with a specific situation? As for those who are included, the question is how they can make their multiple identities come to bear on the issue at hand and how they can effectively use the room for maneuvering that their community of practice or epistemic culture allows them, in contributing to the (transient) transdisciplinary collective. This chapter sets the stage for a close scrutiny of challenges and opportunities in the light of openness and 'closures' of knowledge co-production that the chapters collected together in Part 2 of this book deal with, offering a selection of empirical studies that illuminate pertinent conceptual and practical ways forward to inclusive and diverse forms of transdisciplinarity in a variety of contexts. A reason to do so is the conviction that the complex and persistent problems that are center stage in transdisciplinary work require the use of embodied, tacit and situated experiential knowledge of people's life-world that Ngozi Adichie's quote in the epigraph directs our attention to, which tends to be absent from most academic knowledge production in efforts at promoting societal transformation.

In this chapter we first give an overview of lessons from the literature on what including diverse groups of actors in transdisciplinary work entails in practice. Then we will contemplate the questions *whom to include, when, under what conditions, and to what end,* and look into some of the challenges implied in answering these. In conclusion, we briefly introduce the empirical chapters collected in this second Part of the book for practical illustrations of how such challenges can be dealt with in practice.

[2] Dewey J. (1916). Democracy and Education. Project Gutenberg.

8.2 Dealing with Diversities

Various authors have elaborated the *why*-question posited above, amounting to four 'canonical' arguments for including wide varieties of actors in research and innovation (Schmidt, 2020). First, incorporating multiple perspectives in knowledge production on a problem at hand, it is often argued improves research quality (the *substantive* argument; Lang et al., 2012; Nowotny et al., 2003). Second, legitimacy of and trust in solutions of societal problems will increase, many authors assume, if multiple types of stakeholders contribute substantially to the knowledge that feeds into it (the *instrumental* argument; Owen et al., 2012; Stilgoe et al., 2013; Von Schomberg, 2013). Third, there is a *normative* argument stating that whoever will be affected by the outcomes of research or innovation practices deserves to have a say in these (alternatively dubbed, the democratic argument; Cash et al., 2003; Dryzek, 2002; Habermas, 1981; Jasanoff, 2003). Fourth, the inclusion of a wide variety of actors will result in social learning, enabling participants to come to a mutual understanding of diverse relevant values, problem understandings, interests and the like, which will enable them to act in congruence with one another to resolve a situation (the *actionable knowledge* argument; Hadorn et al., 2006; Innes & Booher, 2004; Klaassen et al., 2018; Loeber et al., 2007).

For these reasons, transdisciplinarity for transformation is about including a diversity of actors, integrating different bodies of knowledge and bringing a diversity of values and perspectives to bear on efforts at ameliorating a problem situation. Transdisciplinarity effectively informs practice because it ties together what Aristotle distinguishes as *episteme*, *praxis* and *poiesis* (Hadorn et al., 2008, p. 31), that is, science, life-world action and production, in a way that informs *phronesis*: the practical wisdom needed to decide on just and effective actions in the face of some complex, contextually defined issue. For that, such knowledge ('wisdom') must enable situated judgment in view of a concrete problem-solving action, and of the ethical choices involved, in a way that does justice simultaneously to the situation at hand and to a more generic understanding as to what is just or wise to do for the community and for humankind (Loeber &Vermeulen, 2007, 2016). While mono- and interdisciplinary work respects the boundaries between the spheres of experiential knowledge holders and practitioners on the one hand and academic experts on the other, transdisciplinary work deliberately transgresses such boundaries—in recognition of the moral and political character of scientific and life-world knowledge.

The development of transformation-oriented knowledge relevant to complex and persistent problems, including the question of how to apply that knowledge in a particular real-life context, often takes place among representatives of a variety of actor groups. In so doing, insights can be unlocked that are vital to finding or co-creating practicable changes for the better, while avoiding exclusion, a priori, of certain groups and the insights these bring to the table, that is, avoiding 'epistemic injustice', defined by Fricker (2007) as 'harming others in their capacity as a knower'.

The recognition of a transdisciplinary approach's distinctive value also means that more (young) researchers need to get acquainted with *doing transdisciplinarity*—something also elaborated in the third Part of this volume. The chapters in this second Part of the book are all concerned with a specific, particularly difficult, aspect to this, namely with *how* to appropriately deal with all sorts of diversities. This comes with challenges concerning the design of transdisciplinary projects and the process of inviting or selecting co-creators. Furthermore, there are challenges in putting transdisciplinarity into practice, enacting inclusivity by making space to genuinely listen to each other and by breaking free from the constraints of one's epistemic culture, to accommodate information from one's sparring partners and their needs and values and associate with these through one or more of one's multiple identities that are less tied in with the dominant epistemic culture.

As for the diversities that deserve to be included, these are many. There is diversity in terms of knowledge and access thereto, diversity in terms of access to resources, distance to and relation with academia, practice, policy and power, diversity in terms of the degree to which one is conventionally allowed to articulate one's needs, diversity in one's capacity to articulate one's needs, desires, problem framings or preferred solutions, diversity in terms of ethnicity, age, gender, sexual orientation—and so on. The various types of differences can exist both between and within actor groups or even individuals and call for a diverse set of approaches to organizing and practicing inclusion. Each of these comes with distinct challenges. The four chapters in Part 2 bring together all have something distinctive to say on the issue of inclusion, and all share a range of lessons for future 'transdisciplinarians' to build on. Moreover, since transdisciplinarity yields context-sensitive knowledge, it brings along additional challenges in regard to knowledge sharing. The question is how to communicate insights across contexts in a way that makes 'experience become available to other experiences', as John Dewey (1916) put it.

One might think that the type of challenge we are singling out here is not necessarily distinctive to transdisciplinarity. And indeed, as was also discussed in Chapter 1 we find that under various labels and in widely diverging contexts, over the past decades, a wide variety of researchers, practitioners and communities have engaged in building bridges between academic disciplines and practices that go beyond both. We come across terms like *co-creation* (Mauser et al., 2013; Taylor et al., 2022), *co-production* (Simon et al., 2018; Turnhout et al., 2020), *participatory action research* (Díaz-Arévalo, 2022; Gorashi & Ponzoni, 2014; Masson et al., 2021), *engaged scholarship* (Franklin, 2022) and *transdisciplinary research* (Klein, 2014; Pohl & Hadorn, 2007)—all used to refer to knowledge production in which disciplinary and academic borders are transgressed. Whether such work is concerned with health system reform (Javadi et al., 2018; Schuitmaker et al., 2021) or nature conservation (Torkar & McGregor, 2012), with the energy transition (Heaslip & Fahy, 2018) or with gender-based violence (Adelman et al., 2012), such efforts tend to be guided by comparable sensitivities and seek to encompass similar values and goals: contributing to positive societal change, in ways that are just and that take into account the needs of those who and that which are most vulnerable and most conventionally ignored—from marginalized groups to threatened wild-life habitats. Of course, this generic characterization leaves plenty of room for variations in all sorts of practical and philosophical details. There are several historical overviews and attempts at analytical description of transdisciplinary research (Balsiger, 2015; Max-Neef, 2005; Popa et al., 2015), so we will not rehearse these here. Rather, we will underscore a small number of observations and lessons to be drawn from past work on transdisciplinary research that we think are relevant for understanding what it means to deal with different actor groups in such work. This includes considering the implications of the fact that all of us are simultaneously member of different professional, personal, cultural or academic communities, and that transdisciplinarity is simultaneously complicated and enriched when it is acknowledged that and how our intersectional positionalities bear on our transdisciplinary practices (Kim, 2023).

We therefore focus on the question on the '*how*' of *inclusive transdisciplinary practices*, and on how to do so appropriately. Assessing the level of 'appropriateness' of inclusion can be determined on the basis of three themes: *(i)* the level of integration of different knowledges (Godemann, 2008), *(ii)* the efficacy or promise of proposed knowledge and solutions to complex persistent problems (Fritz et al., 2019; Lux et al., 2019) and *(iii)* the contribution to (epistemic) justice as shown by its success in bringing solutions into the world that go beyond the reproduction of existing systems and (power) structures (Loorbach, 2007; Turnhout et al., 2020).

Building on the analysis presented by Vermeulen and Witjes (2020), we can see that these themes correspond with the three strands of scholarly work, promoting the different 'flavours' or 'modes' of or approaches to transdisciplinarity (see also Regeer et al., 2024). Each flavor is dedicated to the inclusion of different sets of actors and promotes engaging them differently and at different phases of the process of transdisciplinary research.

First, several scholars associate the 'how' of transdisciplinary research with the *complexity of the problems* to be addressed (e.g., Max-Neef, 2005; Nicolescu, 2014). These scholars consider inclusion and diversity issues in seeking to combine a broad range of disciplinary knowledges of academic experts, as a basis for formulating scenarios and policy options. Initial analytic steps, such as problem analysis, problem-definition and preliminary research need to be completed before bringing in actors from practice, and then these actors may well be industry representatives, policymakers and, possibly, civil society organizations (CSOs). The knowledge that citizens might have to offer does not warrant inviting them (Kua, 2016; Rotmans, 1998; Salem et al., 2018). If research is to lead to real-life solutions, however, some degree of integration of different forms of knowledge appears necessary—at least in order to understand the problem being addressed. Despite this, doing transdisciplinary research in this way, even if it is only modestly inclusive and hardly meets the approach to transdisciplinarity that is central to this book, is already quite hard.

A second group of scholars working on sustainability issues in the domain of transition studies (e.g., Geels, 2012; Smith et al., 2005) stress *problem solving* and 'implementation support, closely linked to local, regional, national and sometimes even supranational policy making' (Vermeulen & Witjes, 2020, p. 16). The relative emphasis these scholars place on the efficacy of transdisciplinarity-based solutions to urgent, complex and persistent problems goes alongside more extensive engagement policies. In other words, although it involves very much the same type of actors who are engaged as in the previously discussed 'flavour' of transdisciplinary research, which largely excludes for instance citizens, in this type of work these actors take part in more of the transdisciplinary process, and hence, often help co-produce the delineation and definition of the problem.

Third, some scholars have the explicit goal of contributing to *increasing (epistemic) justice* in doing transdisciplinary work aimed at transforming existing unfair or unsustainable practices (Rodriguez, 2022; Temper & Del Bene, 2016). Empowering vulnerable, marginalized, or other poorly represented and/or heard groups is part and parcel of transdisciplinary work, and itself an aspect of the transformation(s) to work toward. Engaging citizens or bottom-up networks is a much more conventional part of this type of transdisciplinary research than it is of the other two types and in that sense builds on better-known approaches like Participatory Action Research (see Chapter 1).

This volume presents a range of approaches to and practices of transdisciplinary research. Most of the chapters fall squarely within this third type of scholarly work, and those included in Part 2 address the question of how to deal with differences in the distance to or in the boundedness by science, policy, practice and politics that characterize different groups and individuals, and with the associated differences in the language used to articulate needs, values, insights and knowledges. There is no reason to think that there is only one answer to such *how*-questions. However, the literature draws attention to sensitivities and sensibilities that can help transdisciplinarians think through what does or does not work, what is appropriate and how one might successfully deal with challenges associated with transgressing disciplinary and academic boundaries justly and effectively. The next section briefly discusses some of these challenges.

8.3 Challenges, Proposed Solutions and This Volume's Contributions

Arguably, the focus on 'how' in discussing the three 'flavours' or approaches in transdisciplinarity should also contend with the questions of *whom to include, when, under what conditions, to what end*—questions that all 'transdisciplinarians' will confront at some point. And, of course, also the question how anyone engaged in transdisciplinarity can deal with the different ways in which they are differently bound by different aspects of their positionality, including their ethnic or gender identity, the epistemic culture they were raised in, the community of practice they are member of, and so on, belongs in this list. Answering such questions is deeply political, as any answer will imply including some and excluding other voices (whether they are your own or others'), and relegating different types or degrees of agency and responsibility to groups of differentially interested, affected, distanced, or knowledgeable parties (Chilvers & Kearnes, 2020; Kok, 2021; Soneryd, 2016). We focus on particular challenges that arise and present proposals for dealing with these from each of the different approaches to transdisciplinarity.

8.3.1 Integrating Knowledges in Efforts to Define the Problem

The main challenge in integrating different forms of knowledge concerns how a problem is framed, finding a common currency in delineating a problem, reaching agreement on what values are (most) relevant and the methods that might be appropriate for developing the necessary knowledge and including the languages each actor uses to realize a shared understanding that many would consider meaningful (Schön & Rein, 1994). All transdisciplinary constellations must contend with these issues. For those focusing on the complexity of the problems (the first 'flavour' in transdisciplinary work), this presents the biggest concern. Power differentials among the diverse actors complicate matters considerably. One solution is to set strict 'rules for closure', restricting the problem-definition phase to those who have been involved from the outset, and relegating further engagement to 'end-of-pipe'—arguably at the cost of both the efficacy of proposed solutions and of (epistemic) justice of the research processes and outcomes.

Further complicating issues in integrating knowledge have to do with the composition of the group: how diverse is a transdisciplinary research consortium in terms of disciplinary backgrounds, practical expertise and other forms of knowledge (tacit, experiential, indigenous, *et cetera*)? Communication skills may also differ, as well as reflexive capacities to support collaboration in inter- and transdisciplinary groups. The emerging group dynamics also matters—especially as often groups are unstable and compositions change over time, and group dynamics tends to be influenced by the degree of the of the group's heterogeneity in terms of, e.g., (social, epistemic) status and power (Thomas-Hunt et al., 2003).

Given its grounding in distinct real-world problems and its aim to offer practicable solutions to those, groups conducting transdisciplinary projects almost by definition have to develop not just a shared and common knowledge base, but also one that is uniquely geared to the specific situation. Crucially, exchanges within transdisciplinary collectives should not be seen as forms of knowledge *transfer*, but rather as activities contributing to the expansion of each participant's perspective and perception of problems and solutions (Godemann, 2008; Regeer & Bunders, 2003). Reflexivity, open-mindedness and responsiveness to other perspectives are necessary but not sufficient conditions for successful knowledge integration (Klaassen et al., 2018).

Difficult and demanding as this might seem, of course the different forms of disciplinary and practical knowledge that various contributors bring is first and foremost a resource. Yet, clearly, different forms of knowledge are structured differently, and communication and cooperation across sub-groups with different knowledges, backgrounds, values and needs complicate knowledge integration (Godemann, 2008). The contribution by Brouwers, Egberts and de Hoop (2024, this volume) illustrates how what they dub a *walkshop* (Wickson et al., 2015) offers a particular research method that can help bring to life a shared reality in order to support the construction of a common object and research aim. Similarly, in their chapter on the Frame Reflection Lab, Horn and van der Meij outline a hands-on approach to nurturing reflexivity and transdisciplinary collaboration, and to ways to strengthen relevant communication skills.

8.3.2 Effective Solutions

Issues that result from power differentials among knowledge holders are difficult to brush aside. 'Transdisciplinarians' aiming to create a better understanding of real-world and complex problems, as well as at contributing to instigating transformative dynamics, face the challenge of finding '*ways of working with and around the power relations, which shape and are being shaped by the emerging community' of the transdisciplinary collective*' (Van Breda & Swilling, 2019, pp. 834–5). The intricacies vary and arguably become notably complex and sensitive if transdisciplinary research also aims to contribute to (epistemic) justice (as is the case in the third 'flavour' of transdisciplinarity). Even if we leave aside justice-related issues out of the equation for now, there is a risk that powerful and vested-interest actors use transdisciplinary settings to reproduce the status quo and/or wittingly or unwittingly promote their own self-interest.

This critical risk (see also Stirling, 2008) has been discussed, for instance in relation to participatory experiments in the governance of nuclear waste. Chilvers (2008, p. 1881) discussed the issue, stating that *'[u]nless [the] often tacit power relations are acknowledged, accounted for, and exposed by all involved, but especially vested interests, [transdisciplinary] analytic-deliberative institutions may well undermine public trust, credibility, and legitimacy rather than promote these democratic virtues as is widely claimed'*. As a remedy, there is a need for reflexive and responsive research, as well as governance cultures that authentically support the emancipatory aspirations of transdisciplinarity.

This draws attention to necessary institutional conditions that are, however, seldomly in place. The question is how transdisciplinary research practice can contribute to the normalization of reflexivity and responsiveness in a way that these serve to counter vested interests that reproduce undesirable structures (see also Bourdieu, 1977). The chapter by Ramaswami, Seshadri and Bunders (this volume) shows that phasing transdisciplinary work offers a way forward in this respect, as it enables different actor groups to contribute separately. This makes it possible to make explicit choices about whom to engage at which phase of a project, for what purpose, and together with or apart from which other actors (Lynch et al., 2017; Verwoerd et al., 2023). Thus, separating groups of actors who have more or less power or higher or lower status has proven effective in dealing with this issue (Regeer et al., 2011).

Discussions on power in transdisciplinary processes are inevitably entangled with the debate on how to ensure, without being paternalistic, that vulnerable and marginalized groups are properly engaged. Even if such groups do not pro-actively demand a seat at the table and do not have the same authority or status as other participants—such as corporate representatives, non-governmental organizations (NGOs), policymakers or academic researchers—how can one ensure they fully reap the benefits of transdisciplinary projects? Let us turn to that theme now.

8.3.3 Epistemically Just Research in Transformation Practices

Among the main challenges in transdisciplinary research is to ensure that it is not only its outcomes that serve those who are less likely to articulate their needs, values and perspectives, but also the research process itself. Co-creation by actors among whom there are obvious power imbalances challenges researchers' ability to see the less-privileged groups' knowledges and viewpoints represented in knowledge- and decision-making, without alienating influential stakeholders. As Turnhout and colleagues (2020, p. 16) succinctly put it:

> Literature on participation has demonstrated that elite actors, for example from government, large NGOs, or scientists, have more time and resources available, often initiate these processes, define the scope for participation, have more knowledge and skills, and are, for all these reasons that resonate with social-cultural biases, better able to articulate a contribution that is considered relevant and important. Consequently, elite actors are able to shape these processes to serve their interests. In a co-production context, these power inequalities are further compounded by the strong authority that is attributed to scientific expertise vis-a-vis other knowledge systems.

Ignoring power differentials is not an option, as this risks simply reproducing inequities rather than contributing to transforming the status quo (Cooke & Kothari, 2001). And although, fortunately, there are strategies to circumvent this risk, the difficulty of applying these should not be underestimated. Capacity-building initiatives might be useful to ensure equitable access to resources and information, thus empowering groups, but this can only make a substantial difference when also other proactive measures are taken, including learning how to listen well to people whom one does not easily identify with, creating safe spaces for marginalized groups, fostering partnerships or hybrid forums that enable actors to on-line contribute meaningfully to deliberations (Bruhn et al., 2024; Canel et al., 2022; Fritz & Binder, 2020; Mascarenhas et al., 2021; Stark, 2024, this volume). However, another complicating factor is that not everyone whose knowledge, values and perspectives would be pertinent to resolving a given issue is able and willing to participate in a transdisciplinary project. Some people distrust science, or the public authorities that share responsibility for funding or implementing transdisciplinary research and hence might not wish to participate. The chapters by Holle, Ponzoni and Ghorashi and De Weger, Fraaije, Harambam and Willems (this volume) address these issues. These authors show that fostering a culture of mutual respect, and acknowledging the privileges or disadvantages that characterize certain actors is a starting point for dealing with such challenges, although it does not in itself resolve them.

Power is clearly the central theme in all these challenges (and in the chapters addressing them), namely the intricacies in dealing with power differentials. The different 'flavours' of doing transdisciplinarity show that dealing with power differentials is a complex process in such research, but is essential to achieve equitable outcomes and engage in equitable practices. As these chapters illustrate, there may be many reasons to navigate power differentials in different ways, depending on the circumstances.

8.4 Outlook: Signposting Challenges

The challenges discussed above cannot be resolved but must nevertheless be addressed in practice. 'Transdisciplinarians' who are dedicated to including diverse actors and factors in their work to help ensure it plays a transformative and emancipatory role have to articulate their own response to such challenges, tailored to the particular circumstances in which their project unfolds. There is not and cannot be one single, universally applicable answer to the question what it entails to include all relevant voices when co-creating practicable knowledge for transformation, just like, as Regeer et al. (2024) discussed, more generally there are no one-size-fits-all guidelines to transdisciplinarity.

As we have already seen, there are many reasons for this. First, diversity in what are regarded complex and persistent problems is too large—just think of the commonalities and differences between institutional racism, biodiversity loss, faltering healthcare systems, climate change, child abuse, food insecurity societal polarization and so on. Second, we must consider the possibility that problems are structured differently, calling for the input of different knowledges, different relevant (epistemic) communities, in which there will be different values at play and so on, and so the goals with which various actor groups engage in transdisciplinary co-creation will also differ. Therefore, ways of making knowledge collaboratively, finding meanings and developing solutions will be different from one context to the next. Third, the institutional, epistemic or governance culture one finds oneself in when engaging in transdisciplinary research can vary significantly, being less or more appropriate for the type of reflexivity and responsiveness required to successfully pool knowledge integration and solution formulation and implementation in just and inclusive ways.

This does not mean, though, that we are forever lost in the dark. On the contrary, we hope that structuring some of the challenges as we have outlined here can be of help to everyone who travels the interesting, complex and (hopefully) ultimately rewarding road of transdisciplinarity.

Much as the analysis we have set out in this introductory chapter acts as signpost to challenges that merit close attention, each of the four chapters in Part 2 present some pointers to what including the right voices at the right time, and using the right engagement practices could mean—or: could *for instance* mean. None of the authors makes any claim to be comprehensive, or to give recipes for doing things right in terms of inclusion and diversities—to once again touch upon the theme of Chapter 3 of this book. But these chapters do aim to inspire a wide range of readers, practitioners of transdisciplinary work or budding transdisciplinarians.

In their chapter 'Taking the landscape into conversations', Brouwers, Egberts and de Hoop show how their development and execution of a *walkshop* led to more inclusive conversations on preventing wildfires in the Netherlands. The added value of doing inclusivity becomes clear in the way their walkshop promotes engaging with different knowledges and values, while also allowing for the appropriate recognition of the embodied, situated knowledge that emerges in landscape-human interactions, which more conventional and often sterile formats like round-table discussions or one's everyday post-it session.

The chapter by Ramaswamy, Seshadri and Bunders discusses how transdisciplinarity has enabled systemic transformation in child and youth issues in India, specifically in the area of children's interface with the criminal justice system, in relation to sexual abuse—in particular the substantial power differences in that system and the conventional disregard for children's testimonies. Despite these complexities, the chapter celebrates the transformative potential of transdisciplinary work, which becomes clearly visible in the capacity building it supports and in the policy changes it helped advance.

Similarly focusing on a stigmatized, silenced and marginalized community, the chapter by Holle et al. presents three conditions for co-creative research with refugees in the Netherlands. These are creativity, 'holding space' for change and transformation and the fluid combination of 'decentering and recentering'. Using creative workshops to share stories and content through art practices, the researchers aimed at decentering their own perspectives and alleged leading role to allow the group's desires, knowledges and practices to remain central, recentering when needed—for instance when too much openness risks flowing over into chaos and lack of direction—and incorporating academic insights throughout the process. The chapter illustrates how in transdisciplinary research collectives one can develop ways of working on conducive conditions for fertile co-creation, even when such conditions are not in place.

The final chapter by de Weger, Fraaije, Harambam and Willems presents three different cases, all with different stakeholder groups holding different positions in relation to the interface between science, innovation, policy and practice. The authors focus on the question how to meaningfully engage citizens who experience a distance from, and distrust of, science and government, and who for that reason tend to be excluded in, or choose not to engage in, participatory projects. They discuss pitfalls and (possible) remedies, and once again the authors underscore that 'transdisciplinarians' reflexivity is key, as there are no fail-safe solutions.

Literature

Adelman, M., Haldane, H., & Wies, J. R. (2012). Mobilizing culture as an asset: A transdisciplinary effort to rethink gender violence. *Violence against Women, 18*(6), 691–700. https://doi.org/10.1177/1077801212454121

Balsiger, J. (2015). Transdisciplinarity in the class room? Simulating the co-production of sustainability knowledge. *Futures, 65*, 185–194. https://doi.org/10.1016/j.futures.2014.08.005

Bourdieu, P. (1977). *Outline of a theory of practice*. Cambridge University Press.

Bruhn. (2024). Safe Spaces in Unsafe Environments—Experiences from COP26 About Hosting Inclusive Spaces for Deep Encounters and Reflection. In *Transdisciplinarity for Transformation*. Palgrave Macmillan. (Ahead of Print).

Canel, M. J., Barandiarán, X., & Murphy, A. (2022). What does learning by listening bring to citizen engagement? Lessons from a government program. *Public Relations Review, 48*(1), 102132. https://doi.org/10.1016/j.pubrev.2021.102132

Cash, D. W., Clark, W. C., Alcock, F., Dickson, N. M., et al. (2003). Knowledge systems for sustainable development. *Proceedings of the National Academy of Sciences of the United States of America, 100*(14), 8086–8091. https://doi.org/10.1073/pnas.1231332100

Chilvers, J. (2008). Deliberating competence: Theoretical and practitioner perspectives on effective participatory appraisal practice. *Science, Technology, & Human Values, 33*(2), 155–185. https://doi.org/10.1177/0162243907307594

Chilvers, J., & Kearnes, M. (2020). Remaking participation in science and democracy. *Science, Technology & Human Values, 45*(3), 347–380. https://doi.org/10.1177/0162243919850885

Ciesielski, T. H., Aldrich, M. C., Marsit, C. J., Hiatt, R. A., & Williams, S. M. (2017). Transdisciplinary approaches enhance the production of translational knowledge. *Translational Research, 182*, 123–134. https://doi.org/10.1016/j.trsl.2016.11.002

Code, L. (2006). *Ecological thinking: The politics of epistemic location*. Oxford University Press.

Cooke, B., & Kothari, U. (Eds.). (2001). *Participation: The new tyranny?*. Zed books.

Dewey J. (1916). *Democracy and dducation*. Project Gutenberg.

Díaz-Arévalo, J. M. (2022). In search of the ontology of participation in participatory action research: Orlando Fals-Borda's participatory turn, 1977–1980. *Action Research, 20*(4), 343–362. https://doi.org/10.1177/14767503221103571

Dryzek, J. S. (2002). *Deliberative democracy and beyond: Liberals, critics, contestations*. Oxford University Press.

Fam, D., Palmer, J., Riedy, C., & Mitchell, C. (Eds.). (2016). *Transdisciplinary research and practice for sustainability outcomes*. Taylor & Francis.

Franklin, A. (Ed.). (2022). *Co-creativity and engaged scholarship: Transformative methods in social sustainability research*. Springer Nature. https://doi.org/10.1007/978-3-030-84248-2

Fricker, M. (2007). *Epistemic Injustice: Power and the ethics of knowing*. Oxford University Press.

Fritz, L., Schilling, T., & Binder, C. R. (2019). Participation-effect pathways in transdisciplinary sustainability research: An empirical analysis of researchers' and practitioners' perceptions using a systems approach. *Environmental Science & Policy, 102*, 65–77. https://doi.org/10.1016/j.envsci.2019.08.010

Fritz, L., & Binder, C. R. (2020). Whose knowledge, whose values? An empirical analysis of power in transdisciplinary sustainability research. *European Journal of Futures Research, 8*(1), 1–21. https://doi.org/10.1186/s40309-020-0161-4

Geels, F. W. (2012). A socio-technical analysis of low-carbon transitions: Introducing the multi-level perspective into transport studies. *Journal of Transport Geography, 24*, 471–482. https://doi.org/10.1016/j.jtrangeo.2012.01.021

Ghorashi, H., & Ponzoni, E. (2014). Reviving agency: Taking time and making space for rethinking diversity and inclusion. *European Journal of Social Work, 17*(2), 161–174. https://doi.org/10.1080/13691457.2013.777332

Godemann, J. (2008). Knowledge integration: A key challenge for transdisciplinary cooperation. *Environmental Education Research, 14*(6), 625–641. https://doi.org/10.1080/13504620802469188

Habermas, J. (1981). *Theorie des kommunikativen Handelns* (Vol. 2, pp. 1049–1054). Suhrkamp.

Hadorn, G. H., Biber-Klemm, S., Grossenbacher-Mansuy, W., Hoffmann-Riem, H., Joye, D., Pohl, C., Wiesmann, U., & Zemp, E. (2008). The emergence of transdisciplinarity as a form of research. *Handbook of transdisciplinary research* (pp. 19–39). Springer. https://doi.org/10.1007/978-1-4020-6699-3_2

Hadorn, G. H., Bradley, D., Pohl, C., Rist, S., & Wiesmann, U. (2006). Implications of transdisciplinarity for sustainability research. *Ecological Economics, 60*(1), 119–128.

Heaslip, E., & Fahy, F. (2018). Developing transdisciplinary approaches to community energy transitions: An island case study. *Energy Research & Social Science, 45*, 153–163. https://doi.org/10.1016/j.erss.2018.07.013

Innes, J. E., & Booher, D. E. (2004). Reframing public participation: Strategies for the 21st century. *Planning Theory & Practice, 5*(4), 419–436.

Jasanoff, S. (2003). Breaking the waves in science studies: comment on H.M. Collins and Robert Evans, The third wave of science studies. *Social Studies of Science, 33*(3), 389–400. https://doi.org/10.1177/03063127030333004

Javadi, D., Tran, N., & Ghaffar, A. (2018). Building a workforce for future health systems: Reflections from health policy and systems research. *Health Services Research, 53*, 4024–4033. https://doi.org/10.1111/1475-6773.12978

Jones, T. & Loeber, A. (in review). Responding to power relations in knowledge co-productions: Taking inspiration from original participatory action research as a transformative and liberatory way of learning and doing in EU-funded research.

Kim, S. (2023). Intersectional positionalities across gender, race, ethnicity and immigrant status in qualitative interviews. *International Journal of Qualitative Studies in Education, 1–17*. https://doi.org/10.1080/09518398.2023.2181448

Klaassen, P., Rijnen, M., Vermeulen, S., Kupper, F., & Broerse, J. (2018). Technocracy versus experimental learning in RRI: On making the most of RRI's interpretative flexibility. In *Responsible Research and Innovation* (pp. 77–98). Routledge.

Klein, J. T. (2014). Discourses of transdisciplinarity: Looking back to the future. *Futures, 63*, 68–74. https://doi.org/10.1016/j.futures.2014.08.008

Kok, K. P. W., Gjefsen, M. D., Regeer, B. J., & Broerse, J. E. W. (2021). Unraveling the politics of 'doing inclusion' in transdisciplinarity for sustainable transformation. *Sustainability Science, 16*, 1811–1826. https://doi.org/10.1007/s11625-021-01033-7

Kua, H. W. (2016). A new integrated framework for stakeholder involvement in sustainability policymaking—a multidisciplinary approach. *Sustainable Development, 24*(5), 281–297. https://doi.org/10.1002/sd.1629

Lang, D. J., Wiek, A., Bergmann, M., Stauffacher, M., Martens, P., Moll, P., et al. (2012). Transdisciplinary research in sustainability science: Practice, principles, and challenges. *Sustainability Science, 7*(1), 25–43. https://doi.org/10.1007/s11625-011-0149-x

Latour, B. (1987). *Science in action: How to follow scientists and engineers through society*. Harvard University Press.

Loeber, A. (2004). *Practical wisdom in risk society. Methods and practice of interpretive analysis on questions of sustainable development*. [Doctoral thesis, University of Amsterdam].

Loeber, A. (2007). Designing for *phronèsis*: Experiences with transformative learning on sustainable development. *Critical Policy Analysis [critical Policy Studies], 1*(4), 389–414. https://doi.org/10.1080/19460171.2007.9518528

Loeber, A. V., Mierlo, B., & Leeuwis an, C., & Grin. J. (2007). The practical value of theory: Conceptualizing learning in the pursuit of a sustainable development. In A. Wals & T. van der Leij (Eds.), *Social learning toward a more sustainable world: Principles, perspectives, and praxis* (pp. 83–98). Wageningen Academic Publishers.

Loeber, A., & Vermeulen, T. (2016). Reflexive project management in high-ambition projects: Exploring the competencies for managing innovative sustainable designs. *Social Business, 6*(1), 15–37. https://doi.org/10.1362/204440816X14636485174877

Loorbach, D. (2007). Transition management. In *New mode of governance for sustainable development*. International Books.

Loorbach, D., Frantzeskaki, N., & Avelino, F. (2017). Sustainability transitions research: Transforming science and practice for societal change. *Annual Review of Environment and Resources, 42*, 599–626. https://doi.org/10.1146/annurev-environ-102014-021340

Lux, A., Schäfer, M., Bergmann, M., Jahn, T., Marg, O., Nagy, E., Ransiek, A., & Theiler, L. (2019). Societal effects of transdisciplinary sustainability research—How can they be strengthened during the research process?. Environmental *Science & Policy, 101*, 183–191. https://doi.org/10.1016/j.envsci.2019.08.012

Lynch, D. H., Klaassen, P., & Broerse, J. E. (2017). Unraveling Dutch citizens' perceptions on the bio-based economy: The case of bioplastics, bio-jetfuels

and small-scale bio-refineries. *Industrial Crops and Products, 106*, 130–137. https://doi.org/10.1016/j.indcrop.2016.10.035
Mascarenhas, A., Langemeyer, J., Haase, D., Borgström, S., & Andersson, E. (2021). Assessing the learning process in transdisciplinary research through a novel analytical approach. *Ecology and Society, 26*(4). https://doi.org/10.5751/ES-12631-260419
Masson, J. E., Soustre-Gacougnolle, I., Perrin, M., Schmitt, C., Henaux, M., Jaugey, C., Teillet, E., Lollier, M., Lallemand, J., & Schermesser, F. (2021). Transdisciplinary participatory-action-research from questions to actionable knowledge for sustainable viticulture development. *Humanities and Social Sciences Communications, 8*(1), 1-9. https://doi.org/10.1057/s41599-020-00693-7
Mauser, W., Klepper, G., Rice, M., Schmalzbauer, B. S., Hackmann, H., Leemans, R., & Moore, H. (2013). Transdisciplinary global change research: The co-creation of knowledge for sustainability. *Current Opinion in Environmental Sustainability, 5*(3–4), 420–431. https://doi.org/10.1016/j.cosust.2013.07.001
Max-Neef, M. A. (2005). *Foundations of Transdisciplinarity. Ecological Economics, 53*(1), 5–16. https://doi.org/10.1016/j.ecolecon.2005.01.014
Nicolescu, B. (2014). *Methodology of Transdisciplinarity. World Futures, 70*(3–4), 186–199. https://doi.org/10.1080/02604027.2014.934631
Nilsen, P. (2015). Making sense of implementation theories, models and frameworks. *Implementation Science, 10*(1), article 53. https://doi.org/10.1186/s13012-015-0242-0
Norström, A. V., Cvitanovic, C., Löf, M. F., West, S., Wyborn, C., Balvanera, P., & Österblom, H. et al. (2020). Principles for knowledge co-production in sustainability research. *Nature sustainability, 3*(3), 182–190. https://doi.org/10.1038/s41893-019-0448-2
Nowotny, H., Scott, P., & Gibbons, M. (2003). Introduction: 'Mode 2' revisited: The new production of knowledge. *Minerva, 41*(3), 179–194. https://www.jstor.org/stable/41821245
Owen, R., Macnaghten, P., & Stilgoe, J. (2012). Responsible research and innovation: From science in society to science for society, with society. *Science and Public Policy, 39*(6), 751–760. https://doi.org/10.1093/scipol/scs093
Popa, F., Guillermin, M., & Dedeurwaerdere, T. (2015). A pragmatist approach to transdisciplinarity in sustainability research: From complex systems theory to reflexive science. *Futures, 65*, 45–56. https://doi.org/10.1016/j.futures.2014.02.002
Pohl, C. (2010). From transdisciplinarity to transdisciplinary research. *Transdisciplinary Journal of Engineering & Science, 1*, 65–73. https://doi.org/10.22545/2010/0006

Pohl, C., & Hadorn, G. H. (2007). *Principles for designing transdisciplinary research* (pp. 36–40). Oekom.

Regeer, B. J. et al. (2024). Structuring Design & Evaluation in Transdisciplinarity for Transformation. In *Transdisciplinarity for Transformation*. Palgrave Macmillan. (Ahead of Print).

Regeer, B. J., & Bunders, J. F. (2003). The epistemology of transdisciplinary research: From knowledge integration to communities of practice. *Interdisciplinary Environmental Review, 5*(2), 98–118. https://doi.org/10.1504/IER.2003.053901

Regeer, B. J., Mager, S., & Van Orsouw, Y. (2011). *Licence to grow: innovating sustainable development by connecting values*.

Rodriguez, D. (2022). Transdisciplinarity and epistemic communities: Knowledge decolonisation through university extension programmes. *Geographical Research, 60*(1), 113–125. https://doi.org/10.1111/1745-5871.12524

Rotmans, J. (1998). Methods for IA: The challenges and opportunities ahead. *Environmental Modeling & Assessment, 3*, 155–179. https://doi.org/10.1023/A:1019019024003

Salem, M. A., Shawtari, F., Shamsudin, M. F., & Hussain, H. B. I. (2018). The consequences of integrating stakeholder engagement in sustainable development (environmental perspectives). *Sustainable Development, 26*, 255–268. https://doi.org/10.1002/sd.1699

Schön, D., & Rein, M. (1994). *Frame reflection: Toward the resolution of intractable policy controversies*. Basic Books.

Schuitmaker-Warnaar, T. J., Gunn, C. J., Regeer, B. J., & Broerse, J. E. (2021). Institutionalizing reflexivity for sustainability: Two cases in health care. *Sustainability, 13*(21), 11712. https://doi.org/10.3390/su132111712

Simon, D., Palmer, H., Riise, J., Smit, W., & Valencia, S. (2018). The challenges of transdisciplinary knowledge production: From unilocal to comparative research. *Environment and Urbanization, 30*(2), 481–500. https://doi.org/10.1177/0956247818787177

Smith, A., Stirling, A., & Berkhout, F. (2005). The governance of sustainable socio-technical transitions. *Research Policy, 34*(10), 1491–1510.

Soneryd, L. (2016). Technologies of participation and the making of technologized futures. In J. Chilvers & M. Kearnes (Eds.), *Remaking participation: Science, environment and emergent publics* (pp. 144–161). Routledge.

Stark. (2024). Beyond Interdisciplinary Research: Transdisciplinarity and Transformative Literacy Through Artistic Thinking and Research. In *Transdisciplinarity for Transformation*. Palgrave Macmillan. (Ahead of Print).

Stilgoe, J., Owen, R., & Macnaghten, P. (2013). Developing a framework for responsible innovation. *Research Policy, 42*(9), 1568–1580.

Stirling, A. (2008). "Opening up" and "closing down" power, participation, and pluralism in the social appraisal of technology. *Science, Technology, & Human Values, 33*(2), 262–294.

Temper, L., & Del Bene, D. (2016). Transforming knowledge creation for environmental and epistemic justice. *Current Opinion in Environmental Sustainability, 20*, 41–49. https://doi.org/10.1016/j.cosust.2016.05.004

Thomas-Hunt, M., Ogden, T., & Neale, M. (2003). Who's really sharing? Effects of social and expert status on knowledge exchange within groups. *Management Science, 49*(4), 464–477. https://doi.org/10.1287/mnsc.49.4.464.14425

Taylor, M., Eriksen, S., Vincent, K., Brooks, N., Scoville-Simonds, M., & Schipper, L. (2022). *Putting 'vulnerable groups' at the centre of adaptation interventions by promoting transformative adaptation as a learning process.* Report for Norad. Norwegian University of Life Sciences.

Torkar, G., & McGregor, S. L. (2012). Reframing the conception of nature conservation management by transdisciplinary methodology: From stakeholders to stakesharers. *Journal for Nature Conservation, 20*(2), 65–71. https://doi.org/10.1016/j.jnc.2011.10.002

Turnhout, E., Metze, T., Wyborn, C., Klenk, N., & Louder, E. (2020). The politics of co-production: Participation, power, and transformation. *Current Opinion in Environmental Sustainability, 42*, 15–21. https://doi.org/10.1016/j.cosust.2019.11.009

Van Breda, J., & Swilling, M. (2019). The guiding logics and principles for designing emergent transdisciplinary research processes: Learning experiences and reflections from a transdisciplinary urban case study in Enkanini informal settlement, South Africa. *Sustainability Science, 14*, 823–841.

Von Schomberg, R. (2013). A vision of responsible research and innovation. In R. Owen, J. Bessant & M. Heintz (Eds.), *Responsible innovation: Managing the responsible emergence of science and innovation in society* (pp. 51–74). John Wiley & Sons Ltd. https://doi.org/10.1002/9781118551424.ch3

Vermeulen, W. J., & Witjes, S. (2020). History and mapping of transdisciplinary research on sustainable development issues: Dealing with complex problems in times of urgency. In M. M. Keitsch & W. J. V. Vermeulen (Eds.), *Transdisciplinarity for sustainability: Aligning diverse Practices*. Routledge.

Verwoerd, L., Brouwers, H., Kunseler, E., Regeer, B., & de Hoop, E. (2023). Negotiating space for knowledge co-production. *Science and Public Policy, 50*(1), 59–71. https://doi.org/10.1093/scipol/scac045

Wickson, F., Strand, R., & Kjølberg, K. L. (2015). The walkshop approach to science and technology ethics. *Science and Engineering Ethics, 21*, 241–264. https://doi.org/10.1007/s11948-014-9526-z

Wilde, K., & Hermans, F. (2021). Innovation in the bioeconomy: Perspectives of entrepreneurs on relevant framework conditions. *Journal of Cleaner Production, 314*, article 127979. https://doi.org/10.1016/j.jclepro.2021.127979

CHAPTER 9

Taking the Landscape into Conversations: A Way to Engage (with) Diverse Knowledges and Values

Hilde J. H. Brouwers, Linde R. Egberts, and Evelien de Hoop

9.1 Introduction

Research and practice in sustainability transformations often manifest diverging ideas on the nature and origin of sustainability problem(s), their solutions and the division of roles and responsibilities to address these problems. This diversity is underpinned by a plurality of values, forms of

H. J. H. Brouwers (✉)
Environmental Policy, Wageningen University, Wageningen, The Netherlands
e-mail: hilde.brouwers@wur.nl

L. R. Egberts · E. de Hoop
Vrije Universiteit Amsterdam, Amsterdam, The Netherlands
e-mail: l.egberts@cultureelerfgoed.nl

E. de Hoop
e-mail: e.de.hoop@vu.nl

L. R. Egberts
Knowledge Strategy, Cultural Heritage Agency of the Netherlands, Amersfoort, The Netherlands

B. J. Regeer et al. (eds.), *Transdisciplinarity for Transformation*,
https://doi.org/10.1007/978-3-031-60974-9_9

knowledge and worldviews (Lang et al., 2012; Pohl & Hadorn, 2008; Wickson et al., 2006). Moreover, these diverging ideas are dynamic, both in interaction with each other and with the sustainability challenges they are addressing (Cuppen, 2018). In the field of (predominantly European) sustainability transformations, some scholars have therefore argued that this plurality of views and ways of knowing ought to be embraced, to be able to do justice to the complexity and emergent character of sustainability challenges (Caniglia et al., 2021; Norström et al., 2020). Transdisciplinary research (TDR) aims to build on this recognition, by transcending the boundaries of academia—which are seen as exclusionary to non-academic actors and their knowledges. TDR therefore commonly takes the shape of a collaboration between a range of relevant actors and can be conducted in different places, both in and outside academic settings (Pohl, 2008; Pohl & Hadorn, 2007).

Providing room for these diverse ideas and bringing them into conversation has proven challenging. It has been argued that policy and research contexts may often be oriented towards the identification of a single problem and corresponding solution, thereby reducing the space for a range of actors' diverse ways of articulating and resolving it (e.g. De Hoop & Arora, 2021). Similarly, Cuppen (2018) has highlighted how policy and institutions are often unable to continuously provide space to actors' emergent and dynamic views, again reducing the space for their different ways of articulating and resolving a particular problem. Crucially, scholars in the field of TDR increasingly argue that TDR, too, is frequently conceptualized—even idealized—as a process of integrating diversity into a single problem and solution (Klenk & Meehan, 2015). Such integration takes place during various research stages, including problem framing, stakeholder inclusion and drawing conclusions (Jacobi et al., 2020; Lang et al., 2012; Polk, 2015). Crucially, focusing on integration conceals frictions and paradoxes that are inherent in engaging with diverse and sometimes irreconcilable aims and ways of doing and knowing (Klenk & Meehan, 2015; Wickson et al., 2006). Since such concealment allows dominant norms and values to remain unchallenged, this predominantly tends to favour established interests and actors (Kok et al., 2021; Turnhout et al., 2020).

Second, attempts to address sustainability challenges often rely heavily on cognitive forms of knowledge. A more explicit focus on embodied forms of knowledge—knowledge that emerges through sensory experiences and which informs actions and decisions in a non-cognitive manner—may help to understand stakeholders' actions and worldviews in different ways. Such approaches may allow, for example, for understanding the social dimension of the problem and questioning of accepted norms, rules, policies and practices (Baron, 2020; Bentz et al., 2022; Ingold & Kurttila, 2000; Leichenko & O'Brien, 2020; Pohl et al., 2021). What is important here is that each of these knowledges—including 'cognitive' or 'scientific' knowledges—may be considered partial, specific and situated, only to be seen from a particular vantage point (see also Haraway, 1988; Harding, 1991). Hence, we argue that it is pertinent to render explicit not only which and whose knowledges are involved in TDR, but also how these are entangled with specific material and social sites.

TDR has long been concerned with developing tools and research methods to bring together diverse forms of knowing and valuing. Yet, the challenges outlined above mean that it is important to improve these tools and methods to foreground and build on, rather than erase, diversity—not only with regard to whose voices are included in the process but particularly in relation to how those voices get to express themselves and are heard. In this chapter, we therefore explore a relatively unknown methodology, a transdisciplinary *walkshop*, and reflect on ways in which it may allow diversity to become explicit and to engage with it. A transdisciplinary walkshop combines the act of walking and being at the site where there is a sustainability challenge with a workshop in which multiple stakeholders, including researchers, participate. The method offers the potential to reduce hierarchical relations (Anderson, 2004; Jones et al., 2008; Kinney, 2017; Wickson et al., 2015) among participants and researchers. Moreover, a walkshop may elicit discussions inspired by the material environment, including sights, smells or sounds, which offer data that a room-based setting could not. This may allow for a deeper understanding of how the stakeholders respond not only cognitively to the landscape, but also how they *are* and experience it (Döring & Ratter, 2021). In their sharing of experiences with a walkshop for discussing the ethics of science and technology, Wickson et al. argue that the hiking together, being together for a prolonged amount of time and using the landscape as a prompt for conversation may elicit discussions in which critique can be considered illuminating rather than threatening (Wickson et al., 2015). However, how that illumination occurs exactly—i.e. how diverse perspectives come to the fore—and how this contributes to (sustainable change) remains obscure.

Based on these potential strengths we (the authors) developed our version of the walkshop with the deliberate aim of rendering explicit both diversity in knowing and valuing and the situated emergence of knowledge. In this chapter, we describe this walkshop and reflect the way and extent to which this aim materialized. We organized this walkshop to foster conversations with and between diverse actors with an interest in fire risk-reduction in landscapes under nature management in the Netherlands. These areas are increasingly at risk of uncontrollable fires in the context of climate change and current land- and water-management practices. Such uncontrollable fires have severe consequences for livelihoods, animal life, human health, biodiversity and cultural heritage (De Hoop et al., 2022). This is a rather new problem in the Netherlands, with its moderate northwest European climate. Recent uncontrollable fires, such as the Peel and Meinweg fires in the spring of 2020, led various stakeholders to realize that they are unprepared to handle such fires. These stakeholders—which include various levels of government, land managers and firefighting organizations—are currently attempting to organize themselves and each other around the issue of landscape fire prevention and mitigation. In this context, diversity takes various forms: for example, each of the stakeholders values fires and landscapes under nature management in their own way. Moreover, the increasing risk of uncontrollable fires challenges existing ideas about the appearance, maintenance and indigenous/typical vegetation of a particular landscape.

This issue is also new to Dutch academic institutions, which have historically been concerned mainly with 'wildfires' in regions of the world that have long been at such risk (such as countries in the Mediterranean basin, California or Australia). The increasing risk of uncontrollable fires occurring in Dutch landscapes under nature management has only recently attracted academics' attention. The walkshop we discuss here was part of a pioneering inter- and transdisciplinary research project on this topic, in which scholars in earth sciences, humanities, science and society studies and environmental sciences at the Vrije Universiteit Amsterdam collaborated with these stakeholders.

First we turn to research in the field of geography look at ways of knowing and valuing landscapes, and go on to explore the literature on how walking may make visible the relationships between landscapes and human beings, and may also reshape relationships between researchers and participants. Third, we describe how we developed our transdisciplinary walkshop for the particular case outlined above. Finally, we answer the question that this chapter deals with, namely: how may a transdisciplinary walkshop make landscape–human interactions visible and allow for meaningful engagements with diverse ways of knowing and doing and its potential for transdisciplinary research.

9.2 Theoretical Foundations

This chapter draws on two main bodies of conceptual work. First, we explore more-than-representational approaches and relational approaches to landscapes, which inform both our design of the workshop discussed in this chapter and our analysis of the knowledges that emerged from conducting this walkshop in practice. Second, we review existing literature on walking as an important way in which humans interact with the landscape and that, consequently, walking research methodologies emerged. We also discuss, here, how walking research methodologies may be valuable in the light of transdisciplinary research approaches specifically, given the inherent changing relations between researchers and 'participants' that are central to such approaches.

9.2.1 Living and Walking the Landscape

Geographers have long paid attention to the relations between people's embodied everyday experiences and the socio-material places in which they live (Relph, 1976; Tschakert et al., 2013; Tuan, 1977). In this context, landscapes are seen as integral to human lives, both contributing to and a product of human existence. Experiences in the landscape, and ideas and images about it, are formed through being in the landscape and simultaneously shape it as people determine what is worth preserving or what needs to be changed.

We use the term 'landscape' to refer to physical territories, to forms of governing regions, as well as to symbolic and representational meanings, for example in the form of landscape painting (Olwig, 2019). As any type of representation can be considered as a deliberate act or practice, the materiality and representations of landscapes may be strongly intertwined (Seamon, 2018). Such an approach engages with discourses, representations and ideas about landscapes in relation to the bodily, sensory and emotional experiences of them (Lorimer, 2005). Stressing this intertwined nature are so-called *more-than-representational* and *relational* approaches to landscapes.

More-than representational and relational approaches to landscape have the advantage that they can bring the 'being in the landscape', with all its affective, social and relational aspects, together with representational ways of researching landscapes based on analysing them as 'text', using sources like maps, images, and written text. They allow researchers to grasp how people make sense of and understand the world around them, include body–landscape relationships, in their analysis, and serve to include the role of embodiment, performance and practice in the ways subjectivity is shaped (Harrison, 2000; Macpherson, 2010; Wylie, 2002). More-than-representational approaches enable researchers to come to understand the landscape as a 'lifeworld': an understanding of the inhabitants' social practices and physical activities that co-constitute a landscape's cultural and natural aspects (Döring & Ratter, 2021).

How might landscape be relevant to studying sustainability challenges? In the broader field of research on climate change geographers started to draw attention to how changes in the climate affect embodied engagements with place and landscapes. For instance, Brace and Geoghegan (2011) focus on landscape to understand how climate change may be both an 'intellectual artifact and an embodied and experiential process' (p. 296). Their approach, they argue, makes visible how those who live and work in the landscape feel and sense climatic changes in their everyday life and considers local knowledge to be a valuable means to make sense of climate change where scientific 'facts' (original quotation marks, p. 295) fail to inspire action. In the UK, Köpsel and colleagues analysed the relations between people and place in the face of climate change and distinguished between four different narratives among those responsible for landscape management about what this landscape—in this case Cornwall—means and constitutes. Each narrative implies different directions for how the Cornish landscape ought to be managed and adapted in the face of climate change. Although the authors emphasized the differences between the narratives, they also described how all actors—regardless of the narrative they construct—express feeling attracted to the visual appearance of the Cornish landscape, feeling strongly connected to Cornwall as a place and considering its landscape paramount to the regional identity. As such, they conclude, constructive dialogue on landscape management should make use of this love for the landscape (Köpsel et al., 2017). In Denmark, Nina Baron studied controversies about dyke building and maintenance in the face of climate change. She discusses how these controversies are about much more than finding the 'right' solution based on proper calculations and technical solutions—rather, they are about people's diverse valuations of the landscape and embodied knowledges in relation to the dyke (Baron, 2020). More-than-representational approaches thus allow researchers to bring abstract and cognitive ideas about landscapes together with immediate experiences of and emotions related to landscapes under (climate) transformation. These landscapes in themselves both evoke and are affected by these ideas and effects.

As such, we, too, use a more-than-representational approach to landscape to create space for understanding landscapes and humans as inextricably linked and to create space for stakeholders' diverse experiences in and understandings of the landscape. We acknowledge that the sustainability challenge of fire risk-reduction in the Veluwe is both a cognitive process *and* an embodied and experiential process (in the words of Brace and Geoghegan). Especially when scientific facts cannot inspire action (given the scientific uncertainties in this case and is considered inherent to sustainability challenges) the relations between people and the landscapes they live in and move through may produce a narrative about what aspects are important in the landscape and deserve preservation or require changing. For each actor, this can be a different narrative. In the next section we look at walking research methodologies as a way to study such relations and narratives.

9.2.2 On Walking Research Methodologies

A specific form of perceiving and living the landscape is through walking. In European history, the activity of walking came to be characterized as a form of 'modern corporeal reflexivity' during the Romantic period (Edensor, 2000, p. 82). This means that walking may allow one to become reflexively aware of one's body, senses and self. Walking reveals 'distinctive ways in which we express ourselves physically, simultaneously performing and transmitting meaning while sensually apprehending 'nature' and sustaining wider ideologies about nature and the role of the body in nature' (Edensor, 2000, p. 82). The pattern of how humans walk in a given space creates place-specific 'place ballets'. Such 'place ballets' can be understood as local knowledge, shaped throughout time: people decide (not necessarily (fully) consciously) which landscapes are particularly suitable for walking and create signs and paths within them (Seamon, 1980, cited in Buttimer & Seamon, 2015). Our case-study area, for example, the Veluwe, is an intensely used leisure area, where walking for pleasure has shaped and reshaped the landscape for several centuries (Neefjes, 2018).

We thus consider landscapes and humans as constantly shaping and being shaped by one another. The practice of walking offers dynamic interpretations of both. Walking interviews and methods have often been used to study people's responses to the landscape. In such studies the landscape is considered to be a 'background' or scenery rather than interacting with the human beings who live with them (e.g. Adevi & Grahn, 2012, cited in Macpherson, 2016). In contrast, in more-than-representational and relational approaches to landscape, walking research methods have been used to study interactions *with* the landscape, the dynamic and constant becoming of bodies, minds and landscapes (Döring & Ratter, 2021). Walking, while talking about the landscape, can generate 'thick data of emplaced interactions'—data that is inspired by the context in which this takes place (Döring & Ratter, 2021, p. 320). For example, Döring and Ratter show how smells and sounds become part of their data through their walking interviews. Hence, a walkshop may be used as a research tool to not only gather data on cognitive experiences of the landscape, but also sensory data which may reveal how people experience the landscape in non-cognitive ways.

Particularly relevant for TDR is that walking research methods can also be a means to reshape relationships among participants, including researchers and non-academic actors. These methods may allow for more spontaneous conversation and facilitate interaction and also afford participants a greater degree of control, by placing the emphasis on particular locations, or moving to others. Walking methodologies are thus also seen to help reduce power differentials, allowing participants to speak more freely and create a more even relationship between researchers and participants (Anderson, 2004; Jones et al., 2008; Kinney, 2017). Accordingly, a walkshop may also create a more even relationship among participants when there is distance to political environments and when they are physically away from their respective organizations. However, we acknowledge that walking interviews—like all research methods—require careful consideration and selection as a research method as the potential advantages may very much depend on who the participants are or what are the political tensions: for instance, in particular landscapes such political tensions may also be more prominent than in a room-based interview.

Thus, we explored relations between landscapes and humans, as understanding this relation is pertinent to comprehending how stakeholders know and value landscapes that are prone to fire management and how differences may arise. We furthermore explored walking research methodologies as valuable in understanding these relationships. In the following sections, we explore how these methodologies can contribute to the challenges in TDR that we outlined in the introduction: making diversity in knowing and valuing the landscape explicit and engaging with this. First, however, we turn to our methodology.

9.3 Methodology: Developing and Analysing a Transdisciplinary Walkshop

Here, we elaborate on how we used walking as a research method and tool to develop our walkshop. Moving through and experiencing the landscape can allow stakeholders to use and refer to it in their conversations and discussions (Anderson, 2004; Jones et al., 2008; Kinney, 2017; Wickson et al., 2015). This may contribute, as argued above, to bringing out and understanding how actors variously think about and live in/with the landscape, the different problems they encounter and their diverse views on ways of addressing these problems. A *transdisciplinary* walkshop aims to bring social and scientific practice together and seeks to co-create knowledges that are relevant to the issues they face, while also reflecting on which and whose interests are served in doing so.

The aim of our walkshop emerged from our observations from interviews we had conducted with nine stakeholders of a particular landscape under nature conservation in the Netherlands, namely the Veluwe area. Based on these interviews, we learned that stakeholders had various needs in order to develop more effective tools and measures for fire prevention, which were: (a) revaluation of landscapes and fire in the context of climate change and increasing risks of uncontrollable fires; (b) new, more collaborative and place-based modes of governance; and c) inter- and transdisciplinary knowledge situated within the area's specific characteristics (see also De Hoop et al., 2022). With our walkshop, we sought to contribute to these three needs by bringing stakeholders and researchers together to discuss fire risk-reduction strategies within the landscape, explore possibilities for collaboration and identify knowledge questions.

The design of the walkshop was as follows. First, we welcomed the participants with a lunch at our main venue of the Radio Kootwijk historic building, giving all of us a chance to get to know each other. This, we argue, is relevant to create a more even relationship between researchers and participants, and among participants (Anderson, 2004; Jones et al., 2008). Second, we introduced ourselves and gave a brief overview of the scientific context in which our research took place, by referring to predicted fire risks in research on climate change and current policy-relevant research and research reports (Jones et al., 2022; UNEP, 2022). We also highlighted the findings from our interviews (De Hoop et al., 2022). Third, we split up into three groups (which we determined in advance, to create heterogeneous groups of stakeholders and knowledge from the area) and walked to three specific sites by which we aimed to incite a discussion towards a specific topic, while creating space for stakeholders' own interpretations and understandings of the landscape.

Our theoretical foundations made that we paid in particular attention to the site of the walkshop and the specific locations in which we wanted to facilitate discussions. We chose the Veluwe area and the area around Radio Kootwijk for its suitability for walking, measures already taken, the history of the area (see also De Hoop et al., 2022; Neefjes, 2018) and the practical option to have a lunch location in the building of Radio Kootwijk. To choose sites for conversation, we visited the area with the authors upfront to explore the area. We identified one location as an interesting site for conversation as it provided a good view of the Radio Kootwijk building, which we used as a steppingstone to (try to) instigate a conversation on the history of the area and the desirable future in this landscape given the changing conditions. With this, we aimed to elucidate how human–landscape interactions are informed by perceptions of the past and how this affects what is experienced and valued in the landscape today (which, in turn, affects desirable futures). The second location we chose marked a relatively harsh transition between heathlands and forest, allowing us to provoke a discussion on why such a transition is (not) important and for whom—given various values of biodiversity, cultural value and fire prevention. The third location we chose was a road along which there are some homes and where fire-prevention measures had visibly been taken (felling of conifers and growing deciduous trees), which prompted us to ask why these measures had been taken and what else could or ought to be done to reduce fire risks—here, or elsewhere in the area—and how such measures would relate to each other. Lastly, after the walk and all three groups had covered all three locations, we headed back to the main venue and briefly discussed what stakeholders had taken away from these discussions and what would be the next steps for policy and research.

Our previous interviewees were the first to be invited to the walkshop and were also asked to invite their colleagues. Through this, several additional organizations were reached, such as the relatively new Working Group Fire Prevention (organized by the Ministry of Agriculture, Nature and Food Quality) and the Interprovincial Council (the collaborative organization of the 12 Dutch provinces). In total, the walkshop consists of 20 participants, including firefighting organizations, policymakers, land managers and researchers (see Table 9.1).

Given to the dynamic character of the walkshop we chose not to rely on audio-devices to record our discussions. Instead, the discussions during the walkshop—including those held in break-out groups while walking—were led and followed by at least two researchers actively participating in the discussion and another researcher taking field notes. These notes were written up more extensively within 24 hours (for memory purposes) and edited and complemented by the other researchers who were in the discussions. There were several iterations by all participating researchers within the same week to ensure we ended up with a comprehensive and detailed account. The notes covered what was said as well as the visual and auditory cues that prompted these comments. Photos were also taken at spots that sparked particular interest, to have visual data of what had prompted it. As such, we were able to gather 'thick data' in which the interactions and discussions were related to the landscape and place in which they were situated in (Döring & Ratter, 2021). The data was analysed in relation to how the participants perceived the walkshop, how it made visible the diversity in knowing and valuing the landscape, and how being in the landscape and participating in the walkshop allowed for engaging with this diversity. The analysis was conducted iteratively by the two first authors who checked each other's interpretations.

9.4 Transdisciplinary Walkshop in Practice

In this section we analyse the discussions in our transdisciplinary walkshop as to answer our research question: we answer the question that this chapter deals with, namely: how may a walkshop make landscape–human interactions visible and allow for meaningful engagements with diverse ways of knowing and doing and its potential for transdisciplinary research? As such, we evaluate how the walkshop contributed to engaging

Table 9.1 Overview of the walkshop participants

Type of participant	*Number of participants*	*Function(s) of participants*
Firefighters	6	All were active and working on promoting fire-prevention measures in landscapes Four were active in the Veluwe area; two were from other areas (de Peel and de Meinweg, which experienced major fires in 2020)
Landscape and forest management	4	Two participants were working in their landscape and forest-management organization on developing national policy for prevention of landscape fires Two participants were responsible for local operational decisions and for day-to-day land management in the Veluwe
Provincial government	6	Three participants were from the Province of Gelderland, in which the Veluwe area is located Three participants were from other provinces. All six were policymakers, no politicians were among them
National government	2	Both participants were policymakers at the Ministry of Agriculture, Nature and Food Safety. The national government was not developing policy on fire prevention, but these participants were working on bringing various organizations together to start working on (developing policy for) landscape fire prevention in the Netherlands
Researchers	12	Two were working in the field of fire prevention, one as an independent researcher and the other for the Dutch Institute of Public Safety The ten other researchers were from the Vrije Universiteit, among whom six were involved in the research project and four were interested colleagues helping to facilitate the discussions and note-taking. The disciplines represented were: earth sciences, environmental sciences, humanities, and science and society

with diverse ways of knowing. We first provide a brief overview of the participants' gathering at the start of the walkshop and highlight some interactions that illustrate the scene and the mood of the afternoon.

9.4.1 Getting to Know Each Other in the Landscape

As we mentioned earlier, the participants were welcomed with a simple lunch. During this informal start, many participants expressed their appreciation that the event was dedicated to fire prevention because they felt this issue needed more widespread attention. They also particularly welcomed the participation of so many people with such diverse institutional affiliations: the number of participants seemed to underline the importance of the issue and their diversity was an opportunity to exchange thoughts with people with whom they did not regularly converse. Some participants, especially those who came from outside the Veluwe area, said that it was a relief to talk to other people about this issue, as they were sometimes the only one in their area or organization working on fire prevention. During and after the workshop, participants particularly valued the 'ability to get to know each other'—to get to know stakeholders from other organizations, and in different positions (such as people in executive positions and in more strategic, managerial positions). Participants felt that this 'getting to know each other' was crucial as they thought that although many stakeholders might have heard *of* each other and may know names of who works where, language and priorities were seen to differ enormously between organizations and most interactions between stakeholders had to date been relatively brief—and online during COVID-19. For example, after the walkshop one firefighter said he had learned a lot about land managers' views of the landscape and their considerations regarding land management. Asking how he experienced the workshop, a firefighter who coordinated fire-prevention measures smiled as he explained that 'this afternoon was exactly what was needed for everyone to get closer—physically and experientially—together'. From this, we understand that the walkshop thus helped to facilitate interactions among stakeholders, indeed corresponding to literature on walking research methodologies (Anderson, 2004; Jones et al., 2008; Kinney, 2017).

As we arrived at the venue the anxiety among the firefighters was palpable. As one of them arrived, he almost immediately explained to us that the coming weekend was going to be 'tense' owing to the dry weather and the expected high winds. Although the firefighters acknowledged not minding an occasional fire in nature areas, this weekend would present difficult conditions if there was a fire and they had therefore thoroughly prepared their materials and equipment. The tension here was not only felt at the start, but was brought up repeatedly throughout the discussions: on several occasions, the firefighters referred to the weather forecasts for the coming weekend. This shows how a walkshop can help to allow for spontaneous conversation and incite conversation topics that are at the forefront of stakeholders' minds but not of researchers' (see also Wickson et al., 2015).

At the same time, the firefighters emphasized that the participants in the walkshop were also apprehensive about potentially uncontrollable fires—especially people who were already working on the issue—and found it important that they were invited and made the effort to participate. Such consensus contrasted with their daily discussions as political and social fears about uncontrollable fires was generally limited to moments such as these—where heat, drought and high winds coincide. Once these weather conditions recede, several walkshop participants said that attention dropped off regarding the potential problems of fires in nature areas. This was considered problematic because it means that there is less attention paid to structural and long-term measures, such as redesigning (parts of) the landscape or training in how to deal with fires in nature areas.

9.4.2 Making Diverse Ways in 'Knowing and Valuing the Landscape' Visible

During the walkshop, stakeholders interacted with the landscape for instance by taking particular spots as a prompt for conversation or listening to the sounds in the landscape. As different stakeholders took different cues from the landscape and constructed a different narrative about what is important, there, and why, we consider this as diverse ways of knowing and valuing the landscape coming to the fore. We distinguished affective, operational and policy-based ways of understanding and valuing the landscape.

The *affective* aspect was illustrated in the walkshop approach as participants were able to point out particular trees (e.g. the tree in Fig. 9.1). This happened, for example, in discussions on what type of trees could or should be felled, and for what purposes. In this context, land managers (and a firefighter who was being trained in nature conservation and one who was particularly attached to the area) argued that 'characteristic' pines were not to be felled. They referred to pines that are older and thought to be well-proportioned, for instance with branched trunks or with serpentine-shaped branches. Other trees, for instance oaks, could also be characteristic according to these participants, but the pines were considered more prominent in this area. The way participants introduced these characteristic pines allowed us to see their particular values: what they considered beautiful and authentic in the landscape—and how older trees came to be deemed more valuable. In addition, the focus on these tree types and the value they constituted for the landscape was a topic that we had not expected to feature so prominently in the context of fire prevention. Indeed, the locations at which we stopped for group discussions during the walkshop were not chosen to focus on these trees, nor did we steer the conversations in this direction.

Fig. 9.1 A 'characteristic pine', as indicated by one of the participants. Photo was taken by one of the authors upon this prompt

Another example of affective understanding and valuation of the landscape became visible when one of the participants, walking from location B to the main venue, spoke about how he preferred the heathlands here to others in the vicinity. He argued that the heathlands in this particular area are 'extremely beneficial' for biodiversity as they were relatively varied, with some trees and sand between the heath. Being in the landscape, it was possible to sense this biodiversity, according to this participant: 'just listen to the sound of all these birds, you don't hear that in all places'. Moreover, this participant maintained that the variation in vegetation plus the sand was also beneficial in preventing fires, as fires slow down on sand and may also do so when there are also vegetation types with a higher moisture content than heath. In contrast, he felt less warmly about the heathlands in the vicinity, which have a high density of similar-aged heath but with no other vegetation. Such heathlands are less beneficial for biodiversity or fire prevention, he argued, but because visitors find it attractive when all the plants flower at the same time he referred to them as 'heathlands for tourists'.

Operational ways of knowing and valuing the landscape were made clear when the conversation revolved around the dryness of the vegetation. When one of the researchers provoked a discussion about a non-native tree species in the area (the *Prunus serotina* or black cherry) and whether it would be better appreciated if it had the potential to reduce fire risk, one of the firefighters promptly pulled a moisture-content meter out of his pocket and started measuring the plant's moisture content. Touching the plant, he also remarked on its other characteristics, such as the plumpness of the leaves. Land managers and firefighters both responded that the moisture content is not always important: 'everything burns' if the weather conditions are 'right'. Along this line, participants argued that many trees are not in leaf in the spring—making this a vulnerable period in this area. Moreover, the participants who experienced the Peel fires in 2020 shared what they witnessed during the fire-suppression efforts: the fire brigades encountered situations where fires spread directly over damp and wet areas, even a canal. For them, this also illustrated that moisture levels on the ground at that time were not always relevant in preventing fire. The discussion on dry vegetation further prompted land managers to discuss their operational concerns about how (often) to water new vegetation that could—in the long run—contribute to fire risk-reduction and their considerations about (not) felling trees. For example, they argued that felling certain trees may help to reduce the risk of fires as it reduces the possibility for fire to spread uncontrollably, but that doing so also potentially increases other vegetation's exposure to the sun, thus rendering the area more vulnerable to droughts and fires simultaneously. This discussion illustrates how the same 'source' of information (vegetation dryness) generates diverse reactions (from immediately measuring moisture content to thinking about land management) and expectations of what should be done.

Lastly, the discussion highlighted how *policy*, in this case European policy, may inform some understandings and valuations of the landscape. For instance, on arriving at one of the locations, one participant immediately categorized the heathlands in their particular Natura 2000 type. From this, they inferred that the priority in this area was to 'keep the area open and remove trees'. However, at other points in the discussion some participants referred to the Climate Agreement as applicable to the entire area, which is part of the Dutch climate-mitigation policy and requires the total tree cover in the Netherlands to increase by 10% by 2030. From this, it was assumed that felling trees should be avoided if possible. During the discussions, participants said that although these policies seem restrictive in terms of what is possible for fire-prevention measures, they are sufficiently flexible to allow for such measures, provided that they are mainly on an ad-hoc basis, as there is as yet no policy that integrates these frameworks to ensure that there are structural agreements regarding fire prevention and between different parties such as the fire departments and land managers, and that these are not structurally financed.

In sum, in this section we distinguished various ways of understanding the landscape: affective (e.g. indicating which trees are worth preserving because of their appearance), operational (e.g. how drought is observed and possibly considered relevant to fire prevention) and those based on policy frameworks (e.g. categorizations of Nature 2000 nature types). A more-than-representational or relational approach to landscape helps to distinguish between these various forms of understanding the landscape as it puts the human–landscape interactions centre stage and takes seriously the diverse interactions and resulting understandings. By making visible how stakeholders look at the landscape and where they direct their attention to, we enhance our understanding of what these stakeholders find important and how it shapes their perception of what is worth preserving or what needs to be changed (Tschakert et al., 2013). In addition, our findings underscore that landscapes can be both an 'intellectual artefact', such as a Nature 2000 classification, as well as an experiential process in which the sound of birds and the specific aesthetics of particular trees are experienced and appreciated (see also Brace & Geoghegan, 2011).

9.4.3 *Engaging with Diverse Ways of Knowing and Valuing the Landscape*

In this section we analyse what the walkshop means for how participants engaged with each other's diverse ways of knowing and valuing the landscape. A more-than-representational or relational approach to landscape may draw the attention of researchers to the relations between humans and landscapes and create space for stakeholders' diverse understandings of the landscape, a transdisciplinary walkshop goes a step further and aims to foster such an understanding among stakeholders as to engage explicitly with diversity, which is imperative to addressing sustainability challenges (see Sect. 3.1). We analyse our findings, here, after which we discuss what our contribution in the wider literature (Sect. 3.5).

First, some stakeholders were inclined to *situate distinct measures in their context* and the broader landscape. For instance, when arriving at locations A and B we spoke about the transitions between forest and heathlands, which participants generally described as being as quite harsh in this area. One participant explicitly interpreted this discussion as an illustration of how biodiversity priorities are seen to clash with fire prevention: relatively harsh transitions are likely to prevent the more dangerous tree canopy fires but are less biodiverse. Another participant, one of the land managers, nuanced this by remarking that this does not *have* to be the case. He believed that it is possible to make gradual transitions when land managers and fire brigades examine the area well and take measures to prevent wildfires in other ways, such as vegetation management *in* the forest as opposed to in the transition zone, or measures to improve accessibility for fire brigades. At location C, we spoke about the risks of fires for the local residents and the adjacent village. Here, researchers started the discussion by asking what residents can or should do to protect their homes from fires, for instance, whether they should ensure that the vegetation in their gardens and around their houses is safe. Some participants quickly situated this question in the wider area, by considering the types and heights of trees in the adjacent forest, but also possible escape routes that the residents could use.

Second, engagement with diversity can also be achieved by situating current practices and values in a historical (and future) context, to show the relative position of current norms and values as changing throughout time. Opening a conversation about the eventual appearance of the landscape in the context of a changing climate and increasing fire risks may also create the space to make more radical changes in the landscape to prevent (uncontrollable) fires. However, during our walkshop we observed that there was relatively little space to envisage—or express a desire for—the landscape here to look any different in the future. For instance, at various points in the discussion participants said that it was not possible to reduce the total amount of forests in view of the Dutch Climate Agreement. Although participants from provinces and land managers argued this means that forests may 'move' (i.e. that trees can be uprooted and planted elsewhere, to accommodate both fire-prevention measures and the Climate Agreement), they admitted that—as the Netherlands is considered 'full' and most areas have designated functions—in practice it is easier to maintain the area as it is. Moreover, at location B we (the researchers contributing to the discussions) shared some history about how the landscape had been formed over several centuries. After this, we tried to provoke a discussion about which historic practices of land management are deemed valuable, how these still constitute the landscape, and how we would like this to be in the future given the context of climate change. However, there was little space for this conversation. Land managers did not explicitly acknowledge that particular time periods are taken as the current standard—for instance, how pine trees are seen as typical for the Veluwe whereas most were only planted in the 1800s for the purpose of constructing mines. One land manager said that in an adjacent area they aim to restore sandy dunes, which are, comparably, parts of the landscape that were established only from the 1850s. In sum, we experienced that it was difficult to open up the conversation about what the landscape should look like under changing circumstances such as a changing climate and increased fire risks—and open up to diversity—as particular historical timeframes and current policy frameworks appear to be the leading norms for current land-management practices.

Lastly, as shown in many of the above examples, participants spoke in diverse ways about which fire-prevention measures were valuable and in what setting. Although this diversity was acknowledged and deemed important in the walkshop discussions, there appeared to be little space to engage with it in practice. This tension was particularly evident at locations B and C. Here, the discussion was focused on the transition of particular areas from being dominated by pine trees to deciduous trees (as both locations had such areas). Participants felt that using deciduous trees as a fire-prevention measure was a good practice in general (the rationale is that as leaves and stems of deciduous trees contain more moisture than pine trees, they lower the temperature of a passing fire). Some of the participants with greater knowledge of fire and vegetation voiced concerns regarding the effectiveness of the measure: they pointed out that deciduous trees did were still bare in the spring, which means they do not contain much moisture. They called for more situated analysis of the effectiveness of various measures in an area, including felling trees altogether to create fire breaks. In this walkshop, some participants expressed these concerns. However, in deciding which fire-prevention measures to take, participants argued that there is less space for this as felling trees is not considered a legitimate option given the Dutch Climate Agreement. Notably, in these discussions the Agreement was either challenged due to its limiting appropriate fire measures, or was seen as a given that could not be altered—at least not by this group at this moment.

9.5 Conclusion: The Walkshop for Explicating and Engaging with Diversities

In this section, we draw conclusions with regard to our research question: how may a walkshop make landscape–human interactions visible and allow for meaningful engagements with diverse ways of knowing and doing and its potential for transdisciplinary research? Before doing so, first, we discuss how a more-than-representational or relational approach to landscape helped us in the design and analysis of our walkshop.

As we argued in the introduction, for us, engaging with diverse ways of doing and knowing and including various forms of knowledge is imperative to understanding and addressing sustainability challenges. We chose a more-than-representational or relational approach to landscapes (Lorimer, 2005; Seamon, 2018) as such an approach considers the landscape not as something that 'is' but as something that is constructed through experiences and practices allows for an understanding of the landscape as being lived differently by different stakeholders. Indeed, this approach has previously been used by researchers to allow for an understanding of the landscape as being lived differently by different stakeholders (Baron, 2020; Brace & Geoghegan, 2011; Döring & Ratter, 2021; Köpsel et al., 2017). We, too, found this approach valuable as it allowed us as researchers to engage with diverse ways of doing and knowing and include various forms of knowledge: it allowed us to spark and facilitate discussions in which we aimed to create space for diverse understandings of the landscape, it allowed us to generate data that goes beyond cognitive reasonings but also include affective forms of knowledge and previous experiences in the landscape, and it allowed us to analyse the walkshop and making space for diverse knowledges in this chapter. Although this may not be new for (the aforementioned) researchers in sustainability challenges, walking research methodologies remain relatively uncommon, hence, we find it important to share our positive experiences.

What is new, however, is specifically designing a transdisciplinary walkshop with the aim for sustainability transformation (cf. Döring & Ratter, 2021; Wickson et al., 2015). In the introduction we argued that, ideally, a transdisciplinary walkshop would advance such an understanding among stakeholders, to allow them to place their own values and ways of knowing and doing alongside each other's, in the context of this place and in its historical context. We conclude by reflecting on how the transdisciplinary walkshop allowed enabled this among stakeholders and explore its potential for transformation.

We have shown that our walkshop made visible and explicit how different stakeholders perceive and experience the landscape: affecting, operational and policy-based understandings. We also found three ways in which stakeholders attempted to engage with each other's diverse understandings. First, stakeholders situated measures (and trade-offs) in a wider context by looking at adjacent areas, as to not focus on which value is 'right' or 'wrong' in a specific location but looking at how it adds up in the landscape overall. Second, although we experienced as difficult to foster reflection hereon among stakeholders, a way to engage with diversity is to recognize that most values and understandings of the landscapes are rooted in a particular understanding of what the landscape 'is' and how it 'should be', which depends strongly on the historical context one looks at. Acknowledging the various histories and 'original' landscape elements we find therein makes space for acknowledging that diverse landscape understandings continue to remain (see also Egberts, 2017). Third, we found that stakeholders engaged with diversity and plural possibilities, essentially opening the conversation about the effectiveness of fire-prevention measures by considering under what conditions they would be effective (for instance weather conditions or time of year).

What is more, participants witnessed and experienced what sort of knowledge other stakeholders used to make their arguments and tell their stories about the landscape, for instance when participants measured the moisture content of leaves, indicated particular types of trees or shared specific experiences of something they encountered in the field of fire prevention. Such interactions allowed for an understanding not only of each other's viewpoint, but also of the underlying rationales. This brings about a form of legitimizing other's viewpoints and their reasons for action (see also Habermas, 1985; Wickson et al., 2015). This suggests that the walkshop may be a relevant research method for other TDR projects for transformation, which aim to instigate new forms for participants and researchers to understand others' viewpoints.

Situated knowledge came to the fore when stakeholders considered specific fire-prevention measures mainly *in relation to* other measures, the overall landscape or potential weather conditions. Experiencing the situatedness of fire risks and fire-prevention measures in the landscape and in the context may affect how stakeholders perceive what measures are deemed appropriate (Leino & Peltomaa, 2012). This was shown, for instance, when stakeholders explained how transitions between heathlands and forests may be less harsh in order to promote biodiversity *if* other measures in other areas are taken. Or, how stakeholders expressed concerns regarding the effectiveness of planting deciduous trees as a fire-prevention measure, given that in the vulnerable springtime they do not yet contain sufficient moisture to retard fires.

However, merely bringing diversity and its situatedness to the fore is not the same as productively engaging with diversity regarding (inclusive) transformation. As we argued in the introduction, diversity in TDR has the potential to challenge dominant norms and values regarding transformation (see also Kok et al., 2021; Turnhout et al., 2020). In the walkshop, we saw that some dominant norms—such as the policy framework the Dutch Climate Agreement—were challenged, even though it is currently a given with which our participants have to work.

Engaging with diversity may also entail irreconcilable aims and ways of knowing and doing (Klenk & Meehan, 2015; Wickson et al., 2006). In our walkshop however, we did not witness such irreconcilability, nor did we see much friction between the participants' ways of knowing and doing. All participants made the effort to acknowledge the importance of others' priorities, and of when some of their priorities were detrimental to others (for instance, when fire-prevention measures were detrimental to biodiversity). At times, solutions were suggested in such cases, for instance when one participant argued that a soft gradient would be possible if fire-prevention measures were taken elsewhere in the area. Here, we see how there was empathy regarding each other's viewpoints. Crucially, it is argued that this is crucial to bring about collective action and transformative change (Habermas, 1993, p. 31, cited in Flyvbjerg, 1998). However, the field of fire prevention in the Netherlands is relatively immature: stakeholders are getting to know each other and currently share a common interest in placing the issue more firmly on political and social (and research) agendas. This may have contributed to the open dialogue and a lack of friction during the walkshop. If the interests had been more diverse, the participants may also have been more strategic. This might be at the expense of the level of the openness of the discussion, making it worth developing strategies to address any such strategic tensions.

Acknowledgements This work was funded by the Amsterdam Sustainability Initiative, with project title 'FireScapes: Towards an interdisciplinary understanding of wildfire risk mitigation in the Dutch landscape'. MdR's contributions were supported by the MYRIAD-EU project, which receives funding from the European Union's Horizon 2020 research and innovation programme under Grant Agreement No. 101003276.

The authors also want to acknowledge all researchers who were involved in facilitating the discussions during the walkshop and in taking field notes, as well as the participants who all actively participated and spurred other stakeholders to take part in this walkshop.

References

Anderson, J. (2004). Talking whilst walking: A geographical archaeology of knowledge. *Area, 36*(3), 254–261.

Baron, N. (2020). Flood protection beyond protection against floods: How to make sense of controversies related to the building and maintenance of dikes in Denmark. *Natural Hazards, 103*(1), 967–984.

Bentz, J., & do Carmo, L., Schafenacker, N., Schirok, J., & Corso, S. D. (2022). Creative, embodied practices, and the potentialities for sustainability transformations. *Sustainability Science, 17*(2), 687–699.

Brace, C., & Geoghegan, H. (2011). Human geographies of climate change: Landscape, temporality, and lay knowledges'. *Progress in Human Geography, 35*(3), 284–302.

Buttimer, A., & Seamon, D. (2015). *The human experience of space and place*. Routledge.

Caniglia, G., Luederitz, C., von Wirth, T., Fazey, I., Martin-López, B., Hondrila, K., König, A., von Wehrden, H., Schäpke, N. A., Laubichler, M. D., & Lang, D. J. (2021). A pluralistic and integrated approach to action-oriented knowledge for sustainability. *Nature Sustainability, 4*(2), 93–100

Clark, A., & Emmel, N. (2010). *Using walking interviews*. University of Manchester.

Cuppen, E. (2018). The value of social conflicts. critiquing invited participation in energy projects. *Energy Research & Social Science, 38*, 28–32.

De Hoop, E., & Arora, S. (2021). How policy marginalizes diversity: Politics of knowledge in India's biodiesel promotion. *Science as Culture, 30*(2), 261–286.

De Hoop, E., Brouwers, H. J. H., Buijs, S., Egberts, L., van Gerrevink, M., de Ruiter, M., & Veraverbeke, S. (2022). Multi-stakeholder analysis of fire risk reduction in a densely populated area in the Netherlands: A case-study in the Veluwe area. *Environmental Research Letters, 17*(9), 095011.

Döring, M., & Ratter, B. (2021). 'I show you my coast…'—a relational study of coastscapes in the North Frisian Wadden Sea. *Maritime Studies, 20*(3), 317–327.

Edensor, T. (2000). Walking in the British countryside: Reflexivity, embodied practices and ways to escape. *Body & Society, 6*(3&4), 81–106.

Egberts, L. (2017). *Chosen Legacies Heritage in Regional Identity*. Routledge.

Flyvbjerg, B. (1998). Habermas and Foucault: Thinkers for civil society? *British Journal of Sociology, 49*(2), 210–233.

Habermas, J. (1985). *The theory of communicative action: Volume 1: Reason and the rationalization of society*. Beacon Press.

Haraway, D. (1988). Situated knowledges: The science question in feminism and the privilege of partial perspective. *Feminist Studies, 14*(3), 575–599.

Harding, S. (1991). *Whose science, whose knowledge?* Open University Press.

Harrison, P. (2000). Making sense: Embodiment and the sensibilities of the everyday'. *Environment and Planning d: Society and Space, 18*(4), 497–517.

Ingold, T., & Kurttila, T. (2000). Perceiving the environment in Finnish Lapland. *Body & Society, 6*(3–4), 183–196.

Jacobi, J., Llanque, A., Bieri, S., Birachi, E., Cochard, R., Chauvin, N. D., Diebold, C., et al. (2020). Utilization of research knowledge in sustainable development pathways: Insights from a transdisciplinary research-for-development programme. *Environmental Science & Policy, 103*, 21–29.

Jones, M. W., Abatzoglou, J. T., Veraverbeke, S., Andela, N., Lasslop, G., Forkel, M., & Smith, A. J. P., et al. (2022). Global and regional trends and drivers of fire under climate change. *Reviews of Geophysics*, e2020RG000726.

Jones, P., Bunce, G., Evans, J., Gibbs, H., & Hein, J. R. (2008). Exploring space and place with walking interviews. *Journal of Research Practice*, *4*(2), article D2.

Kinney, P. (2017). Walking interviews. *Social Research. Update*, *67*, 1–4.

Klenk, N., & Meehan, K. (2015). Climate change and transdisciplinary science: Problematizing the integration imperative. *Environmental Science & Policy, 54*, 160–167.

Kok, K. P. W., Gjefsen, M. D., Regeer, B. J., & Broerse, J. E. W. (2021). Unraveling the politics of 'doing inclusion' in transdisciplinarity for sustainable transformation. *Sustainability Science, 16*(6), 1811–1826.

Köpsel, V., Walsh, C., & Leyshon, C. (2017). Landscape narratives in practice: Implications for climate change adaptation. *The Geographical Journal, 183*(2), 175–186.

Lang, D. J., Wiek, A., Bergmann, M., Stauffacher, M., Martens, P., Moll, P., Swilling, M., et al. (2012). Transdisciplinary research in sustainability science: Practice, principles, and challenges. *Sustainability Science, 7*(1), 25–43.

Leichenko, R., & O'Brien, K. (2020). Teaching climate change in the Anthropocene: An integrative approach. *Anthropocene, 30*, 100241.

Leino, H., & Peltomaa, J. (2012). Situated knowledge–situated legitimacy: Consequences of citizen participation in local environmental governance. *Policy and Society, 31*(2), 159–168.

Lorimer, H. (2005). Cultural geography: The busyness of 'being more-than-representational.' *Progress in Human Geography, 29*(1), 83–94.

Macpherson, H. (2010). Non-representational approaches to body–landscape relations. *Geography Compass, 4*(1), 1–13.

Macpherson, H. (2016). Walking methods in landscape research: Moving bodies, spaces of disclosure and rapport. *Landscape Research, 41*(4), 425–432.

Neefjes, J. W. P. M. (2018). *Landschapsbiografie van de Veluwe: historisch-landschappelijke karakteristieken en hun ontstaan*. Rijksdienst voor het Cultureel Erfgoed.

Norström, A. V., Cvitanovic, C., Lof, M. F., West, S., Wyborn, C., & Osterblom, H. et al. (2020). Principles for knowledge co-production in sustainability research. *Nature Sustainability, 3*, 182–190. https://doi.org/10.1038/s41893-019-0448-2

Olwig, K. R. (2019). *The meanings of landscape: Essays on place, space, environment and justice*. Routledge.

Pohl, C. (2008). From science to policy through transdisciplinary research. *Environmental Science & Policy, 11*(1), 46–53.

Pohl, C., & Hadorn, G. H. (2007). *Principles for designing transdisciplinary research*. oekom Munich.

Pohl, C., & Hadorn, G. H., et al. (2008). Core terms in transdisciplinary research. In G. H. Hadorn, C. Pohl, & H. Hoffmann-Riem (Eds.), *Handbook of transdisciplinary research* (pp. 427–432). Springer.

Pohl, C., Klein, J. T., Hoffmann, S., Mitchell, C., & Fam, D. (2021). Conceptualising transdisciplinary integration as a multidimensional interactive process. *Environmental Science & Policy, 118*, 18–26.

Polk, M. (2015). Transdisciplinary co-production: Designing and testing a transdisciplinary research framework for societal problem solving. *Futures, 65*, 110–122.

Relph, E. (1976). *Place and placelessness*. Pion.

Seamon, D. (2018). *Life takes place: Phenomenology, lifeworlds, and place making*. Routledge.

Tschakert, P., Tutu, R., & Alcaro, A. (2013). Embodied experiences of environmental and climatic changes in landscapes of everyday life in Ghana. *Emotion, Space and Society, 7*, 13–25.

Tuan, Y.-F. (1977). *Space and place: The perspective of experience*. University of Minnesota Press.

Turnhout, E., Metze, T., Wyborn, C., Klenk, N., & Louder, E. (2020). The politics of co-production: Participation, power, and transformation. *Current Opinion in Environmental Sustainability, 42*, 15–21.

UNEP. (2022). *Spreading like wildfire—the rising threat of extraordinary landscape fires. A UNEP rapid response assessment*. United Nations Environment Programme.

Wickson, F., Carew, A. L., & Russell, A. W. (2006). Transdisciplinary research: Characteristics, quandaries and quality. *Futures, 38*(9), 1046–1059.

Wickson, F., Strand, R., & Kjølberg, K. L. (2015). The walkshop approach to science and technology ethics. *Science and Engineering Ethics, 21*, 241–264.

Wylie, J. (2002). An essay on ascending Glastonbury Tor. *Geoforum, 33*(4), 441–454.

CHAPTER 10

Transdisciplinary Approaches in Assimilating Power and Diversity to Address Children's Interface with the Law: A Case Study from India

Sheila Ramaswamy, Shekhar Seshadri, and Joske Bunders

10.1 Introduction

Many children worldwide interact with the justice system, which presents a host of legal and practical problems (Marsil et al., 2002). The substantial power differentials in children's interface with the criminal justice system are skewed towards professionals from the judicial, protection and mental health fields. Since legal and judicial systems tend to be particularly patriarchal, paternalistic and hierarchical, they are poorly equipped

S. Ramaswamy (✉) · S. Seshadri
National Institute of Mental Health and Neurosciences, Bengaluru, India
e-mail: sheila@nimhanschildprotect.in

S. Seshadri
e-mail: docshekharseshadri@gmail.com

J. Bunders
Vrije Universiteit Amsterdam, Amsterdam, The Netherlands
e-mail: j.g.f.bunders-aelen@vu.nl

B. J. Regeer et al. (eds.), *Transdisciplinarity for Transformation*,
https://doi.org/10.1007/978-3-031-60974-9_10

to grasp the need for democratic and more fluid interactions in order to include marginalized and vulnerable witnesses, such as children. Furthermore, legal and judicial personnel are seldom adequately trained in child development and child interviewing techniques (Birnbaum et al., 2013), resulting in a lack of clarity on the best ways to help children in their interface with the law. There are similar professional boundaries among mental health service providers, who are skilled in the treatment and care of childhood trauma and child behaviour, but may fail to adequately consider the subjective realities of children's lives, specifically their interaction with the legal and protection systems, and the often negative impacts on such children (Ramaswamy et al., 2021; Ramaswamy, Seshadri et al., 2022a, 2022b).

While legal and mental health systems have the duty to protect and uphold the rights of child witnesses, many of these professionals do not acknowledge these children as important players; yet these often deal with increasingly complex problems, characterized by tensions and conflicting demands that challenge their professional judgement, leading to handling such problems inconsistently (Olsvik & Saus, 2022), and in ways that are not always in children's best interests. Consequently, the need to perceive children as social actors and co-constructors of knowledge about their experiences of the world (Brady et al., 2015) is often overlooked, thus contributing to the enormous power differentials between children and professional bodies. This results in children having to contend with difficult legal and judicial processes with little support from the various relevant professionals, which exacerbates their helplessness within the criminal justice system.

The complex problem of children's interface with the law is therefore not easily resolved from the perspective of a single discipline; it calls for crossing borders between the traditions of academic knowledge, as well as a deliberate attempt to co-create knowledge with relevant non-academic actors, not least the children concerned. The judicial system is charged with obtaining child witness testimony that is valid and reliable—which requires children to have competencies ranging from language, communication and comprehension to observation, memory and recall, as well as an appreciation for telling the truth (Pantell et al., 2017). However, India's adversarial justice system is known to have negative impacts on children's language, cognitive and socio-emotional competencies to testify (Eades, 2012; Zajac & Hayne, 2003), as well as causing children to experience secondary trauma and re-traumatization during court processes (Caprioli & Crenshaw, 2017) (Whitcomb, 2003). The lack of adequate support to help children navigate the justice systems, to know that legal systems offer them an opportunity to speak, and to participate in dialogue and discourse that may determine decisions that affect their lives, reflect the contextual influences and power imbalances that shape these adult–child interactions.

Underpinning these power differentials is the concept of epistemic injustice. In her work on the subject, the philosopher Miranda Fricker identifies two types of epistemic injustice: testimonial injustice—when some bias on the part of the listener discredits a speaker's testimony or report, in the absence of further corroboration; and hermeneutical injustice—when there is a lack of a (systemic) interpretative framework to allow for the speaker's experiences to be accurately understood (Fricker, 2007). In the context of children's interface with the law, there are both types of epistemic injustice, as children's testimonies tend to be discredited on the basis of their age and developmental capacities, and the failure to create interpretative frameworks that take account of their capacities, competencies, life experiences and subjective realities.

Consequently, the concept of epistemic injustice centres around questions of whether children (as opposed to adults) are heard, and believed, and how adults are biased in ways that render children more vulnerable to such injustice (Carel & Györffy, 2014) in the absence of methods and techniques to integrate different orientations and perspectives (e.g. the need to establish facts versus protecting children's rights and wellbeing). In such a situation, the challenge is to fuse the knowledge and concerns of children and judicial personnel by adapting processes to support children in 'telling the truth' (Magnusson et al., 2021), in order to arrive at solutions that ensure children's rights and promote mental health and wellbeing in their involvement in legal processes, while also allowing the legal and protection professionals to adhere to laws and due process. The tension, as reflected in the concept of boundary dynamics (Regeer & Bunders, 2009), is to work within traditional disciplinary borders while also considering practical realities that call for crossing the divisions between science and society.

Furthermore, despite a considerable body of scientific knowledge on the challenges children experience in their interface with the law, the 'linear model of expertise' (Beck, 2011) neither adequately defines nor makes for the translation of science to policy (Hegger et al., 2012). Contrary to assumptions that science and power can be kept distinct

(Jasanoff & Wynne, 1998) or that science will propel policy change (Pielke, 2005), scientific agendas are also socially constructed, and scientific research needs to be transformed and contextualized (Latour, 1987). Thus, knowledge co-evolution or co-production is a joint venture between science and society (Jasanoff, 2004; Jasanoff & Martello, 2004).

In principle and practice, for a justice system to address issues of protection and inclusion in its interface with children, the questions and complexities of its design and function need to be addressed through a process of deliberation and co-creation with different knowledge brokers and actors working at the interface of child protection, mental health and law. At present, because legal and judicial, as well as mental health systems, tend to function in a mono-disciplinary fashion, often overwhelmed by numerous responsibilities and agendas, these institutions tend to find it difficult to grasp the need for boundary-transgressive interactions and collaborations across disciplinary borders (Olsvik & Saus, 2022). Other reasons for professionals and organizations to remain within their respective domains include the fear of the unknown, or that the priorities and epistemological perspectives of another discipline might dilute, impinge on or even overpower their own; or perhaps because of the insecurity experienced in leaving one's own expertise and entering a zone of uncertainty and 'not-knowing' (Gaim & Wåhlin, 2016; Lotz-Sisitka et al., 2015; Nicolescu, 2006).

It is in this context that the case study presented in this chapter uses Transdisciplinary Research (TDR) as a way to understand how to facilitate a process of multi-stakeholder deliberation and knowledge co-creation to find more inclusive ways to ensure children's interface with the law. This is because TDR seeks to enable systemic transformations to address complex problems, by enabling the integration of knowledges from diverse stakeholders who interact according to varying interests, power relations and agreements (Regeer & Bunders, 2009). Various methodologies have been developed to facilitate such processes, and to overcome the numerous challenges in bringing stakeholders together in processes of deliberation and integration.

The study draws on the work of SAMVAD (Support, Advocacy & Mental Health Interventions for children in Vulnerable circumstances and Distress) to understand how TDR methodologies work. It reviews SAMVADs' attempts to address the complex problems of child protection and mental health and law—both with children, and in intersectoral collaboration between service providers and stakeholders from the domains of child mental health and law with the specific aim to elicit lessons on how to overcome systematic and structural barriers and power asymmetries to bringing stakeholders together in a process of knowledge co-creation.

10.2 Imperatives for TDR Approaches in Working Towards Child-Inclusive Legal Systems

> Transdisciplinarity is a reflexive, integrative, method-driven scientific principle aiming at the solution or transition of societal problems and concurrently of related scientific problems by differentiating and integrating knowledge from various scientific and societal bodies of knowledge. (Lang et al., 2012, pp. 26–27)

One important incentive for using TDR approaches is the realization that some social problems are too complex or 'wicked' to solve on the basis of one discipline or through science alone (Adler et al., 2018). These problems are so interconnected and multifaceted that they can only be resolved through processes of deliberation, frame-reflection and consideration of various challenges, consequences, potential solutions and the possible tensions among different stakeholders (Lawrence et al., 2022). Another important aspect of TDR is that of learning and co-creating knowledge in transformational processes (Kok et al., 2021), meaning that an important step in the TDR process is to understand how various stakeholders experience the current situation and their ideas for working towards a better future in any complex arena, (Gaziulusoy & Ryan, 2017), in this case, more child-inclusive legal systems.

TDR approaches are particularly appropriate for child and youth studies (Mitchell & Moore, 2018) and recent efforts range from shaping education on climate change (Cutter-Mackenzie & Rousell, 2019), to how the COVID-19 pandemic opened up new spaces and opportunities for children's participation (Faria et al., 2021), early intervention services for infants and young children in community settings (Bell et al., 2010), decision-making regarding children and technology (Straker et al., 2022), paediatric and child health-related issues (Reaman, 2004), and co-producing data with children by understanding their experiences and viewpoints through physical movements in their living environment (Camponovo et al., 2023). These examples tend, however, to be limited to collaborating with children and/or with stakeholders in children's immediate environment, such as parents, teachers and health professionals. Moreover, although such studies present strong arguments for the need to apply transdisciplinary methodologies in child-related work, they fail to take fully into account the critical importance of involving (sufficiently) diverse stakeholders in TDR.

Stakeholders are defined as those who can affect or be affected by a decision or action (Freeman, 1984). Their active involvement and participation, reflecting their interests and knowledge, and integrated into solution-oriented research, is essential to transdisciplinarity and its effective assimilation into policy and practice (Bracken et al., 2015; Hurni & Wiesmann, 2014; Lang et al., 2012). Hence, while parents, teachers, health workers and children are critical stakeholders, and bring transdisciplinary perspectives to children's issues, they may well find it hard to translate the resulting recommendations into practice. This is because they are less likely to have the power to bring about the necessary policy changes to achieve sustainable, large-scale systemic transformation. Children's problems thus risk remaining only partially addressed, because even transdisciplinary approaches tend to remain rooted in a specific discipline (Collins, 2017; Nicolescu, 2018), or do not achieve the level of social change to make meaningful impacts on children's lives.

TDR is especially relevant to children and their problems because one of the normative objectives of diversity among stakeholders in TDR is that everyone who is affected should have a voice in formulating research questions, and also in their implementation and outcomes (Arnstein, 1969; Fiorino, 1990; Hage et al., 2010; Pohl & Hirsch Hadorn, 2007; Stirling, 2008). However, a major challenge in relation to the co-production of knowledge is the asymmetric power among diverse stakeholders; particularly when children are a key stakeholder group, this may adversely affect emerging concerns and perspectives on a given issue. This power imbalance highlights tensions such as the types of conflicts in seeking transdisciplinary solutions, based on differing values, conflicting interests and dissimilar knowledges, and how these affect the fusion of diverse sources of knowledge (Siebenhüner, 2018). In TDR, the diversity of stakeholders is reflected not only in the varied domains and disciplines, or academic and non-academic characteristics, but also by asymmetries in power and influence.

TDR methodologies are designed to provide insights on how to forge collaborations among various stakeholders, and facilitate processes of shaping and improving practices, learning, evaluation and adaptation. Besides being agenda-driven and focused on integrating various types of knowledge, they help to expand and enrich distinct disciplines, and extend the landscape of possible practices of the relevant parties (Guzmán Ruiz et al., 2017); and also gather 'heterogeneous sets of relationships between epistemic ends and epistemic means' (Weichselgartner & Truffer, 2015). This may allow for knowledge from less powerful groups, such as children, to be integrated into policy and practice.

10.3 Challenges in TDR Processes

While much literature supports using TDR approaches for projects that are aimed at social, systemic and policy transformation (such as the current case), their facilitation poses several challenges. Thus, in order to ensure that the objectives of TDR research, entailing inclusive co-creation of improved practices and systems, it is necessary to explore strategies to better understand and overcome these challenges (Gaziulusoy et al., 2016). While some barriers pertain to institutions—such as universities, publishers, research funds, etc.—there are also numerous methodological and epistemological, as well as social and political, challenges. Methodological and epistemological barriers relate to the research content, and the difficulty of integrating knowledge across disciplines, different epistemological orientations and framings of the problems (Brouwer et al., 2018). The social and political level entails different actors transcending their professional domain and accepting/engaging with the knowledge of others, which can lead to conflicts. TDR process facilitation also means recognizing practical versus academic and discipline-based knowledge, and how to integrate these so as to generate solutions to practical problems, by creating new knowledge that transcends disciplinary boundaries—essentially how knowledge can be exchanged and integrated to achieve common goals (Godemann, 2008). The third level concerns the difficulty of acquiring funding for TDR studies and publishing in prestigious journals (Hessels et al., 2018).

Related challenges include the need to address the separation between science and society (i.e. academics versus practitioners), biases in the engagement and influence of various stakeholder groups and the tensions between academic, policy and social outputs and outcomes (Jahn et al., 2022).

A further critical difficulty in TDR concerns the psychological barriers of fear and uncertainty for academics and other professionals. The two modes of the self and its existence, i.e. egocentrism and socio-centrism, can hinder people from considering the possibility that their construction of reality and experiences may be biased or misleading; consequently, they develop little or no (conscious) awareness of the potential for alternative ways to (re)conceptualize these perspectives (Naidoo, 2015). Addressing such challenges calls for flexibility and openness among researchers and other actors to overcome initial reluctance and inertia, leave their 'comfort' zone and engage with a meaningful joint problem- and solution-framing process (Pohl et al., 2021).

Another key challenge in facilitating TDR processes with diverse actors is that not everyone speaks the same language, which can lead to mutual misunderstandings. For instance, legal personnel, mental health professionals and children use very different languages. The effectiveness of TDR depends on whether group members (including the TDR team) can engage in self-reflexivity, collectively reflecting on and adapting their ways of working in the face of uncertainty and a lack of clarity (Schippers et al., 2015). According to Pohl et al. (2021), actors in the real world need to reflect upon their perception and significance of the problem, as well as that of others—and this calls for the relatively difficult task of relativizing one's own perspective as well as accepting the view points of others.

Finally, as mentioned above, the power differentials among diverse actors involved in working towards system change make it hard to strike a balance between the knowledge of those who hold power and those who are marginalized (Bunders et al., 2010)—in our case, for instance, between judicial officers and children—and the relatively little knowledge exchange that traditionally occurred between various actors and across disciplines. It was thus expected that facilitating a TDR process towards enhancing the child–law interface would face these process obstacles. This chapter explores how these were overcome, and what can be learned from them.

10.4 Study Context

The chapter describes child and law work in India over a timespan of five years, based on the work of SAMVAD, which is a national initiative and integrated resource for child protection, mental health and psychosocial care. Established by the Ministry of Women and Child Development, it is based in the Department of Child and Adolescent Psychiatry at the National Institute of Mental Health and Neurosciences (NIMHANS) in Bangalore, a tertiary-level mental health institution under the Ministry of Health and Family Welfare.

Various stakeholders of child protection and mental health services, particularly those working within the government and judicial systems, observed that children's subjective realities, and consequently, their rights, mental health and wellbeing, are often neglected both because of the systems' hierarchical workings, and because various systems and stakeholders work in isolation. In recognition of the critical nature of convergence and transdisciplinary approaches for systemic transformations, SAMVAD was set up to address complex problems of child protection and mental health, particularly through intersectoral collaboration among service providers and stakeholders from the fields of (child) welfare, health, education and the law (Ramaswamy, Vijay Sagar et al., 2022b). While SAMVAD's core expertise is mental health, given the increasingly complex child protection and mental health needs and contexts—such as in families and communities, pre-schools and schools, child care institutions and courts of law—it recognizes the critical importance of understanding core 'truths' in and about other domains, including their (systemic) objectives, perspectives and mandates regarding children, to gain an in-depth understanding of how to reform and transform the systems that are responsible for or govern children's lives. This calls for the convergence of multiple stakeholders in ways that transcend a given discipline (such as mental health or law) to develop effective solutions to these complex concerns and problems.

Initially a state-level community-based child mental health initiative, SAMVAD has pioneered work in the field of children and the law, particularly in the contexts of child sexual abuse and children in conflict with the law, to enable them to have a voice in legal proceedings. These activities have ranged from developing training curriculums integrating child mental health into legal and judicial procedures and engagement in judicial education programmes through the training academies across the country to extending psychosocial, mental health and legal support to individual children, and assisting investigating agencies under Supreme Court direction in cases of child abuse. It has possibly been the only agency to engage in research activities with the judiciary, and to implement a transdisciplinary review of the child sexual abuse law in India. It is against this backdrop that we adopted several transdisciplinary methods and initiatives to integrate the domains of child protection, mental health and the law, in order to find solutions to children's problems, particularly in relation to their marginalization and powerlessness in their interface with the law.

The chapter reflects on various phases of SAMVAD's work and the specific case it presents enabled us to understand how transdisciplinary methods could be used in the very complex and sensitive context described above, which involves diverse stakeholder groups characterized by large power differentials. It describes working with two primary groups of stakeholders at opposite ends of the power spectrum, along with a third:

(i) At one end are children (child witnesses), who lack the developmental and mental health capacities to engage with a criminal justice system that was not designed to accommodate their needs; yet, they must attend, and be heard, in legal and quasi-legal spaces, such as in courts and state juvenile justice systems, to engage effectively in legal processes, which are governed by relatively non-negotiable rules and procedures.
(ii) At the other end are the judicial personnel, whose role is to ensure the rule of law.
(iii) Towards the 'powerful', albeit less influential, end of the spectrum are child protection officers and mental health service providers, who variously facilitate children's legal engagement, ensure protection of their rights and safety and support their mental health and wellbeing.

10.5 Methodology

Our work focuses explicitly on processes in order to understand how and why phenomena occur and evolve, since it is impossible to separate the research process and its outcomes (Garfield, 1990), and hence the methods of inquiry and the content that they generate are an intrinsic part of the results. We use a case-study approach to understand the effectiveness of transdisciplinarity, based on the transdisciplinary field practice methodologies we developed.

Crucial to this, in this context, is to acknowledge that in transdisciplinarity it is hard if not impossible to discern what belongs to the phenomenon being investigated and what to the context in which it arises. We therefore adopted a qualitative case-study approach which, as Yin explains (Yin, 1984, p. 23), is a form of 'empirical inquiry that investigates a contemporary phenomenon within its real-life context; when the boundaries between phenomenon and context are not clearly evident; and in which multiple sources of evidence are used'. This methodology enabled us to understand the contextual nuances of children's engagement with legal processes in courts and other settings, necessitating interface with the law and its systems, facilitated by mental health service providers and by legal and judicial personnel.

10.5.1 Data Collection and Analysis

Our transdisciplinary research used a mixed-methods design (predominantly qualitative) to understand the complexities pertaining to child and law issues. These methods ranged from deliberations and integrated review consultations with judicial personnel, to field practice methods for use in direct interventions with children.

- **Participant Observation**

We used participant observation in the context of child sexual abuse to understand how children responded in court, based on the transdisciplinary field practice methodologies we had developed to use with them. The support personnel recorded their observations of the court or trial process, including social interactions and communication between the child witnesses with the court officers.

Participant observation was also used in training and teaching, during which we observed the types of responses the training sessions elicited from mental health professionals and judicial personnel, including how far they generated discussion, (divergent) viewpoints, dialogues and ultimate convergence and the relevance to challenges and dilemmas in their own practice.

- **Focus Group Discussions**

We used focus group discussions (FGDs) in training and capacity-building initiatives, deliberations, and in the review of consultation on the law with mental health professionals and judicial personnel. We conducted about 35 FGDs with judicial groups (over 1,000 participant judges) and some 10 FGDs with mental health professionals (around 500 mental health service providers). At the end of each FGD, the participants' opinions were elicited in discussion and in feedback forms. These focused on the issues and learning methods they had found most helpful, and the relevance to their practice of the transdisciplinary collaboration and the issues discussed. Process-related feedback on the methodologies used in specific sessions was also obtained throughout the consultations and training programmes.

10.5.2 Outcomes of Our Transdisciplinary Methods and Initiatives Used by Other Agencies

Following the development and use of our transdisciplinary methods, we examined high-impact references to our work related to child law. Since our methodologies are in the public domain, we wanted to see how far our work had influenced other agencies' training programmes and field practice, and government policy.

We then analysed patterns, challenges and successes related to the use of transdisciplinary methods and processes for evidence of effective knowledge integration and collaboration, to understand the impact of transdisciplinary approaches on addressing complex issues. For instance, based on the dialogue and responses in the stakeholder groups and their feedback, we sought to understand how far this knowledge had been assimilated in relation to understanding how to support child witnesses in adversarial justice systems: for instance, in the case of mental health service providers, we followed up on their case work and interventions with sexually abused children to understand whether they were implementing the child–law interface methodologies we had shared with them; and whether they were disseminating such knowledge and practice among their trainees/students. In the case of judicial personnel, we examined High Court judgements, for example, to understand whether child-oriented adjudicatory methods and processes were adopted.

10.6 Findings

10.6.1 Practical Solutions for Systemic Transformation in Child-Law Interfaces

10.6.1.1 Practice Methodologies for Understanding Children's Perspectives and Assisting Them in Their Interface with the Law

Based on our direct interactions and interventions with children in contact with the law, we developed three methodologies to use with them, with a view to (i) eliciting and understanding their developmental capacities and mental health issues; (ii) enabling them to navigate the complexities of the law and judicial systems; (iii) minimizing the harmful and traumatizing impact of the legal system, and ensuring children's best interests by being aware of their developmental and protection concerns, and their mental health status, and also of their rights to participation and agency. These methodologies, depending on the stage of the legal procedure, were designed to be implemented by various stakeholders in mental health, protection and law.

These practice methodologies are transdisciplinary in nature because the psycho-legal assessments and interventions they encompass are primarily located in the field of mental health, but are applied in the legal context. Through directly interviewing children in therapeutic and forensic contexts, using creative and child-friendly methods, they seek to understand and assess children's psychosocial vulnerabilities, mental health issues and their cognitive capacities; the implications of how these psychological assessments and interventions play out in the legal arena by equipping children to provide testimony in court, or recommendations regarding juvenile transfer and rehabilitation. The methodologies, which relied almost completely on our direct engagement with children and are based on child rights, developmental and mental health frameworks, ensured children's participation in legal processes. Thus, they represent a major effort to alter the power imbalance by ensuring that the marginalized and vulnerable have a voice in processes that will affect their life.

10.6.1.2 Development and Implementation of Transdisciplinary Capacity-Building Workshops for Multiple Stakeholders

A vital part of implementing TDR processes entailed integrating the various ideas proposed by stakeholder groups, analysing the recurring themes and patterns and drawing out themes and content to feed into subsequent rounds of training and capacity building. Thus, the SAMVAD team continually engaged in analysis, which enabled it to select and use stories or cases from real-life experiences and interactions between children and which reflected child development, ethical, practical and systemic tensions and dilemmas. These themes and stories helped to refine, revise and sometimes completely change the contents, methodology and structure of the existing training curriculums and programmes for judicial personnel and mental health service providers.

Thus, building on our transdisciplinary approaches and what we learned from them, we developed India's first-ever transdisciplinary professional training courses on child and law issues. Courses for judicial personnel aimed to facilitate an understanding of the psychosocial and mental health contexts and vulnerabilities of sexually abused children, and of children in conflict with the law, to enable them to integrate knowledge of abuse, trauma and other adverse circumstances into how they interview children and pass judgements. Others were designed to introduce mental health and legal dimensions of forensics regarding sexually abused children and children in conflict with law, focus on clinical and psycholegal assessments, and therapeutic and court-support interventions, and were geared to secondary and tertiary mental health service providers and paediatricians. The training courses promote intersectional methods to alleviate the negative impacts of children's interface with the criminal justice system. The transdisciplinary element lies in content that encompasses themes such as linkages between child rights, protection, mental health and legal procedures, capacity assessments for children in contact with the law, forensic interviewing methods for children, pre-trial and in-trial interventions for supporting children and provision of expert witness testimony (see Fig. 10.1).

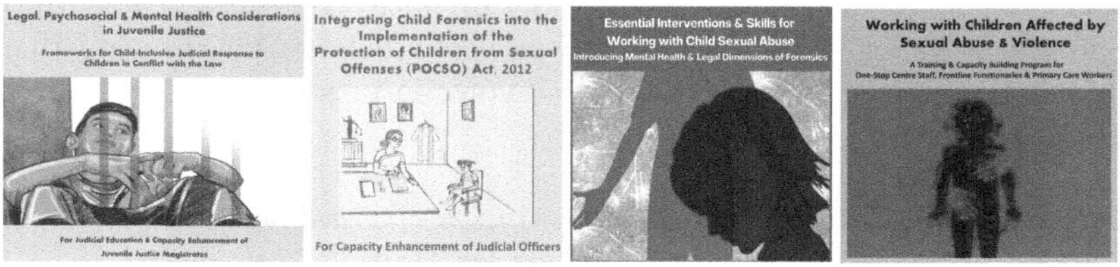

Fig. 10.1 Training curriclums developed for key stakeholders

These have been used in training and capacity-building workshops for the stakeholder groups at state and national levels. Evaluations of the training workshops indicate that participants feel that these truly attempt to address the 'grey areas' and dilemmas that they face in their clinics and courtrooms regarding children's contact with the law. The quotes in Box 10.2 reflect various ways in which participants internalized the transdisciplinary approaches and methods used to prompt reflection and dialogue in the training and education programmes—in particular, they highlight the importance of the methodologies, the room for discussion on dilemmas and difficult issues in field practice, and their changed perspectives. In addition to helping to navigate tensions in child and law work, the workshops resulted in the co-production of knowledge as participants shared experiences and challenges. Furthermore, knowledge co-produced in one group of stakeholders (such as judges) was shared at a subsequent workshop with another group of stakeholders (such as mental health professionals); thus different stakeholder groups had access to each other's perspectives and dilemmas through the facilitators, who are also the transdisciplinary researchers. Such exchange of knowledge and viewpoints helps to transcend disciplinary barriers and allows for new information to flow between different stakeholders—thus helping to iron out, to some extent, power imbalances between adult stakeholders.

10.6.1.3 Policy Transformation

A tremendously important outcome, and a testimony to the need for transdisciplinary approaches to resolve problems relating to children's interface with the law, was a landmark judgement passed by the Supreme Court of India, namely the Barun Chandra Thakur v. Master Bholu & Anr (*Barun Chandra Thakur v. Master Bholu & Anr. (2022 SCC OnLine SC 870)*, 2022). In this case, a 17-year-old came into conflict with the law for allegedly killing a younger child, and the Supreme Court drew on the practice methodologies for children SAMVAD had developed on juvenile transfer, to emphasize the importance of mental health assessments, and the need to use rehabilitative and treatment-based approaches to children in conflict with the law, in recognition of their (neuro)developmental, mental health and psychosocial vulnerabilities. Thus, not only did this judgement recognize the need to adopt transdisciplinary methodologies to address difficult legal questions that address the tensions of child rights, protection, law, mental health, rehabilitation and public safety, but in doing so, it moved the child law, protection and mental health systems towards systemic change by directing all relevant parties to adopt the transdisciplinary practice methodology that SAMVAD had developed.

The Supreme Court direction is thus an example of how the use of transdisciplinary methodologies in shifting the perspectives of powerful and influential stakeholders could result in systemic changes in policy and practice. The Supreme Court's direction is binding on all Juvenile Justice Boards, Special Courts, mental health and child protection service providers across the country. Compliance is likely to result in use of standardized methods and the transformation of child and law work, both of which may well lead to large-scale and sustainable change in decisions pertaining to child justice.

10.6.1.4 Generation of New Scientific Knowledge

The use of TDR approaches, through methodological innovations and generation of diverse perspectives, gave rise to insights and new scientific knowledge that informed both policy and practice in the context of child–law interface. For example, guidelines for mandatory reporting of child sexual abuse and proformas for decisions based on child rights in juvenile transfer were developed on the basis of scientific and empirical evidence; and the newly developed methods are therefore scientific ways to approach child work in the legal context. We developed a repository of materials, such as assessment proformas, activity books for children and training manuals and demo videos, all of which are available on the SAMVAD website.

Box 10.1: Research and Information Dissemination: List of Publications

Ramaswamy, S., Seshadri, S., & Bunders-Aelen, J. (2023). Transdisciplinary training for forensic mental health in child sexual abuse in India. *The Lancet Psychiatry*, *10*(5), 317–318.

Ramaswamy, S., Kommu, J.V.S., & Seshadri, S. (2023). Support, advocacy, and mental health interventions for children in vulnerable circumstances and distress: a unique public child mental health initiative in India. *World Social Psychiatry*, *5*(2), 166–169.

Ramaswamy, S., Devgun, M., Seshadri, S., & Bunders-Aelen, J. (2023). "The child needs to tell it to me in words": Barriers and Facilitators to Witness Competencies in Child Sexual Abuse Trials. *The International Journal of Children's Rights*, *31*(2), 403–443.

Ramaswamy, S., Devgun, M., Seshadri, S., & Bunders-Aelen, J. (2023). Balancing the law with children's rights to participation and decision-making: Practice guidelines for mandatory reporting processes in child sexual abuse. *Asian Journal of Psychiatry*, *81*, 103464.

Ramaswamy, S., Sagar, J. V., & Seshadri, S. (2022). A transdisciplinary public health model for child and adolescent mental healthcare in low-and middle-income countries. *The Lancet Regional Health-Southeast Asia*, 3.

Ramaswamy, S., Seshadri, S., & Bunders-Aelen, J. (2022). Shifting landscapes of global child mental health: Imperatives for transdisciplinary approaches. *Asian Journal of Psychiatry*, *69*, 103002.

Ramaswamy, S., Ashok, S., & Seshadri, S. (2022). Mental health issues and challenges of children in Karnataka. In M.R. Narayana (Ed.), *Public finances for development of children in Karnataka: policy issues and challenges*. (In Press by FPI-UNICEF).

Ramaswamy, S.. Seshadri, S., & Bunders-Aelen, J. (2021). Navigating juvenile transfer laws: the application of vulnerability, mental health, and rights frameworks in psycholegal assessments of children in conflict with the law. In S.O. Okapo (Ed.), *Innovations in global mental health* (pp. 1–30). Springer.

Ramaswamy, S., Seshadri, S., & Bunders-Aelen, J. (2021). Building a research agenda for mental health assessments in resolving legal dilemmas on adolescent sexual consent. *Asian Journal of Psychiatry*, *66*, 102907.

To ensure that our methods are widely accessible to practitioners of child work, we selected journals that either have a global impact but are largely mono-disciplinary, to reach readers who may not be familiar with transdisciplinarity (e.g. *The Lancet Psychiatry*); or are widely read by practitioners in a given country or cultural context (e.g. the *Asian Journal of Psychiatry*). These publications (see Box 10.1) reflect how TDR may not only lead to creating collaborative knowledge and social transformation, but also contribute critically to generating new knowledge and publishing of scientific research.

10.6.2 Development of TDR Methodological Approaches and Strategies

10.6.2.1 Engagement of Homogeneous Groups

The recognizable phases in the process start with a negotiation of how all stakeholders perceived and related to the current and established (procedural) interactions between children and the legal system. Given that the research process is strongly influenced by relational features across the stakeholder networks, such as power asymmetries, interpersonal trust and cultural differences, we deliberately worked separately with each stakeholder group in view of their different agendas and priorities with regard to children's interface with the law. For example, judicial officers' mandate is to implement the law, and to apply impartial justice in ways that allow children to be protected; child mental health service providers' agendas are more keenly focused on ensuring children's mental wellbeing—which could often be at odds with the law, whose priority is to elicit accurate evidence. In sum, the reasons for establishing homogeneous groups, and why they worked were:

- They helped create a safe and empathic space in which members of homogeneous stakeholder groups would feel free to share varied and difficult problems and dilemmas with colleagues, who are likely to validate rather than counter their experiences.
- They avoided direct confrontation between stakeholder groups with significantly different perspectives (no matter how 'child-unfriendly' it may seem) as presenting more critical or 'radical' approaches might result in defensiveness and insecurity, and less openness and inclination to consider new perspectives.
- They allowed for forums to introduce new knowledge and information from child mental health and/or the law in ways that were applicable to all present; and to offer alternative perspectives, including possible solutions for the stakeholders to consider, for them gradually to absorb these perspectives, methods or skills into their field or practice (again common to all group members).

Moreover, our experience with heterogeneous groups was largely unsuccessful because of the different interests and agendas, as well as linguistic and semantic difficulties; for instance, not only do children not speak the same language as the technical jargon that adult stakeholders use, but even among the latter the language of the law is very different from that of mental health. Therefore, while our stakeholder groups were diverse, and we ensured that knowledge was transferred from the actors of one group to those of another. In seeking to effect transformations in stakeholder interactions and prevailing discourses, the methods were used simultaneously: information from one was fed into another, to understand another (group of) stakeholders' responses. For example, we presented judicial officers with knowledge from child mental health (through deliberations and research studies), to identify the legal constraints to assimilating these, and used the resulting understanding in refining the practice guidelines and methodologies we developed to use with children. We did this by presenting children (using simple language and age-appropriate life examples and analogies) with the provisions of the law concerning child sexual abuse and juvenile justice—and used their responses, questions and concerns to inform the practice guidelines and methodologies.

10.6.2.2 Co-construction of Knowledge in Multi-stakeholder Processes of TDR

- ***Use of Creative and Experiential Methodologies***

While we engaged in a more 'formal' implementation of research methodologies, such as survey questionnaires, schedules for FGDs and participant observation, our TDR methodologies extended to include a plethora of creative and experiential methodologies. The methods of engagement included various creative, participatory methodologies such as art, film, role plays, simulation games and case analysis to enable participants to orient themselves to children's perspectives and reflect deeply on their struggles in engaging with the law. These activity-based methods were designed to inquire about children's experiences or the field practice of (adult) stakeholders, to elicit perspectives, opinions and dilemmas for deeper reflection and further dialogue in order to arrive at solutions. Thus, the (meta) processing that followed these methods applied an inquiry-based approach to consultations, training and education programmes, to allow for co-creation of new knowledge and alternative perspectives. This approach was also based on our belief that individual and systemic transformation can rarely be coerced, and that with hierarchical institutions such as the judiciary, or even with proprietorial disciplines such as mental health, change needs to be incremental rather than radical.

- ***Implementation of Deliberations and Consultations***

In one of the judicial education programmes, the research team explained that although the judicial academy's (well-intentioned) titling of it as 'Sensitization of Judges…' on the child law and its processes, our objective was not to 'sensitize' them. Rather, we sought to facilitate a deeper examination of the implementation of child laws and procedures through a 'developmental and mental health lens' to enable judges to consider skills and strategies that may enable them to navigate the many challenges of working with child witnesses. The judges expressed relief that there was at last some acknowledgement that they were not 'ignorant of or insensitive to' children in their courtrooms.

Box 10.2: Deliberations
A deliberation is a collaborative process of discussing contested issues by considering various perspectives with the aim of forming opinions and guiding judgment; while deliberative practice can take various forms such as discussions, role-playing exercises, or formal debates, it incorporates sustained and appropriate arguments, with a view to exploring diverse perspectives and informing various decisions. Given the formal and hierarchical structure of the judiciary, we opted for a form of structured debate on three motions, with arguments advanced in favour of, and against, each. A panel of experts from mental health and legal domains, i.e. judges from the High Court and Supreme Court, moderated by child mental health professionals from a national tertiary-care mental health institution, also providing their own perspectives on issues in order to ultimately influence the officers'. Thus, the panellists served a strategic purpose in including child development and mental health agendas, as well as strongly presenting child-centred perspectives in what judicial officers typically perceive to be core legal work.

Such a response is indicative of the approaches that emphasize openness and reflexivity: that they can create a sense of joint engagement in a change process (rather than of being passive recipients of others' change efforts), and therefore help break with underlying assumptions about the 'us versus them' type of thinking that serves to exacerbate bias and feelings of 'otherness'. In this case, judicial officers were sending a strong message that they were also caring about children—as much as mental health professionals were, perhaps—thus, humanizing the practice of law. In turn, this helped shift the perspective of mental health professionals, who started to develop ideas about other actors in the system, and that judges may not be as uninterested or unable to think about children as was imagined.

This anecdote also underscores the effectiveness of our transdisciplinary approach to integrating child-centred perspectives that straddle the domains of child rights and protection, mental health and law, with the goal of initiating systemic change, albeit incrementally. The aim was first to understand the thinking and perspectives, including the challenges, of judicial personnel working at the district level in matters concerning children, and then to integrate child mental health perspectives into their approaches and methods. As part of the TDR objective of (co)production of knowledge, we implemented deliberations and review consultations and a quantitative research study (see below), to facilitate spaces where new knowledge could emerge from a joint sense of urgency and need to change—and set the stage for the requisite education and training programmes.

Deliberations were conducted with judicial personnel on how key aspects of the child sexual abuse and juvenile justice legislation helped to establish an ongoing dialogue with judicial officers from across the country, and better understand their current thinking on child law, including its interpretation and consequent (gaps and challenges in) implementation. The deliberations not only elicited judicial views and understandings of child law, but also enabled the co-production of knowledge through an exchange of perspectives between judicial and mental health domains. In allowing for judicial systems to shift from didactic training to learning and education that was based on sharing and reflection between colleagues and with leaders, the use of transdisciplinary methods also helped create a more equal platform for dialogue in an essentially very hierarchical system. Furthermore, since they comprised High Court and Supreme Court judges they also served as platforms for strategic advocacy, critical in working with stakeholders who are bound in role and function to organizations that are inherently based on hierarchy and positional power.

An integrated consultation on the Indian child sexual abuse law was conducted, with the participation of judges of the Supreme Court and of the State High Courts, senior police personnel, child mental health professionals and legal academics. This consultation enabled the use of transdisciplinary approaches to ensure the integration of diverse professional and psychosocial perspectives in children's interface with the law. With a focus on complex and controversial questions regarding the criminal justice system and its applicability to children (such as mandatory reporting of child sexual abuse, and the age of consent), such consultations, by reframing concerns about the law regarding children, generated two key outcomes: (i) convergence among stakeholders on possible solutions to solutions to the complex problem of children's interface with the law; and (ii) agreements on the need for implementation to be in line with procedural and child-inclusive justice, by taking into account children's (developmental and mental health) competencies, and consequently, for courts to work in partnership with mental health professionals in eliciting evidence and providing other assistance to child witnesses. Thus, the use of transdisciplinary methodologies, through convergence and co-production of knowledge between these two key groups of adult stakeholders, enhanced the prospects for large-scale, sustainable impact on the application of justice, and for transformations in child policy and practice.

10.6.2.3 Employment of Non-linear Pathways in TDR

This case study can be understood as a non-linear pathway towards a more child-inclusive legal system in India (see Fig. 10.2), with (i) some distinguishable phases and steps, including deliberation and reflection on the current state and where we want to go; (ii) steps towards creating alternative practices and policy by integrating various knowledge and ideas; and (iii) reflecting on the impact of these new practices. These steps were part and parcel of a context in which some of SAMVAD's existing work (such as training professionals on child-sensitivity), and that were adapted to fit the objectives of a joint TDR process.

As shown in Fig. 10.2, the transdisciplinary practice and research approach, in loops or spiralling cycles, deepened our understanding of the issues by surfacing or identifying more potential problems and solutions. This continuous process of exchanging perspectives and responses between stakeholder groups helped to identify (possible) ways to integrate very diverse priorities, and enabled an analytical process in which

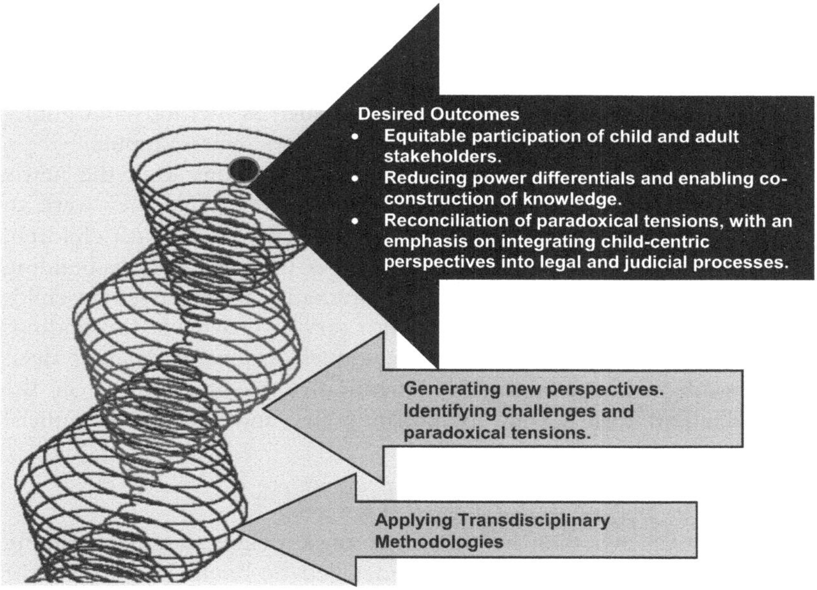

Fig. 10.2 Transdisciplinary practice and research processes towards more child-inclusive legal systems (*Source* Authors used an image of W.B. Yeats's 'A Vision', which pertains to his gyre theory—to depict the spiralling cycles of the transdisciplinary research processes)

the stakeholders co-constructed knowledge. Finally, this iterative engagement generated 'widening circles' of thought until saturation was reached in the sense of being enough to yield feasible solutions to address a major part of the problem. In keeping with our goal of developing practice-oriented solutions to contribute to social transformation, at every spiral of the ever-widening circle of knowledge and understanding, we engaged in reflection to ensure that successive rounds of social action and transdisciplinary methods were indeed leading us (closer) to our desired outcome for children.

These adaptations included, for instance, a more expansive network of stakeholders who were considered part of the research process, including a more deliberate attempt to: (i) better understand and include children's perspectives and experiences; (ii) (re) conceptualize training content and methodologies facilitation to allow for enhanced co-creation of knowledge and improved field practices; and (iii) ensure a stronger focus on designing and presenting new methods for use in the procedures governing children's interface with the law (such as methods and guidelines for use with children), based on stakeholders' diverse inputs.

In fact, what greatly contributed to the adaptations were the newly generated scientific knowledge and methodologies. While they were an end in themselves, the guidelines and proformas for use with children, activity books and training videos were also used in capacity-building initiatives to generate further new and alternative perspectives on child-centred legal work among mental health service providers; the findings from the deliberations with judicial personnel on juvenile transfer decisions provided the basis for the quantitative research study on the issue—which in turn served to inform policy and practice in judicial systems.

10.6.2.4 *De-briefing and Reflexive Monitoring*

Every initiative, whether in the form of a deliberation or training programme, was followed up with a detailed de-briefing in which the SAMVAD team members participated. These de-briefings focused on challenges pertaining to participant stakeholders who were less open or willing to engage in TDR processes, how a given group overall seemed to be experiencing and receiving the processes and methodologies. Such reflexive monitoring also led to iterative decisions on changes and additions in terms of content and methodology in order to enhance engagement or to move towards possible solutions. Given that content and process factors are highly interconnected and influence each other, when the complexity of the problem and content threaten to overwhelm people, there is a tendency to cling to professional boundaries with greater tenacity, and thus be less open to alternative perspectives and viewpoints. Reflexive de-briefing and monitoring helped us identify these blockages, and analytically examine their underlying reasons, as well as introducing new methodologies and approaches.

10.6.2.5 *Communicating Research to Policymakers and Those Responsible for Implementing the Law*

In an effort to address the divides between researchers and academics, and policymakers and practitioners, and also to influence law and policy, we always shared our new scientific knowledge—our research findings—with policymakers and practitioners. Indeed, one of the key outcomes of our transdisciplinary research towards systemic transformation came from a relatively unexpected quarter, namely the quantitative study on judicial decision-making. Our dissemination strategy led the academic, law and policy communities, who work according to diverse rules, timelines and mandates, to converge on solutions related to children's interface with the law. The two-pronged strategy included a brief presentation of research findings, including simply elucidating the implications for policy, practice and training (avoiding academic jargon), and the inclusion of judges from the Supreme Court of India to discuss the significance of the study, particularly in terms of child law practice.

Although our research team made no specific efforts to advocate for changes in policy and practice, the discussions in the research dissemination forum, particularly among judicial officers, moved naturally towards advocacy in favour of child-centred inter-operations and implementation of the law. This, and the apex Court's recognition of the importance of TDR approaches, was reflected in the remarks made by Judges and Members of the Juvenile Justice Committee of the Supreme Court of India in the course of the (country-wide) dissemination of the study findings[1]:

> '...There is an imperative for judicial personnel to work with mental health professionals and non-judicial members of the juvenile justice boards, in transdisciplinary ways, that adhere to constitutional safeguards, and knowledge from legal and non-legal disciplines...the means for ensuring [child] accountability should be grounded in child and adolescent psychology...-transferring a child or adolescent to adult court must be invoked only as a measure of last resort, in exceptional circumstances...
>
> This study will contribute to judicial education and decision-making on juvenile transfer...
>
> The nature and quality of judicial education will ultimately determine the ease and fairness with which judicial transfer decisions are made. I hope the study will scientifically guide judicial discretion and limit the propensity for arbitrariness...I would like to appeal to all the Directors of the Judicial Academies to have a separate training program and curricula on

[1] Online event, dissemination of findings from Report on the study titled 'Factors impacting Judicial Decision Making on Juvenile Transfer in India', 28 September 2023, available at: https://www.youtube.com/watch?v=qMbMTPJHv58

Sections 16, 16 and 18 of the [Juvenile Justice] Act...with the appropriate resource persons...'

—Judge and Chairperson, Juvenile Justice Committee, Supreme Court of India

'...The entire discussion on Section 15 and transfer of [juvenile] cases is a very, very important one...from the point of view of criminal jurisprudence, from [that of] child rights...outside assistance [for the legal system] must be taken whenever available, there is no reason to be hesitant about it [in transfer decision-making]...'

—(Former) Judge and Chairperson, Juvenile Justice Committee, Supreme Court of India

TDR findings can, if disseminated in ways that are dispassionate and impartial, be very powerful in themselves: allowing stakeholders to judge their implications may propel them to advocate in support of vulnerable groups. Such dissemination forums play a critical role in enhancing communication and interaction among researchers, policymakers and the judiciary, and consequently in a gradual increase in the potential impact of research on social attitudes, practice and policy with regard to children and vulnerable groups.

10.7 Emerging Challenges and Roadblocks in Our TDR Journey

Despite our many (successful) efforts to implement TDR approaches in the child–law interface, and its role in helping to address various challenges, there remained some outstanding concerns that need to be addressed in any future work in this area.

10.7.1 The Impact of Knowledge Integration on Power Imbalances

10.7.1.1 Integrating Children's Thinking and Worldview

Our transdisciplinary methodology involved the participation of marginalized and vulnerable children whose perceptions were at the centre of our solutions for their interface with the law. Given children's thinking and worldviews, depending on their developmental abilities, it was sometimes challenging to involve them in dialogues on the issue. For instance, it is difficult for them to understand the procedural workings of the court, let alone the legal philosophies and ethics that underpin them.

Thus, although we sought to create an equitable platform for stakeholders' engagement, the inherent capacities of individual stakeholder groups might make this particularly difficult. Consequently, we had to adopt different forms of communication, such as simplified language and games, stories and play in engaging with children, to help them understand legal perspectives and engage with judicial systems.

10.7.1.2 Hegemonies That Hinder

The law and how it is applied are largely inflexible, including court procedures and judicial officers' almost absolute belief in the adversarial system of justice. These issues do not align with children's worldviews and ways of thinking. Judicial officers also exercise authority over other stakeholders by virtue of the systemic power vested in them, namely child protection functionaries, child rights activists, mental health professionals, and of course over children themselves. Consequently, a child's trauma might find limited space in the courtroom as might a mental health professional's advocacy to consider the child's developmental capacities and mental health status in the operations of the law. Since the application of the law is largely non-negotiable, the challenge was to work within the hegemony of the judiciary, and to acknowledge some degree of powerlessness in our endeavours to integrate scientific knowledge in legal practice.

However, the transdisciplinary methodologies of deliberations, consultations and training interventions enabled a deeper level of qualitative interactions with judges. We gradually understood that judges, though apparently powerful, also work within a larger hegemonic, paternalistic, even patriarchal system—none of which may be empowering for an individual judge attempting to adjudicate in the case of an individual child. Transdisciplinary methodologies made for closer and more open discussions with judges, helping us to realize that despite their seeming inflexibility, they were often naturally inclined to integrating multiple domains and stakeholders in the dispensation of (social) justice, including that of children. These methodologies also helped to coalesce different types of knowledge and practice, juxtaposing these to create room for a shift in perspectives, in ways that 'ordinary' social science research may not easily be able to do. The resulting insights enabled us to find alternative solutions that were feasible from a legal perspective to assist children in their interface with the law.

10.7.2 Communication and Collaboration Barriers Among Diverse Stakeholders

10.7.2.1 Breaking Out of the Mono-Disciplinary Mode: Why Our Domains Are So Dear to Us

In our experience, each group of professionals or stakeholders tends to remain within their disciplinary boundaries to pursue their core roles and functions. This was particularly challenging for any type of work that sought to provide holistic and integrated support to children, and find solutions to their problems, by converging diverse stakeholder perspectives and professional expertise.

The underlying issue points to psychological rather than knowledge- and skill-related factors. These may pertain to the anxiety and insecurity we experience when we move out of our comfort zones, in which we feel competent and confident that we have the right knowledge and skills. TDR entails moving beyond these disciplinary confines—and stakeholders may fear that the priorities and objectives of another discipline could impinge on their own, possibly uncovering tensions (such as those between the professional disciplines of mental health and law) that they do not feel equipped to address.

The collaborative spirit of transdisciplinarity, which goes between, across and beyond disciplines, and which informed our judicial deliberations and consultations, and our forensic training and capacity-building curricula and programmes, helped various stakeholders overcome disciplinary barriers. These initiatives deliberately integrated child-centred knowledge from both legal and mental health perspectives, allowing for intensive professional exchanges that enabled stakeholders to learn about an unfamiliar subject.

10.7.2.2 Restricting the (Researcher's) Competent Self

A preference for mono-disciplinarity is related to competence. We often feel the need to assert ourselves to prove our (superior) knowledge and skills, especially in our primary domain. Competence is cherished and helpful, often coming from profound socio-cultural and family values and beliefs. It can therefore be daunting for researchers to have to discard this competence and embrace transdisciplinarity—to accept a lack of expertise in order to be able to receive new knowledge, participate in different ways of thinking and accept alternative perspectives. To replace the ensuing anxiety with the realization that transdisciplinarity can be liberating, because it affords a space unencumbered by the burden of competence, is a process of developing psychological readiness. It takes time to be able to move away from what Chimamanda Ngozi Adichie calls 'the danger of a single story', to experience the joy of curiosity and wonder, as we immerse ourselves in the subjective realities and challenges in others' lives. But only then can we be effortlessly empathic in our approach and interactions with our research respondents and internalize their stories and perspectives—be they children, judges or administrators—so as devise solutions that are located in their particular life experiences.

10.7.2.3 Human Resource Development for Transdisciplinary Research

The methodological approach we developed relied heavily on facilitation skills. The research team had to embrace a transdisciplinary ethos and have a range of technical and psychological skills in order to facilitate transdisciplinary engagement. These included (i) technical knowledge in their main area of expertise (i.e. law or mental health) and related communication and teaching skills; (ii) a basic understanding of other relevant disciplines, and a willingness to learn about them, in order to connect their knowledge in discussions and workshops to the main issue; and (iii) psychological skills, particularly being open to accepting and reflecting on new perspectives, and freely allowing for dissent.

Given the general preference for mono-disciplinarity, and psychological orientations that often favour individualism, competence and competition, it is not easy to find and retain professionals with the necessary skills and attitude, or create a team of transdisciplinary researchers. Building such a team is time-consuming, needing hours of careful mentorship to instil the requisite skills and work methods, and ensure that individuals' goals are largely aligned—for only then can teaching and learning initiatives with external stakeholder groups effectively engage in practising transdisciplinary methodologies.

10.7.3 Effecting Social Transformation

10.7.3.1 Challenges in Translating Research into Policy

One of the challenges in our TDR for transformation, in terms of assimilating diversity from a professional domain and disciplinary perspective, was integrating scientific knowledge with the implementation of child rights and the development of policy and law. Initially it was difficult for scientific knowledge and global evidence on best practices in relation to children in conflict with law and sexually abused children to find traction in law and human rights communities. The evidence on adolescent neurodevelopment and the negative consequences of using retributive (versus rehabilitative) approaches to help change children's behaviour, for example, were not readily accepted as considerations for child-centred implementation of the law.

A possible reason for this is the divide between academics or researchers and practitioners (including professionals or service providers and policymakers). Although these groups should inform each other in mutually beneficial ways, they tend to be somewhat isolated and separate. This is because the respective stakeholder groups work with different timeframes, use different language and communication styles and have different professional goals. For instance, researchers often have longer timeframes, and use language that practitioners and policymakers do not readily understand. This may lead practitioners to question the relevance of research as being 'too theoretical' and 'far removed from field realities'.

Given the research–practice divide, we endeavoured in our deliberations, consultations and training programmes to present scientific knowledge and evidence on child mental health and law-related issues, adapting the information to the relevant professional backgrounds, roles and functions. We presented our research findings in ways that made knowledge and information accessible and relevant to their contexts of practice, also suggesting how the findings could be used to implement existing child laws and policies more effectively as a way to find better solutions to children's problems. Again, this approach was gentle and non-confrontational, presenting alternative perspectives and solutions before law and policy audiences, to steer them towards incremental (rather than radical) changes in policy and practice—often the only way forward in strongly hierarchical systems.

10.7.3.2 Constraints of Communities and Culture

Ultimately, no matter how much child rights activists and policymakers may promote issues of children's rights and best interests, or how committed child protection and mental health service providers are to ensuring children's safety and wellbeing, and how conscientious judicial personnel are about interpreting and implementing laws in a child-centred manner, children are subject to the ambiguities, inconsistencies and biases of their own families and communities. In the Indian context, the socio-cultural taboos and stigma that surround sexuality and abuse, and public and community prejudices towards children in conflict with the law, are exceedingly hard to overcome.

We found that despite our child-centred methodologies, unless the children's families and communities are willing to support them by permitting a child victim to testify in court, or following through with rehabilitation and reintegration processes, our aspirations for children to be able to access justice, and safety and wellbeing, come to naught. The question, therefore, particularly in the light of issues concerning children who in most situations require their family's assent, is how to address families and communities in ways that will enable them to participate in transforming (their) children's lives.

10.8 Discussion

This case study sought to understand how TDR methodologies may be applied to effectively address challenges in convening stakeholders to co-create a more child-inclusive legal system in a context characterized by considerable power asymmetries—and to examine their outcomes and challenges. We described SAMVAD's experience in India of bringing together various stakeholders in order to transform the nature of child–law interface. One of the key objectives for involving diverse stakeholders in TDR is to promote democratic ideologies by emphasizing processes of inclusion, improving outcomes through eliciting relevant knowledge and risk perceptions from different stakeholders, and consequently, to generate effective and valid decisions and solutions to problems (Fiorino, 1990; Stirling, 2008). According to the notion of 'knowledge democracy', dominant and non-dominant social actors have equal access and ability to participate in resolving social problems, so that both are heard in the decision-making processes leading to research and policy mandates—and which can be achieved with the use of TDR methods (Bunders et al., 2010). The democratic or normative argument for inclusion also asserts that those affected by research and innovation outcomes should be involved in the process (Kok et al., 2021), particularly through democratic methods of participation and deliberation (Jasanoff, 2003; Latour, 2004). The issue of deliberative democracy may also be viewed through the lens of epistemic justice (Catala, 2015), so that we are better able to identify such situations, and failures of inclusion, within deliberative spaces, and to take measures to construct spaces that are more inclusive, thus promoting democratic ideologies (Dieleman, 2015).

From the perspective of epistemic injustice, when knowledge systems reflect social prejudices and biases, they result in discrediting and misinterpreting marginalized groups, thus further excluding them (Bhakuni & Abimbola, 2021). The age and developmental lens is one way to view children and child-related research; equally important is how adults perceive children and their role and influence in society (Klyve, 2019). Our interaction with others is essentially based 'on the categories that we spontaneously place them in' (Gilbert, 2009, p. 93). The 'very things that make children children' is what contributes to their 'otherness'; and there is a need to acknowledge this otherness in social science research and in society more broadly, by listening to their voices and perceiving the world through their eyes and thus bringing their perspectives into practice (Jones, 2001). Our (adult) failure to do so is what lies at the core of power asymmetries, such as in children's interface with the law. In our case, prevalent assumptions about children's capacities to shape decisions about their lives potentially interfere with their participation in knowledge co-production with regard to legal processes. But our engagement with children and inclusion of their subjective experiences and worldviews reflects how our transdisciplinary approaches actively addressed issues of epistemic injustice, and indeed its consequences for the very groups that were marginalized and powerless when they are unable to contribute to knowledge and decision-making.

Our transdisciplinary work, contending with tensions between children's perspectives and the roles and functions of judicial personnel and mental health service providers, are reflective of the Strumińska-Kutra's analytical framework (Strumińska-Kutra, 2016) on some of the major power-related tensions that pervade collaborative approaches to research, namely tensions between participatory processes and socio-cultural and institutional contexts that seldom favour dialogue and more egalitarian methods of discourse (such as in case of the hegemonic nature of judicial and even mental health systems); and tensions within the stakeholder communities, which may include highly heterogeneous groups (as in the case of children and judges), who vary considerably in their values, interests and capacities. Given that these power-related tensions are complex and dynamic, the processes of (research) engagement may be abstruse and confusing (Strumińska-Kutra & Scholl, 2022)—as we described in our roadblocks, particularly those relating to the preference for mono-disciplinarity, community-level, cultural and institutional hegemonies, and the impediments they create for the epistemic participation of child stakeholders, as well as those regarding the researchers' skills, competencies and psychological readiness to embrace transdisciplinarity.

Addressing these tensions therefore calls for a 'flexible repertoire of responses' (Strumińska-Kutra, 2016), as strongly reflected in our methodology. This included enormous amounts of open dialogue and discussion, albeit structured by the use of creative methodologies (such as art, film, games, etc.), in order to stimulate alternative thinking and discourse in relation to children. Specifically acknowledging power dynamics within a given social problem also entails developing a language and tools that make it easier to discuss power, values and interests, both in the research and in policy development (Strumińska-Kutra & Scholl, 2022), particularly in institutional settings that are governed historically and politically by power inequities (Forester et al., 2023). Our engagement with various stakeholders was constantly alert to language and semantics, not only because it entailed direct involvement with children (and the need to find a language and methodology to engage them in court processes), but also with adult stakeholders: for instance, we were conscious of having to demystify medical and psychological jargon and terminology in discussion with judicial personnel, and to make the provisions and concepts of child law more relatable for mental health professionals. Moreover, our creative, non-didactic, empathic and non-judgemental approaches to deploying any methodology enabled us to navigate the frequently rigid and hierarchical spaces, to discuss complex power-related issues, gently shifting adult stakeholder groups towards more child-centred worldviews.

The diverse, layered, non-linear implementation of transdisciplinary methodologies helped us at every stage to identify and consolidate the emerging tensions between child protection, law and mental health in the context of children's interface with law, including perspectives and viewpoints that conflicted with each other and with our own. In other words, the 'widening circles' created by cycles of implementation of various methods with different stakeholders (shown in Fig. 10.1) helped us move towards synthesizing tensions arising from competing demands, as reflected in Herbert Simon's concept of 'Design Thinking' (Simon, 1996, p. 111): 'everyone designs who devises courses of action aimed at changing existing situations into preferred ones'. We were then able to channel these tensions back into the interplay between science, law and child mental health in ways that allowed for creation of platforms for (continuing) transdisciplinary dialogues. Such platforms sometimes appeared to take on a life of their own, such as in case of judges, thus moving beyond existing hegemonies within the judiciary; conversely, judicial isolation, leading to weak interaction and conversation among judges, along with having to conform to strict (social) codes of conduct, can result in superficial dialogue (Zimmerman, 2000). Transdisciplinary methodologies allowed for critical exchanges within the judicial stakeholder group, helping them to move beyond their usual restraint and defence of the status quo to reflect on their perceived professional ethos and social role, as well as their varied motivations of social activism and public service (Gomes et al., 2016).

In addition to democracy and epistemic justice, the substantive argument for research and innovation ensuring diversity and inclusion is that the co-production of knowledge by science and society leads to 'better' outcomes (Kok et al., 2021); they are outcomes that helping to lead to more practicable solutions to real-world challenges, because of greater integration of multiple stakeholder perspectives, knowledge and values (Lang et al., 2012). By using transdisciplinary methodologies to attempt to shift the underlying power dynamics, and thereby enabling the more powerful groups to perceive issues through a child-centred lens, we were also able to effect policy change—as illustrated by the Supreme Court direction in the Barun Chandra Thakur case on the more child-centred implementation of juvenile transfer laws. We recognize that individual shifts may not always lead to large-scale change, or at least not immediately; but by opening doors to alternative ways of thinking and acting, we have at the very least set the stage for further changes in (adult–child) power dynamics, so that marginalized groups such as children have a voice in systems that make decisions for them—and that there is every possibility that policy and law changes may follow. In relation to outcomes, it is worth highlighting that our work, by assimilating diversity and leading to 'better outcomes', also generated new scientific knowledge, based on 'evidence-based patient information' approaches (Bunge et al., 2010), and resulting in the development of scientific methodologies for engagement with children in the context of the law.

In conclusion, underpinning our central research question of resolving tensions in the context of children's interface with the law, by reducing power asymmetries between children and the judicial system, is the issue of inclusion. Transdisciplinary approaches, by transforming underlying power relationships, helped to create a more equitable platform for children's voices in legal processes. In sum, our work concurs with the TDR literature that discusses the importance of forging collaborations to ensure 'reciprocity of expertise' among various stakeholders, and reducing power hierarchies to develop effective research and solutions to complex problems (Kareem et al., 2022).

However, while TDR may begin with wanting to understand and assimilate knowledge from different disciplines and stakeholders, being tossed into the swirling currents of diversity and power is then almost inevitable. Indeed, transdisciplinary researchers may often find themselves in the 'eye of the storm'—frequently the tensions and dilemmas that are generated by power and diversity. In this storm, our focus was always on children—their rights and best interests continually helped us steer our efforts in favour of the marginalized and powerless.

Finally, while we have endeavoured in this chapter to describe transdisciplinary approaches and methodologies, tensions and challenges and ways forward, perhaps the quintessential experience of transdisciplinary approaches can never be articulated exactly as we would like: the emotions of curiosity, awe and wonder that pervade our initiatives and activities, the excitement of (scientific) discovery in the process of knowledge co-production, the joy of the 'oneness' that emerges in deliberations, the new-found empathic bonds created when we (dare to) '*cross disciplinary boundaries, expand epistemological horizons, transgress stubborn research and education routines and hegemonic powers, and transcend mono-cultural practices in order to create new forms of human activity and new social systems that are more sustainable and socially just*' (Lotz-Sisitka et al., 2015 p. 74).

References

Adler, C., Hirsch Hadorn, G., Breu, T., Wiesmann, U., & Pohl, C. (2018). Conceptualizing the transfer of knowledge across cases in transdisciplinary research. *Sustainability Science, 13*(1), 179–190. https://doi.org/10.1007/s11625-017-0444-2

Arnstein, S. R. (1969). A ladder of citizen participation. *Journal of the American Institute of Planners, 35*(4), 216–224. https://doi.org/10.1080/01944366908977225

Beck, S. (2011). Moving beyond the linear model of expertise? IPCC and the test of adaptation. *Regional Environmental Change, 11*(2), 297–306. https://doi.org/10.1007/s10113-010-0136-2

Bell, A., Corfield, M., Davies, J., & Richardson, N. (2010). Collaborative transdisciplinary intervention in early years—Putting theory into practice. *Child: Care, Health and Development, 36*(1), 142–148. https://doi.org/10.1111/j.1365-2214.2009.01027.x

Bhakuni, H., & Abimbola, S. (2021). Epistemic injustice in academic global health. *The Lancet Global Health, 9*(10), e1465–e1470. https://doi.org/10.1016/S2214-109X(21)00301-6

Birnbaum, R., Cyr, F., & McColley, D. (2013). Children's voices in family court: Guidelines for judges meeting children. *Family Law Quarterly, 47*, 379–408.

Bracken, L. J., Bulkeley, H. A., & Whitman, G. (2015). Transdisciplinary research: Understanding the stakeholder perspective. *Journal of Environmental Planning and Management, 58*(7), 1291–1308. https://doi.org/10.1080/09640568.2014.921596

Brady, G., Lowe, P., & Lauritzen, S. O. (2015). Connecting a sociology of childhood perspective with the study of child health, illness and wellbeing: Introduction. In *Children, Health and Well-being* (pp. 1–12). https://doi.org/10.1002/9781119069522.ch1. John Wiley & Sons.

Brouwer, S., Büscher, C., & Hessels, L. K. (2018). Towards transdisciplinarity: A water research programme in transition. *Science and Public Policy, 45*(2), 211–220. https://doi.org/10.1093/scipol/scx058

Bunders, J. F. G., Broerse, J. E. W., Keil, F., Pohl, C., Scholz, R. W., & Zweekhorst, M. B. M. (2010). How can transdisciplinary research contribute to knowledge democracy? In R. J. in 't Veld (Ed.), *Knowledge democracy* (pp. 125–152). Springer.

Bunge, M., Mühlhauser, I., & Steckelberg, A. (2010). What constitutes evidence-based patient information? Overview of discussed criteria. *Patient Education and Counseling, 78*(3), 316–328. https://doi.org/10.1016/j.pec.2009.10.029

Camponovo, S., Monnet, N., Moody, Z., & Darbellay, F. (2023). Research with children from a transdisciplinary perspective: Coproduction of knowledge by walking. *Children's Geographies, 21*(1), 163–176. https://doi.org/10.1080/14733285.2021.2017405

Caprioli, S., & Crenshaw, D. A. (2017). The culture of silencing child victims of sexual abuse: Implications for child witnesses in court. *Journal of Humanistic Psychology, 57*(2), 190–209. https://doi.org/10.1177/0022167815604442

Carel, H., & Györffy, G. (2014). Seen but not heard: Children and epistemic injustice. *The Lancet, 384*(9950), 1256–1257. https://doi.org/10.1016/S0140-6736(14)61759-1

Catala, A. (2015). Democracy, trust, and epistemic justice. *The Monist, 98*(4), 424–440. https://doi.org/10.1093/monist/onv022

Collins, T. M. (2017). A child's right to participate: Implications for international child protection. *The International Journal of Human Rights, 21*(1), 14–46. https://doi.org/10.1080/13642987.2016.1248122

Cutter-Mackenzie, A., & Rousell, D. (2019). Education for what? Shaping the field of climate change education with children and young people as co-researchers. *Children's Geographies, 17*(1), 90–104. https://doi.org/10.1080/14733285.2018.1467556

Dieleman, S. (2015). Epistemic Justice and democratic legitimacy. *Hypatia, 30*(4), 794–810. https://doi.org/10.1111/hypa.12173

Eades, D. (2012). The social consequences of language ideologies in courtroom cross-examination. *Language in Society, 41*(4), 471–497. https://doi.org/10.1017/S0047404512000474

Faria, C. G.-D., Bendo, D., & Mitchell, R. C. (2021). "Post-covid" childhood-s?: Transdisciplinary reflections on participatory praxis in the lives of 21st century children. *The International Journal of Children's Rights, 29*(2), 371–399. https://doi.org/10.1163/15718182-29020007

Fiorino, D. J. (1990). Citizen participation and environmental risk: A survey of institutional mechanisms. *Science, Technology, & Human Values, 15*(2), 226–243. https://doi.org/10.1177/016224399001500204

Forester, J., Verloo, N., & Laws, D. (2023). Creative discretion and the structure of context-responsive improvising. *Journal of Urban Affairs, 45*(6), 1145–1162. https://doi.org/10.1080/07352166.2021.1901589

Freeman, R. E. (1984). *Stakeholder management: Framework and philosophy*. Mansfield.

Fricker, M. (2007). *Epistemic injustice: Power and the ethics of knowing*. Oxford University Press.

Gaim, M., & Wåhlin, N. (2016). In search of a creative space: A conceptual framework of synthesizing paradoxical tensions. *Scandinavian Journal of Management, 32*(1), 33–44. https://doi.org/10.1016/j.scaman.2015.12.002

Garfield, S. L. (1990). Issues and methods in psychotherapy process research. *Journal of Consulting and Clinical Psychology, 58*(3), 273–280. https://doi.org/10.1037/0022-006X.58.3.273

Gaziulusoy, A. İ, & Ryan, C. (2017). Roles of design in sustainability transitions projects: A case study of Visions and Pathways 2040 project from Australia. *Journal of Cleaner Production, 162*, 1297–1307. https://doi.org/10.1016/j.jclepro.2017.06.122

Gaziulusoy, A. I., Ryan, C., McGrail, S., Chandler, P., & Twomey, P. (2016). Identifying and addressing challenges faced by transdisciplinary research teams in climate change research. *Journal of Cleaner Production, 123*, 55–64. https://doi.org/10.1016/j.jclepro.2015.08.049

Gilbert, M. A. (2009). Defeating bigenderism: Changing gender assumptions in the twenty-first century. *Hypatia, 24*(3), 93–112. https://doi.org/10.1111/j.1527-2001.2009.01047.x

Godemann, J. (2008). Knowledge integration: A key challenge for transdisciplinary cooperation. *Environmental Education Research, 14*(6), 625–641. https://doi.org/10.1080/13504620802469188

Gomes, A. O., Guimaraes, T. A., & de Souza, E. C. L. (2016). Judicial work and judges' motivation: The perceptions of Brazilian state judges. *Law & Policy, 38*(2), 162–176. https://doi.org/10.1111/lapo.12050

Guzmán Ruiz, A., Dobbie, M., & Brown, R. R. (2017). Toward multifunctional landscapes in Australian cities: What disciplinary dynamics and practitioner strategies inform transdisciplinary practice? *Urban Forestry & Urban Greening, 27*, 15–23. https://doi.org/10.1016/j.ufug.2017.06.015

Hage, M., Leroy, P., & Petersen, A. C. (2010). Stakeholder participation in environmental knowledge production. *Futures, 42*(3), 254–264. https://doi.org/10.1016/j.futures.2009.11.011

Hegger, D., Lamers, M., Van Zeijl-Rozema, A., & Dieperink, C. (2012). Conceptualising joint knowledge production in regional climate change adaptation projects: Success conditions and levers for action. *Environmental Science & Policy, 18*, 52–65. https://doi.org/10.1016/j.envsci.2012.01.002

Hessels, L. K., De Jong, S. P. L., & Brouwer, S. (2018). Collaboration between heterogeneous practitioners in sustainability research: A comparative analysis of three transdisciplinary programmes. *Sustainability, 10*(12), Article 12. https://doi.org/10.3390/su10124760

Hurni, H., & Wiesmann, U. (2014). Transdisciplinarity in practice. experience from a concept-based research programme addressing global change and sustainable development. *GAIA—Ecological Perspectives for Science and Society, 23*(3), 275–277. https://doi.org/10.14512/gaia.23.3.15

Jahn, S., Newig, J., Lang, D. J., Kahle, J., & Bergmann, M. (2022). Demarcating transdisciplinary research in sustainability science—Five clusters of research modes based on evidence from 59 research projects. *Sustainable Development, 30*(2), 343–357. https://doi.org/10.1002/sd.2278

Jasanoff, S. (2003). Breaking the waves in science studies: Comment on H.M. Collins and Robert Evans, 'The Third Wave of Science Studies'. *Social Studies of Science, 33*(3), 389–400. https://doi.org/10.1177/03063127030333004

Jasanoff, S. (2004). *States of knowledge: The co-production of science and the social order.* Routledge.

Jasanoff, S., & Martello, M. (2004). *Earthly politics: Local and global in environmental governance*. MIT Press.

Jasanoff, S., & Wynne, B. (1998). Science and decisionmaking. Human choice and climate change. In S. Rayner & E. L. Malone (Eds.), *Human choice and climate change. 1: The societal framework*. Batelle Press.

Jones, O. (2001). "Before the dark of reason": Some ethical and epistemological considerations on the otherness of children. *Ethics, Place & Environment, 4*(2), 173–178. https://doi.org/10.1080/13668790123774

Kareem, B., McClure, A., Walubwa, J., Koranteng, K., Mukwaya, P. I., & Taylor, A. (2022). Power dynamics in transdisciplinary research for sustainable urban transitions. *Environmental Science & Policy, 131*, 135–142. https://doi.org/10.1016/j.envsci.2022.02.001

Klyve, G. P. (2019). Whose knowledge? Epistemic injustice and challenges in hearing children's voices. *Voices: A World Forum for Music Therapy, 19*(3), Article 3. https://doi.org/10.15845/voices.v19i3.2834

Kok, K. P. W., Gjefsen, M. D., Regeer, B. J., & Broerse, J. E. W. (2021). Unraveling the politics of 'doing inclusion' in transdisciplinarity for sustainable transformation. *Sustainability Science, 16*(6), 1811–1826. https://doi.org/10.1007/s11625-021-01033-7

Lang, D. J., Wiek, A., Bergmann, M., Stauffacher, M., Martens, P., Moll, P., Swilling, M., & Thomas, C. J. (2012). Transdisciplinary research in sustainability science: Practice, principles, and challenges. *Sustainability Science, 7*(1), 25–43. https://doi.org/10.1007/s11625-011-0149-x

Latour, B. (1987). *Science in action: How to follow scientists and engineers through society*. Harvard University Press.

Latour, B. (2004). *Politics of nature*. Harvard University Press.

Lawrence, M. G., Williams, S., Nanz, P., & Renn, O. (2022). Characteristics, potentials, and challenges of transdisciplinary research. *One Earth, 5*(1), 44–61. https://doi.org/10.1016/j.oneear.2021.12.010

Lotz-Sisitka, H., Wals, A. E., Kronlid, D., & McGarry, D. (2015). Transformative, transgressive social learning: Rethinking higher education pedagogy in times of systemic global dysfunction. *Current Opinion in Environmental Sustainability, 16*, 73–80. https://doi.org/10.1016/j.cosust.2015.07.018

Magnusson, M., Joleby, M., Luke, T. J., Ask, K., & Lefsaker Sakrisvold, M. (2021). Swedish and Norwegian police interviewers' goals, tactics, and emotions when interviewing suspects of child sexual abuse. *Frontiers in Psychology, 12*. https://doi.org/10.3389/fpsyg.2021.606774

Marsil, D. F., Montoya, J., Ross, D., & Graham, L. (2002). Child witness policy: Law interfacing with social science. *Law and Contemporary Problems, 65*(1), 209. https://doi.org/10.2307/1192371

Mitchell, R. C., & Moore, S. A. (2018). Transdisciplinary child and youth studies: Critical praxis, global perspectives. *World Futures, 74*(7–8), Article 7–8. https://doi.org/10.1080/02604027.2018.1485435

Naidoo, R. (2015). Beyond the academic's dilemma: Transdisciplinary and existential perspectives of re-enchantment. *The Journal for Transdisciplinary Research in Southern Africa, 11*(2). https://doi.org/10.4102/td.v11i2.70

Nicolescu, B. (2006). Transdisciplinarity—Past, present and future. In B. Haverkort & C. Reijntjes (Eds.), *Moving worldviews: Reshaping sciences, policies and practices for endogenous sustainable development*. ETC/Compas.

Nicolescu, B. (2018). The transdisciplinary evolution of the university condition for sustainable development. In D. Fam, L. Neuhauser, & P. Gibbs (Eds.), *Transdisciplinary theory, practice and education: The art of collaborative research and collective learning* (pp. 73–81). Springer International Publishing. https://doi.org/10.1007/978-3-319-93743-4_6

Olsvik, B. S., & Saus, M. (2022). Coping with paradoxes: Norwegian child welfare leaders managing complexity. *Child Care in Practice, 28*(3), 464–481. https://doi.org/10.1080/13575279.2020.1776683

Pantell, R. H., Yogman, M., Gambon, T., Lavin, A., Mattson, G., Rafferty, J. R., & Committee on Psychosocial Aspects of Child and Family Health. (2017). The child witness in the courtroom. *Pediatrics, 139*(3).

Pielke, R. A. (2005). Misdefining "climate change": Consequences for science and action. *Environmental Science & Policy, 8*(6), 548–561. https://doi.org/10.1016/j.envsci.2005.06.013

Pohl, C., & Hirsch Hadorn, G. (2007). *Principles for designing transdisciplinary research*. oekom verlag. https://doi.org/10.14512/9783962388638

Pohl, C., Klein, J. T., Hoffmann, S., Mitchell, C., & Fam, D. (2021). Conceptualising transdisciplinary integration as a multidimensional interactive process. *Environmental Science & Policy, 118*, 18–26. https://doi.org/10.1016/j.envsci.2020.12.005

Ramaswamy, S., Seshadri, S., & Bunders-Aelen, J. (2021). Building a research agenda for mental health assessments in resolving legal dilemmas on adolescent sexual consent. *Asian Journal of Psychiatry, 66*, 102907. https://doi.org/10.1016/j.ajp.2021.102907

Ramaswamy, S., Seshadri, S., & Bunders-Aelen, J. (2022a). Shifting landscapes of global child mental health: Imperatives for transdisciplinary approaches. *Asian Journal of Psychiatry, 69*, 103002. https://doi.org/10.1016/j.ajp.2021.103002

Ramaswamy, S., Vijay Sagar, J., & Seshadri, S. (2022b). A transdisciplinary public health model for child and adolescent mental healthcare in low- and middle-income countries. *Lancet South East Asia*. https://doi.org/10.1016/j.lansea.2022.100024

Reaman, G. H. (2004). Pediatric Cancer research from past successes through collaboration to future transdisciplinary research. *Journal of Pediatric Oncology Nursing, 21*(3), 123–127. https://doi.org/10.1177/1043454204264406

Regeer, B., & Bunders, J. (2009). *Knowledge co-creation: Interaction between science and society*. Advisory Council for Research on Spatial Planning, Nature and the Environment/Consultative Committee of Sector Councils in the Netherlands.

Schippers, M. C., West, M. A., & Dawson, J. F. (2015). Team reflexivity and innovation: The moderating role of team context. *Journal of Management, 41*(3), 769–788. https://doi.org/10.1177/0149206312441210

Siebenhüner, B. (2018). Conflicts in transdisciplinary research: Reviewing literature and analysing a case of climate adaptation in Northwestern Germany. *Ecological Economics, 154*, 117–127. https://doi.org/10.1016/j.ecolecon.2018.07.011

Simon, H. A. (1996). *The sciences of the artificial*. MIT Press.

Stirling, A. (2008). "Opening up" and "closing down": Power, participation, and pluralism in the social appraisal of technology. *Science, Technology, & Human Values, 33*(2), 262–294. https://doi.org/10.1177/0162243907311265

Straker, L., Beynon, A., Smith, S., Johnson, D., Wyeth, P., Sefton-Green, J., & Kervin, L. (2022). *Towards a transdisciplinary approach to evidence-based decision making regarding digital technology use with, by and for children. Digital Child Working Paper 2022-01.* ARC Centre of Excellence for the Digital Child, Brisbane, Australia. https://www.digitalchild.org.au/wp-content/uploads/2022/07/Straker-et-al-2022-Towards-a-transdisciplinary-approach-Digital-Child-Working-Paper-2022-01.pdf

Strumińska-Kutra, M. (2016). Engaged scholarship: Steering between the risks of paternalism, opportunism, and paralysis. *Organization, 23*(6), 864–883. https://doi.org/10.1177/1350508416631163

Strumińska-Kutra, M., & Scholl, C. (2022). Taking power seriously: Towards a power-sensitive approach for transdisciplinary action research. *Futures, 135*, 102881. https://doi.org/10.1016/j.futures.2021.102881

Weichselgartner, J., & Truffer, B. (2015). From knowledge co-production to transdisciplinary research: lessons from the quest to produce socially robust knowledge. In B. Werlen (Ed.), *Global sustainability: Cultural perspectives and challenges for transdisciplinary integrated research* (pp. 89–106). Springer International Publishing. https://doi.org/10.1007/978-3-319-16477-9_5

Whitcomb, D. (2003). Legal interventions for child victims. *Journal of Traumatic Stress, 16*(2), 149–157. https://doi.org/10.1023/A:1022895106297

Yin, R. K. (1984). *Case study research: Design and method*. Sage Publications.

Zajac, R., & Hayne, H. (2003). I don't think that's what really happened: The effect of cross-examination on the accuracy of children's reports. *Journal of Experimental Psychology: Applied, 9*(3), 187–195. https://doi.org/10.1037/1076-898X.9.3.187

Zimmerman, I. M. (2000). Isolation in the judicial career. *Court Review, 36*, 1–6.

CHAPTER 11

Conditions for Transformative Engaged Scholarship in Co-creation with Queer Refugees

Fabian Holle, Elena Ponzoni, and Halleh Ghorashi

11.1 Introduction

This chapter focuses on conditions for co-creative research with regard to experiences and diverse forms of knowledge that are silenced by stigmatizing, negative labelling and (subtle) kinds of exclusion that must be countered. We reflect on three core conditions that emerged in LIMBO: a creative research collaboration between queer refugee[1] community

[1] We use the term 'queer refugees' as an umbrella term to include people who identify as LGBTQIA+ (Lesbian, Gay, Bisexual, Transgender, Queer, Intersex, Asexual plus other sexual identities) (Jagose, 1996) and who are refugees or have experienced forced migration.

F. Holle (✉) · E. Ponzoni · H. Ghorashi
VU Amsterdam, Amsterdam, The Netherlands
e-mail: f.y.holle@vu.nl

E. Ponzoni
e-mail: e.ponzoni@vu.nl

H. Ghorashi
e-mail: h.ghorashi@vu.nl

B. J. Regeer et al. (eds.), *Transdisciplinarity for Transformation*,
https://doi.org/10.1007/978-3-031-60974-9_11

organizers and artists, the Amsterdam art institute Framer Framed and the Refugee Academy (expertise lab based at the Institute for Societal Resilience at VU Amsterdam). LIMBO is part of the *Engaged Scholarship and Narratives of Change* project, which investigates specific forms of transdisciplinary cooperation between academic institutions and society, which we refer to here as transformative engaged scholarship.

Transformative engaged scholarship refers to critical forms of academic work with the aim of enabling more inclusive practices both in society and in academic institutions. It is grounded in the belief that academia has an important role to play in expanding the collective imagination and facilitating the inclusion of marginalized groups. Other than understandings of engaged scholarship that hinge on researchers' advisory and consulting roles in existing policy or professional frames, which are familiar in Science and Technology Studies (Zuiderent-Jerak, 2015), this form of engaged scholarship involves critical reflection on power dynamics in knowledge production and creating spaces for mutual learning with different social actors, particularly those from marginalized communities (Meekosha et al., 2013). Transdisciplinary work from this perspective involves investing in challenging normalized structures of exclusion by incorporating diverse forms of knowledge, making power structures more visible and promoting shared responsibility for change (Medina, 2013), through processes of knowledge co-creation.

In the Dutch welfare context, one instance of a normalized form of exclusion is the image of refugees as deviating from the norm. These images are informed by hegemonic discourses based on gendered, racialized and cultural hierarchies of difference that reproduce structures of inequality in everyday practices (Young, 2007). In these normalized structures, members of non-privileged groups—such as refugees—are depicted as both completely different and 'lacking' and also, increasingly, as dangerous and unwanted (Ghorashi, 2020). Transformative engaged scholarship, seen as transdisciplinary work, addresses this normalized form of power and recognizes the agency and voices of marginalized individuals and communities, but also goes beyond merely documenting such issues. In our view, enabling change starts with scholars acknowledging power relationships (including their own position) and making efforts to establish sustainable conditions for both epistemic and social justice, recognizing and valuing experiential and community knowledge as meaningful ways of existing and understanding in the world. This chapter describes such conditions that are relevant to co-creative research in the context of LIMBO, a collaboratively designed community to enable non-academic participants to co-own the process of knowledge production.

11.1.1 LIMBO: Co-creative Engagement in the Margins

LIMBO is an alternative space for knowledge creation outside dominant structures for people who experience the intersection of queerness and refugeeness. It facilitates twice-monthly creative workshops by and for those identifying as queer refugees, and who face intersecting dimensions of (blatant and/or subtle) exclusion in the Dutch context. LIMBO has aimed to create space for sharing stories and content through art, such as poetry, drawing, creative writing, collage and photography. Reflecting on this project, which was initiated and co-facilitated by the first author (hereafter FH), enables us to highlight the transformative potential of co-creative spaces and to investigate the premises and conditions that can lead to creating these. We have reconstructed the methodological opportunities and challenges through reflective dialogues among ourselves, thinking through the skills, aptitudes, choices, processes and relations that led the organizers of LIMBO to navigate the complexity of this creative and collaborative journey. In addition to seeing the transformative value of engaged scholarship, the experience of LIMBO made us more aware of academics' vulnerability when engaging with communities. Hence, our chapter is guided by the following questions: What is the role of academia when community-engaged research decentres the hierarchical position of academic knowledge to enable epistemic justice and multiplicity? What are the challenges and opportunities?

Decentring involves researchers stepping back from their (personal, academic) perspectives in their leadership and researcher roles to allow the group's desires, needs, stories and practices to remain central, in order to become part of the group and let others take the lead (Braidotti, 2018). This means that, like most participatory research, the researcher shares power over the research process with the group (Caretta & Riaño, 2016). For instance, the researcher does not depart from a fixed research frame when approaching the research participants but approaches the group with a flexible framework with the aim of adjusting it throughout the process of co-creation. It thus involves researchers being open to different perspectives and values within the group and being willing to set aside their own preconceived notions and opinions, allowing for epistemic justice through a multiplicity of knowledges (Caretta & Riaño, 2016). Recentring is stepping in when needed, moving towards a productive direction by including academic insights. De/recentring can be seen as acts of situational humility (Edmondson, 2018), which are at the core of our approach.

In an earlier co-creation project led by FH (see Fig. 11.1), it became evident that decentring presents specific challenges and opportunities. We learned that it is important to find a balance between decentring and recentring. In Trans-Clash, the group of participants[2] was in control of designing the project and the researcher was almost entirely decentred. However, this process led to a somewhat limited outcome in FH's perception. The relatively small group of six members wanted to create a performance event and co-author an application to fund a small tour. FH took the lead in co-writing this funding application, which served as an opportunity to learn what was important to this group and why. However, FH aimed to reach and include the larger queer refugee community as part of their PhD research. They[3] wanted to create a project that was meaningful to many/most queer refugees in the hope of attracting more (diverse) queer refugees by using the snowball technique (Tracy, 2013). FH also deemed specific elements based on theory and experience to be important, such as holding space and abstaining from pressure to work towards creative results. Because most members of the Trans-Clash group did not give priority to these values and since FH had almost completely decentred themselves, they could only follow the direction of the group. After one co-organized performance event that was warmly received by the audience, the funding body rejected the application. LIMBO was then co-created in order to navigate democratic decision-making while at the same time including insights and knowledge based on theory and the researcher's experience.

The dynamic interplay of decentring and recentring is relevant not just in the project's design but throughout the entire process. The following sections critically unpack this as one of three conditions that appeared in LIMBO as crucial avenues in working towards transformative impact. The first condition concerns the use of creativity, more specifically using art practices as a research method. The second condition concerns the

[2] The Trans-Clash group and the LIMBO group are distinct.

[3] FH uses they/them pronouns.

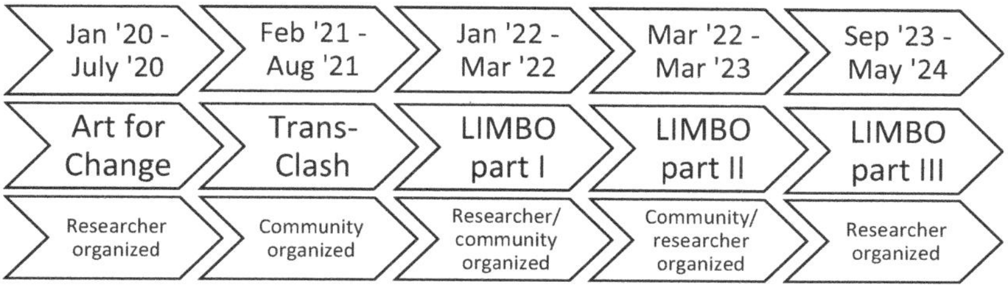

Fig. 11.1 FH's co-creation timeline about decentring-recentring in the project's design

notion of 'holding space' for all kinds of intimate and often uncomfortable feelings and conversations. The third is the ability of decision-makers to decentre and recentre.

11.2 Creativity

During the first co-creation project Art for Change (Fig. 11.1), the potential of creativity to cross language barriers became evident. FH was researching the role of art in the lives of queer refugees in the context of the social distancing and lockdowns during the COVID-19 pandemic. The entire research team was suddenly forced to reflect on what community engagement means in times of social isolation. What role can academics play? An online platform was created with the intention of continuing meaningful and mutually beneficial academic community engagement with queer refugee artists (Holle et al., 2021). There, we found that art practices are experienced as liberating because they are not restricted to verbal language (Karimi, 2020; O'Neill et al., 2002). For people with a refugee experience, for whom Dutch or English is often a third or fourth language, art offers a 'means of imagining beyond the limits of language (or playing creatively with language) via more sensorial, visual or emotional levels' (O'Neill et al., 2002 in Holle et al., 2021).

> I think poetry, it surpasses just words, I think they're more like illustrations and they're like colours. They are stories. I was quite impressed with what a lot of people came with. [...] It was very mind opening to me to learn

about new cultures, new people, their ways of life, how they got to the Netherlands. (Musa)[4]

Art practices are playful, more embodied, forms of sharing and expression and for that reason might be easier to understand, feel or relate to, than (academic) texts (Dewhurst, 2010; Oliveira, 2019). For this reason, creative expression can be an important way to increase epistemic multiplicity, which means recognizing and valuing the simultaneous presence of different forms and sources of knowledge. This project significantly deepened our understanding of creativity (a broader frame for art practices) as a valuable methodological approach. Despite this importance, creativity in the neoliberal context is commonly defined interchangeably with innovation—referring mainly to bigger, better, higher, faster or more (Jeanes, 2006). This is a rather hierarchical perception as it focuses on linear growth. As an alternative to this understanding, we embrace the horizontal value of liminality inherent to creativity and creative practices (Turner, 1979). When the focus is on the process and not on the final results, it removes the pressure to perform, making the experience more liberating. Creativity through the lens of liminality enables tuning into oneself in playful ways and encourages (self-)discovery and (self-) expression.

A focus on liminality is relevant because (queer) refugees often find themselves in a liminal state arising from displacement and exclusion (Manjikian, 2014). In the context of (forced) migration, liminality refers to the state of transition, which can be experienced as temporary or permanent, depending on both individual and social perceptions: when does a person 'belong' in their new society? Being in a liminal state can be challenging for refugees, as it can involve a loss of identity, culture and community (Turner, 1969, 1979). In many cases, refugees cannot return to their own country and must navigate the challenges of building a new life in a foreign land. This liminal state can also have an impact on mental health and well-being, as refugees may experience feelings of isolation, marginalization, loss, uncertainty and feeling cut off from social structures (Elferink & Emmen, 2017).

Despite its difficulties, liminality also involves specific sources of situated knowledge. Edward Said (1994) argued that those living in exile occupy a position of in-betweenness, which makes it impossible to fully

[4] Pseudonyms used for all the quotes from LIMBO members.

assimilate into specific social structures. It involves both a sense of permanent non-belonging and having the potential for creativity and originality. Said uses the metaphor of exile (or, in contemporary terms, refugees), to show that juxtaposing the old and new cultural and structural contexts enables reflexivity. This 'forced reflexivity' allows exiles to hold an insider–outsider position that challenges assumed positions in either context (Ghorashi, 2018). In a similar way, queerness can also be seen as a liminal condition that fosters reflexivity, particularly when 'queer' refers to activism on challenging dominant binary social norms (Jagose, 1996). Hence, both queerness and refugeeness involve a potential for the critical thinking necessary to stretch the boundaries of what is considered the norm. The question is how engaged scholars can highlight this other side of liminality, which is in contrast to dominant perceptions of those in liminal positions as 'lacking' (Ghorashi, 2020). To broaden this singular perspective of lack, the name LIMBO—referring to liminality—aims to boost this positive source of creativity and originality.

Liminality has been described as having transformative potential at the individual, group and public levels (Turner, 1979). At an individual level, liminality is described as 'a rite of passage', which refers to a state or a journey in which one lets go (or is forced to let go) of one identity while transitioning to another (Turner, 1979; Van Gennep [1909] 2019). Examples are adolescence, when one is no longer a child but not yet fully adult; migration, where one is no longer part of the former society, but not yet fully settled in the new one; or being in the process of losing a loved one, where one is transitioning to an identity without that other person. These are all forms of identity (re/de)construction (Beech, 2011). Creativity in the form of art practices can enable such identity (re/de)construction (Holle et al., 2021).

At a group level, liminality in creativity triggers 'plural reflexivity' (Turner, 1979, p. 465). In contrast to making art in solitude, plural reflexivity is a form of sharing reflections and art in a group—strengthening not only one's own but also the group's identity. It can enable a sense of belonging because plural reflexivity refers to 'ways in which communities think about, see and present themselves through signs, symbols, music, dance and visual arts, thereby strengthening a sense of community' (Holle et al., 2021, p. 4).

At a public level, artists and activists use creativity and reflexivity to enable social change. This is what Turner refers to as 'public liminality' or 'public reflexivity' (Turner, 1979). Activists use art or creative expression to 'make us wonder about what we are doing, to rupture a stream of thought' by using 'non-discursive means' such as 'pictures, song, poetic imagery, and expressions of mockery and longing performed in rowdy and even playful ways aimed not at commanding assent but disturbing complacency' (Young, 2001, p. 687). This is in line with what Toni Cade Bambara says: 'The role of the artist is to make the revolution irresistible' (cited in Brown, 2019).

11.2.1 Playfully Unsettling Normalized Positioning

Essentially, creativity in LIMBO concerns the possibility of unsettling and liberating oneself from normalized or conditioned positioning. With this in mind, we realized the importance of icebreakers, games and/or movement when Kenza Badi gave a workshop about clowning. Clowning is another way of playfully questioning and resisting normalized positioning. Badi's workshop was mentioned by several members as one of the highlights:

> It was very funny to me, because for you to be able to do this kind of weird things, it really, really, really, really relieved a lot of stress for me. Because sometimes you don't need to feel that I'm too big to do this, this is for small kids, but when you try to be weird like a baby or all those things you see that stress is still relieved. (Faadin)

Badi's workshop idea relates to Adrienne Maree Brown's understanding of pleasure activism, which she defines as 'the work we do to reclaim our whole, happy, and satisfiable selves from the impacts, delusions, and limitations of oppression and/or supremacy' (Brown, 2019, p. 13).

In LIMBO, there was only one clowning workshop, but elements of playfully unsettling remained. Over time it became evident that moving, dancing and playing games should become part of the LIMBO routine. We borrowed several games from other contexts and developed some that focus on collectivity. Such games have the potential of learning to leave one's comfort zone in a relatively safe setting. By constantly switching games, creative practices, workshop facilitators and the changing environment of the LIMBO homebase, participants essentially practise being comfortable with change. The group is continuously invited to practise something unknown, and in effect, never have the opportunity to rely on stability or reproduction of any kind. The only constant is the practice of sharing reflections. One thing that we learned in LIMBO is to avoid typical introduction rounds: 'My name is ..., I was born in ..., I studied ..., I am working as, etcetera'. Such routines can produce a pressure to 'properly' introduce oneself. Instead, in LIMBO the members get to know each other through (name) games and (reflections on) creative practices.

11.2.2 Liminal Space

In addition, liminality is also described in the literature in terms of space, which refers to 'transitory in-between spaces' through which people pass, such as hallways, airports or streets (Shortt, 2015). They are situated between 'dominant' spaces that have clearer social codes on how to behave, even though those codes are often internalized and taken for granted. Think, for example, of classrooms, dining rooms or conference halls. Ways to behave in spaces with such clear purposes are culturally shaped and conditioned, because 'the practices within them are interwoven with social expectation, routines and norms' (Shortt, 2015, p. 634). Whereas dwelling in liminal spaces, as opposed to passing through them, may allow for new possibilities and a sense of freedom on how to behave. We might think of dwelling with colleagues around office areas that are not designed for hanging out. For instance, running into somebody in a hallway and spontaneously staying there, or smokers gathering around an exit. These experiences may open up different types of conversations and allow a freedom to act outside one's conditioned behaviour according to social codes (Shortt, 2015).

LIMBO is situated in a creative surrounding that can be seen as a liminal space (Shortt, 2015). Its homebase is an art institution that holds changing exhibitions and periods of in-betweenness when new exhibitions are constructed. The experience of some members in LIMBO is one of being temporarily outside their routine of going to spaces that are goal-oriented and filled with expectations, such as taking Dutch classes or meetings with case managers. Meetings with case managers, lawyers or the IND (Dutch Immigration and Naturalization Service) are examples of obligations that are paired with social codes that are culturally conditioned.

Many queer refugees in LIMBO were conditioned to behave 'properly' in their former contexts because of the risk of violence or prosecution. In their new context, when undergoing the asylum process, some are at risk of the IND not believing their claim for asylum based on gender or sexuality (Renkens, 2018; Spijkerboer, 2013). Many LIMBO members were denied their asylum claim because the IND did not believe them. In most cases, this was due to not giving enough detail. While in their former life they were forced to hide or suppress their sexuality, they now have to be open and somehow prove they are queer. FH heard many life story details in private conversations, which are often extremely violent and traumatic, full of examples of torture and other brutalities. These stories are difficult to share, particularly with institutions that approach them with scepticism. In LIMBO almost all participants mentioned feeling relief during the workshops.

> I was feeling relieved and started opening up freely, because LIMBO is a safe place and family that makes me feel myself. (Elea)
>
> Before it was so difficult to say I'm gay, but now I feel confident to share I'm gay. I'm feeling relief through these kinds of workshops. (Dex)

Participants mention this feeling of relief in connection to feeling safe with people who have similar experiences as well as the playful and fun character of the workshops, whose main goal is to hold space for whatever they want to share. However, participants also mentioned that if LIMBO had been situated in a more institutional setting, like a classroom, it could have been more anxiety-inducing because they would have felt more out of place. Hence, in line with Shortt (2015), we argue that dwelling in a creative liminal space makes it easier to shake off internalized expectations of how to behave, which adds to momentarily feeling liberated from such dominant norms and expectations. This relief then creates space to tune into oneself, which is a form of (re/de)conditioning when practised and repeated over the longer term.

11.2.3 Transformative Potential of Creativity

The focus on epistemic multiplicity and justice brought us to strive for three levels of transformation: individual, group and public, as described by Turner (1979). At the individual level, the aim is to encourage tuning into oneself without feeling any pressure to arrive somewhere or produce something. Rather than trying to make something that others may or may not believe as true, categorize as good or bad, valid or invalid, our focus is on safely tuning into oneself, fostering reflexivity. Reflections by LIMBO members are often profound and reveal relationships with their own and other's ideas, values, histories and dreams. Over time, in sharing and reflecting about their creative works, we have witnessed changes.

> I get to learn every day. Every time I come, I learned something new. Yeah, the other time I learned how to write a poem. In the other day, maybe to train my mind [...] I also wanted to say thank you for having us have this opportunity and make my handwriting, my painting on that wall. I left Uganda. And now to enjoy this whole experience, it helps me with my mind. I am struggling to put my mind to keep fighting. I don't know why I'm fighting but it has helped me to know more about myself. It's like I can talk, I can just use my mind and your paper, without even speaking what's actually going through my mind. I'm actually a very quiet person, but it has helped me to open up. I actually love you guys. Yeah, it has been an amazing experience for me. (Jenny)

Jenny refers to constantly learning something different, apparently safely stepping outside her comfort zone. She mentions not knowing why she is fighting but learning more about herself in LIMBO. She appreciates 'talking' without speaking, referring to art being an intuitive and liberating language, because creative practices make it safer and easier to express and construct notions about oneself in relation to the other and to their experiences in the new society.

At a group level, the mutual learning experience is central to LIMBO. By sharing stories and art in a group setting, participants recognize the similarities and differences, which creates bonds. Everyone shares a refugee experience; some are currently going through the asylum procedure, while others did so up to eight years ago. With regard to queerness, experiences are often more implicitly shared and negotiated. For example, how a person does not adhere to gender norms in the way they dress. Although the workshop facilitators almost never explicitly work with themes of gender or sexuality, queerness is negotiated simply through sharing a space with various queer displaced bodies. It fuels feelings of kinship, which became evident in the fact that most participants refer to LIMBO as family.

> LIMBO is a family, we all come together, there are no insecurities when I'm in LIMBO. I know that I'm surrounded by people who love me, who have the same sexuality with me. (Mira)

Over time, we observed members expressing themselves more comfortably and confidently, which is also recognized and acknowledged by others in the group and further encouraged, strengthening the sense of collectivity and kinship.

The transformative potential at the public level is enabled through LIMBOs public presentations, events, co-created booklets and articles. These outings all serve as a platform to share their stories and art and are forms of public reflexivity (Turner, 1979). Outside audiences are invited to engage and reflect on the creative output presented. Queerness plays a role because it is often seen by LIMBO members as being inherently activist in terms of breaking or blurring dominant social (gender, sexual, relationship) norms (Jagose, 1996).

> I am a queer from Uganda and it's my first time to open up for everyone [tearing up]. My first time and I'm very happy in LIMBO. LIMBO has helped me to get confidence, and it has helped me find a family which is the same as me. I'm happy, I'm very, very grateful for LIMBO. I will live with all of it in my life, thank you. (Elea)

Elea stated this at a public LIMBO event in which she found confidence to speak in front of more than 200 people. Public reflexivity took place because visitors referred to this moment as 'powerful in its vulnerability'. Elea experienced that her voice matters for others and for herself. She also refers to 'family which is the same as me', referring to the developed kinship in LIMBO, feeling connected by recognizing similarities in difference.

11.3 Holding Space

Aminata Cairo (2021) argues that it is important to hold space for one's full range of human emotion and experience, without immediately trying to (re)solve them. As discussed above, we argue that creative practices are a useful tool to hold space for the full spectrum of experiences without striving for, or arriving at, results. In the context of facilitating groups, Heather Plett defines holding space as 'what we do when we walk alongside a person or group on a journey through liminal space. We do this without making them feel inadequate, without trying to fix them, and without trying to impact the outcome. We open our hearts, offer unconditional support, and let go of judgment and control' (Plett, 2020, p. 18).

For Plett (2020), holding space is about transformation to witness and guide from a place of humility, for which a safer environment needs to be created. One metaphor for a liminal journey is transforming from being a caterpillar into becoming a butterfly, letting go of one identity to become another (Plett, 2020). She goes on to say that the journey often feels like a labyrinth, 'moving you close to the destination and then, with a simple turn, taking you far away from where you think you're meant to be' (Plett, 2020, p. 29). The journey is thus never linear but rather complex, messy, layered and challenging, including a full range of emotions from shock and grief to joy and relief.

In LIMBO, we hold space for queer refugee narratives through creative practices. Plett (2020) metaphorically describes those on a liminal journey as crystals with sharp edges. Because of the sharp edges and variety of shapes and sizes, she suggests the idea of a bowl or container with a soft protective inside layer, a safer space for exploration. There needs to be a sense of comfort and protection so that our sharp edges do not hurt ourselves or others along the way. In LIMBO, we co-create this safer space by constantly practising consent, and listing intentions for how we want to be together that specific day. The safer space is based on three principles: autonomy, safety to fail and simple instructions (Plett, 2020). Plett describes additional principles in depth, but in the context of co-creation with queer refugees, these three are the most relevant.

11.3.1 Autonomy

The first principle is autonomy, which is defined as 'the quality or state of being self-governing; self-directing freedom and especially moral independence' (Plett, 2020, p. 52). We encourage everyone to trust their own intuition and wisdom and do so by focusing on consent. In addition to an information and consent letter, we practise consent continuously. At the start of each workshop, we remind everyone to check in with themselves and others to feel whether it feels good to do something together or share one's story. We encourage everyone to make their own decisions, whether to share a story, whether to keep it light and easy, heavy and deeply personal or anything in between. In this way we embrace diversity by making different decisions and sharing different experiences.

The first LIMBO workshop was about consent as an ongoing practice, checking in with oneself and the others, that everyone (enthusiastically) commits to whatever they are consenting to. We imagined that a consent workshop would contribute to feelings of safety and autonomy. However, there are university ethics and guidelines regarding consent forms that research participants are obligated to sign. Such forms usually include content, activities such as interviews, agreements on using pseudonyms, authorship and potential risks. Although this can have an important function in protecting participants' rights, this way of obtaining consent is not always appropriate for all research contexts. In LIMBO, which is an ongoing journey where participants learn to hold space for each other as part of collective research, signing consent forms gradually seemed to be out of context. Often the issues cited in those documents are rather abstract, and people may change their mind when experiences become concrete. How can someone consent to something when they don't yet know what they are consenting to? Moreover, we learned that signing formal documents triggered anxiety, memories of bad experiences and difficulty understanding Dutch bureaucracy. Signing documents can thus be scary or complicated for refugees, especially for undocumented migrants. We hoped that, through a playful workshop, we would make a conversation about consent more meaningful and fun.

Workshop facilitator Maha Youssef addressed issues of consent by practising with rope. One person would tie another person's wrists or ankles so that they would physically and emotionally feel constraint. This constraint became a measuring tool for the person to check internally and read their levels of (dis)comfort. It was stressed that consent can be evoked for no apparent reason, or for reasons one is unwilling to share. In this way, the responsibility is shared for deciding whether something feels (mutually) good. Youssef borrowed rope play exercises from the Bondage, Discipline (or Domination), Sadism (or Submission), Masochism (BDSM) community, where there is extensive knowledge about consent in practice (see, for instance Bennett, 2018; Fanghanel, 2020).

While consent in LIMBO seldom concerns physical touch, it is mainly directed to decisions about sharing stories and creative works. These (art) narratives are shared inside the safer space among ourselves and sometimes in public events. For some, this step from private to public depends on the time and space whether they choose to share stories and different layers of meaning connected to the work. As part of how we approach consent as an ongoing practice, as opposed to completing a single consent form at the start, we continuously emphasize that everyone should share if, where, or when, they are ready. All choices are encouraged, although the authors believe there is power in sharing, because it may be inspiring, helpful or courageous. It might connect people, trigger reflection, introspection or bring clarity for the person and/or the receiver.

11.3.2 Safety to Fail

The second principle concerns a focus on learning and growing. In LIMBO this translates into a continuous effort to support everyone to feel safe enough to fail. We encourage taking risks and learn from 'mistakes' without fear, judgement or shame. In terms of art practices, the term 'mistakes' does not really apply. Any drawing or text can be silly or serious, technically 'good' or 'bad'. One does not have to be good at art or sharing stories. It's about learning and discovering together and finding ways to tell our stories. In this sense, group members are all researchers because they engage in some form of internal, external or relational inquiry. This relates to what we discussed about creativity; through playfulness and dwelling in between, as opposed to striving for results, implies that making mistakes, failing, or stumbling is fine or even encouraged. Because LIMBO members' daily challenges are often intense and overwhelming, we remind ourselves in the workshops to laugh, act in silly ways, dance and enjoy. In 'The Queer Art of Failure' (Halberstam, 2011), it is argued that living up to social norms is overrated and that failure should be celebrated because it points out taken-for-granted forms of exclusion. If the idea is to make the revolution irresistible, queer people, particularly those who are displaced and/or of colour, must (re)learn to joyously embrace failure of living up to social (gender, beauty, relationship, heterosexual, institutional) norms.

In practising being safe to fail, we work with the term 'intentions' for how we choose to be together in that moment. In contrast to terms as rules or guidelines, the term 'intentions' is more active as something for which to strive. Rules are more static because they can be broken or bent. Guidelines may limit one's freedom to act because they suggest certain forms of behaviour. Using intentions is thus both more liberating and also makes it safer to fail. For example, one might unintentionally 'break the rule' of addressing someone with their proper pronouns. Rules that are established together, and unintentionally breaking them, may then trigger feelings of shame, or feelings of (sometimes constant) failure. In using intentions, 'mistakes' are less likely to be seen as problematic, and one can move on more quickly without feeling shame or failure. Moreover, queer activists are often actively breaking rules in terms of dominant norms or standards. Collectively creating intentions is thus a more liberating and inclusive practice because it does not suggest behaviour.

In LIMBO, intentions are co-created and constantly developed and revisited in every meeting. They are collectively listed on a large piece of paper and placed in a prominent spot. In every gathering the list is checked and revised in case anyone wants to add or change something. This is particularly salient when current affairs affect (individuals in) the group. For example, when something occurs in their country of origin—such as war, revolution, natural disaster—they might feel worried about friends and families, feel vulnerable, overwhelmed or easily triggered. The collective remembering, revisiting and listing of intentions becomes an appropriate moment to tune in with oneself and share specific desires or needs—for example, 'don't ask me anything today', 'I just want to sit in a corner' or 'hugs are welcome'.

11.3.3 Simple Instructions

The third principle does not require much clarification. It concerns keeping it simple in terms of tasks and information. A liminal journey can be overwhelming, given that norms, values and ways of being in the world no longer apply. In addition, most participants struggle with stress, depressive thoughts or other mental health issues. Therefore, we have experienced the importance of keeping information to a minimum to create space for a mutual learning experience and limit feelings of being overwhelmed. This is another reason why the focus is not on learning artistic techniques or skills, but rather on reflecting on one's context through creativity.

11.3.4 Belonging

Holding space is important in the context of LIMBO because the aim is to provide a safer, supportive and non-judgemental environment for refugees, who are often approached as a weak group in need of help. This dominant image of refugees as lacking inhibits their chances of participating and belonging in society, despite various policies aimed at improving their integration (Ghorashi, 2020). According to Brené Brown (2017), belonging is the opposite of fitting in. Therefore, LIMBO offers an alternative approach, rather focusing on belonging in enabling refugees to express themselves freely.

> And when I found you [FH] and found LIMBO, I start giving my heart, I start giving my energy, sharing myself, being vulnerable. I open myself in the exhibition to show them how SAFE I fucking feel. [...] I found people who also I feel comfortable with to speak up and to be safe to express myself. Maybe my ideas about it [are] very big for some people, but I don't want you to understand. I want to be able to talk. I want to be able to express. I want to be! And if I cannot speak it, I can draw it, I can dance it, and create it for you. Just give me the space, give me the time [laughs]. (Neo)

Neo's quote highlights the transformative impact of LIMBO in providing a space where they feel comfortable and safe to share their thoughts and ideas. In this space, they can be vulnerable and express themself without judgement or the need for others to fully understand them. LIMBO gives them the opportunity to be heard, whether through words, art or dance. This sense of freedom is crucial for Neo, as it allows them to break free from the limiting perspective of being seen as lacking and instead to embrace their own agency and creativity.

The significance of creating a safer space for queer refugees became more evident during a specific workshop focused on holding space in the context of clubbing. The workshop took place at the start of a semi-public clubbing LIMBO event that was organized by queer refugee artists for approximately 300 invited guests. To ensure the safety of all attendees, 25 volunteers, mostly LIMBO members, were asked to hold space in short shifts. This preparation included a workshop led by two experienced individuals in organizing queer safer clubbing events (for more context on these events, listen to an interview with one of the organizers here: Van Driel, 2020). The participants were asked to share memories of when they felt truly safe. It was remarkable that a majority of LIMBO members expressed only feeling safe within LIMBO, highlighting the importance of the space in providing a sense of belonging and safety for queer refugees.

11.3.5 Challenges of Holding Space

Holding space for guests at semi-public events is of different nature to doing so in the LIMBO workshops. This is because in public events it is not entirely possible to work 'without trying to impact the outcome [...], and let[ting] go of judgment and control' (Plett, 2020, p. 18). The reason is that events are end results in which safety needs to be organized for both the public and the hosts. This makes it important to have clear judgement while trying to control a situation into an outcome that is positive and safe.

In contrast to events, during the workshops it is possible to dwell in between and let go of controlling an outcome. In an ideal situation, the responsibilities of holding space are shared within the group. LIMBO members' shared queer refugee experience and similar challenges faced in the Dutch context, potentially make the group members themselves ideal to hold space for each other's humanity in all its complexities, ambiguities and challenges. When done successfully, diversity and inclusion are practised by recognizing and supporting similarity and difference at the same time. It has the potential to counter victimizing narratives (Ghorashi, 2018), which is an important aim of research collaborations and co-creation with refugees generally.

However, holding space is intense emotional labour. It becomes increasingly difficult when the narratives are complex, painful, full of trauma and triggers. There are constant challenges refugees face in asylum centres, shelters and their inability to be with their family, finding suitable jobs or places where they feel safe. Under such conditions of ongoing crises, mistakes in holding space are bound to be made even if they're done with the best intentions. In contrast to creativity where mistakes are encouraged, mistakes in holding space can break trust from (individuals in) the group that has been built over time. In view of this, space holders should have proper training, time and resources. Unfortunately, LIMBO does not have the capacity, skills or budget to provide such training, time or resources.

Moreover, if one is holding space for queer refugee narratives, the sharers' needs, emotions and transformation need to be prioritized. This implies that, in addition to proper training, the space holders' needs have to be met elsewhere (Plett, 2020). In the case of LIMBO members as space holders, they often have no support system outside LIMBO—so many members see LIMBO as their family and sometimes jokingly call the organizers 'mom'. The concept of chosen family is not uncommon in queer culture. Particularly in queer ballroom culture, there are 'houses' in which the 'legendary children' are led by the 'mother' of that house (Bailey, 2011). In research methodologies where care is central, especially in the context of queer culture, it is important to remember to care, while keeping a healthy distance and not taking on a 'mother role'. Excessive intimacy can hinder the ability to properly hold space. Plett (2020) argues that holding space is not recommended 'when close is too close', simply because the greater attachment makes it harder 'to let go of the outcome – because the outcome has a direct impact on us' (p. 103). This is why healthy personal, as well as structural (space and time), boundaries are important, because one needs to protect oneself and the group while holding space—particularly because '[m]any people, especially those in crisis mode, are unaware of how much emotional labor they ask of others' (Plett, 2020, p. 87).

11.4 De/Recentring

Both decentring and recentring appeared in LIMBO as important conditions for co-creation, as they involve leadership in terms of democratic decision-making and sharing responsibilities. Co-creation entails the ability of those leading the research process (who are generally the researchers, but also applies to collaborators or co-creators leading parts of the process) to move in and outside the centre. Decentring involves temporarily letting go of the leading position to become part of the group (Braidotti, 2018). Recentring involves making critical decisions when clear direction is needed. For example, in case of conflicting desires or needs when there are too many captains trying to steer the ship in different directions. By decentring and recentring, the researcher can actively navigate diverse values and expectations in co-creation projects, which may differ greatly within a group as well as between institutions and communities.

Decentring is described in the literature as essential to 'intimate scholarship'. Transformative engaged scholarship can be seen as a form of intimate scholarship, when the researcher is part of the study and therefore also being researched (Hamilton & Pinnegar, 2014). The 'researcher being researched' does not necessarily mean by someone else but could also be in the form of rigorous self-reflexivity (Caretta & Riaño, 2016). By allowing the group to take the lead, the researcher can enable the co-creation process to be truly collaborative. Overall, decentring enables the group to take ownership of their own stories and experiences and ensures that the co-creation process is respectful, inclusive and reflective of the group's needs and desires. It allows for a more open and transformative co-creation process that can lead to deeper understanding of and support for the research itself.

In contrast, recentring is important because it involves stepping in as a leader and making decisions when necessary. Additionally, recentring helps make academic knowledge and experience more visible when embracing epistemic multiplicity. While decentring is about stepping back and allowing the group to lead, the notion of recentring recognizes there are times when the researcher's expertise and perspective are valuable in guiding the co-creation process. While the researcher should prioritize the voices and experiences of the participating co-creators, they must also ensure that the general direction and main purpose of the project stays on track. Furthermore, recentring allows the researcher to address potential risky situations or conflicts that arise. The researcher may have to intervene in conflict and guide the group towards resolutions that align with the project's overall goals. Rather than maintaining the complete control over the co-creative research, decentring-recentring is a balanced and dynamic process that requires the researcher to recognize when their expertise and leading role are needed. It is a delicate balance between stepping back and stepping in order to support and facilitate a collaborative and transformative co-creation experience.

11.4.1 What Are the Challenges of Decentring-Recentring?

In the context of LIMBO, this careful navigation of decentring-recentring proved to be challenging. In the first period (LIMBO part I in Fig. 11.1), FH drove the organizing with input from community members with whom they had previously worked. The experience of Trans-Clash was still fresh, which made balancing community needs and theory and former research insights relatively easy. Seven community-led workshops were organized, and the project was supposed to finish with a final presentation in the form of an exhibition at the art institution. The project generated a lot of enthusiasm. Also, the snowball effect worked because the group had grown from six participants in the first workshop to a group of 20 at the public presentation that attracted more than 200 visitors. Given the group's enthusiasm and the feelings of hope, safety and joy it generated, we felt that we needed to continue.

> I felt safe enough to share this with this group and I felt there was the sense of hope being built, and from the conversation that I had with a

lot of people they really did want this initiative to continue because it represents hope for them. (Musa)

Almost all members asked for LIMBO to continue and expressed concern that it might end. We also received an abundance of positive reactions from the visitors at the public event and saw how LIMBO had a positive transformative influence. As researchers and the art institution, we were equally excited and therefore decided to continue. In terms of research, it meant that we could see what is entailed in making such a project durational and sustainable.

However, in continuing LIMBO (part II in Fig. 11.1), challenges concerning time and responsibilities arose along the way that needed immediate solutions. FH spent many hours co-organizing and co-facilitating workshops, which proved too much alongside other academic commitments. Given the emotional labour of holding space, we sought to shift this responsibility to people in the group. The idea was that the researcher would progressively fully decentre, so that LIMBO members could step in and take on these responsibilities. However, as LIMBO continued to expand to more than 50 participants and started to be recognized by the wider Dutch LGBTQIA+ community and refugee organizations, the requests for collaborations and responsibilities grew. This was received with enthusiasm by the organizers and group members. It fostered feelings of finally belonging in the Dutch context in ways that suit them. For some, the hope that LIMBO represented became a goal of making it into a permanent job—especially for those who had been struggling for many years to find work in the Netherlands and longed to move out of the uncertainty and challenges of living in their continuous liminal state. In LIMBO, they felt that their expertise of in-betweenness was finally recognized and that they could use this knowledge and creativity to reach a more stable situation, both financially and emotionally. While FH tried to manage expectations, they also didn't want to kill or put brakes on hopes and dreams.

In reality, FH failed to decentre. From their perspective, the initial focus had shifted from holding space for workshop participants to now also guiding and accommodating the needs and desires of organizers and workshop leaders. This was not only more work, but the work itself was also of a different nature. Holding space is not controlling the outcome (Plett, 2020), which is more in line with a task of a researcher, to observe and analyse without intervening. In contrast, organizing is meant to control an outcome in terms of content, scheduling, safety measures and so forth. We communicated boundaries in what we as organizers could do and what was beyond our capacity. However, in trying to keep up with the growing enthusiasm and requests for more events, the boundaries became blurred. We were spending increasingly more time listening to ideas for new initiatives and collaborations with other organizations. Although we kept explaining that we could not support 'outside LIMBO projects', we were continuing to give support in terms of listening and sometimes advising.

This was a gradual process, which made it difficult to start establishing clear boundaries. Given the pressure of growth expectations within the group, the organizers and FH in particular were slowly but steadily becoming overwhelmed with all this growth and enthusiasm. The stakes for the organizers also became higher and conflicting desires among LIMBO members affected our core task of holding space together as a team. This meant that in situations of conflict between LIMBO members, FH needed to be involved in resolution or mediation. The increase in communication and guidance, while trying to hold space for everyone's ideas and dreams, became too much and FH was forced to temporarily step back due to symptoms of burnout.

During the necessary break for healing and reflection, a few things became evident. For LIMBO part III (Fig. 11.1), FH had to recentre and establish clear boundaries, particularly in terms of time and space. We realized that for the project to be sustainable, we could only hold space during contained meetings. We had to shift back to the original idea of simply holding space for queer refugee narratives, without trying to guide the growth of community organizers. In addition, rather than different people organizing different workshops, which was the case in the first two periods, we found two experienced people with more 'relational distance' to facilitate all the workshops for the entire last period. These facilitators have a (forced) migration background, but are not relationally entangled with the group. Our aim thus shifted from co-organizing and co-holding space, to more 'experienced outsiders' holding space. We believe that this is better because of the greater emotional distance, which makes it easier to 'let go of controlling an outcome' (Plett, 2020), in an attempt to remove (result-oriented) stress.

Because FH was forced to decide on this re-evaluation and new direction and was thus not co-created with the group, it posed the question of horizontal versus hierarchical decision-making. If decisions are made in a top-down fashion, does it still align with a democratic approach where the group's input and voices are considered? The answer to this question relates to the context and general aim of the project. If its aim exceeds holding space for narratives to co-organizing a long-term project, then the necessary resources and support need to be in place. In fact, LIMBO is inspired by Phoebe Kisubi Mbasalaki's research in South Africa as part of the Global Grace project (GlobalGRACE, 2018), where there was extensive support in terms of therapists, conflict mediators and so on. In LIMBO, we realized later the need to keep it smaller and simpler because we had been seduced by the enthusiasm and growth. Fortunately, the group expressed a lot of care and understanding of this decision to go back to the core essence, as most members desired first and foremost a space to express their narratives safely and creatively.

11.5 Conclusion

This chapter has delved into the potential of transformative engaged scholarship in co-creation with queer refugees through the lens of the LIMBO project. Our exploration was guided by the principles of transformative and critical academic work, which seek to challenge established power dynamics, biases and discriminatory practices while fostering inclusivity and equity. We have shown how this requires an act of decentring to create the necessary space for co-creation in knowledges that are often excluded in the process of academic knowledge production. Even when voices of disadvantaged communities are included in research there is often a lack of meaningful engagement with such perspectives as integral part of knowledge production. This means that a selective inclusion of those voices do not challenge the normalized assumptions inherent in much research. This is partly to do with the fact that acts of decentring make scholars vulnerable in the process of knowledge production. We also aimed to show how such vulnerability can become a strength when this form of scholarship contributes to transformation at the level of individual co-creators and of the group as a whole. Transformative academic engagement goes beyond mere description and documentation in recognizing agency and creativity, while following the perspectives of marginalized communities as leaders in the process. Within the context of this co-creative project, three pivotal conditions emerged that underpinned this transformative process: creativity, holding space and the dynamic interplay of de/recentring.

Creativity in LIMBO is seen as the possibility of unsettling and liberating oneself from normalized positioning. The liminal space of LIMBO allows individuals to temporarily suspend social expectations and explore their identities within the new (Dutch) context. The playful and creative atmosphere of LIMBO provides a sense of relief from daily stress and struggles, promoting joy and tuning into what feels good. Creativity allows individuals to move away from the pressure to fit into social norms, in reference to queerness and refugeeness. The transformative potential of LIMBO is evident at the individual, group and public levels. At the individual level, LIMBO encourages tuning into oneself without the pressure to produce specific outcomes. At a group level, the mutual learning experience fosters bonds and a sense of collectivity and kinship among the participants. At the public level, LIMBO's public presentations and co-created publications serve as platforms to share stories and art, enabling public reflexivity.

Holding space is a crucial element in the context of LIMBO as it provides a safer and supportive environment for queer refugees. This non-judgemental space allows them to freely express themselves without feeling inadequate. However, holding space also involves intense emotional labour, especially when narratives are complex and traumatic. It is essential to prioritize the needs and emotions of the sharers and provide proper training and/or support for the space holders.

Decentring and recentring are other important aspects in the co-creation process, involving stepping in and back as leaders when needed and making the value of academic knowledge more visible. While holding space is aligned with decentring as it encourages autonomy and decision-making, recentring requires making critical decisions that may challenge certain values or may not fully align with the ideas of (everyone in) the group. The challenges of recentring and decentring in LIMBO are thus multifaceted and require careful navigation. In LIMBO, we were seduced by the enthusiasm, growth and visible transformations that occurred, but realized that we had to keep it simple and go back to the essential core of holding space for queer refugee narratives.

Overall, transformative engaged scholarship in co-creation with queer refugees in LIMBO offers a space for the agency of queer refugees, enabling them to challenge social norms and create their own narratives on their own terms. At the same time, it is important to recognize the challenges and complexities in maintaining these conditions and to continuously adapt and evolve the co-creation process to best serve the needs and desires of the participants, while honouring the organizers' emotional, practical and structural capacity and boundaries.

Acknowledgements We would like to thank all LIMBO members for co-creating this project with creativity, trust and openness. We thank Framer Framed for co-organizing, hosting and facilitating this project. We are grateful for the financial support of the Dutch Research Council (NWO) [016.Vici.185.084] as well as Fonds Cultuurparticipatie for funding 'Through Art We Care', which LIMBO is part of.

References

Bailey, M. M. (2011). Gender/racial realness: Theorizing the gender system in ballroom culture. *Feminist Studies, 37*(2), 36–386. https://doi.org/10.1353/fem.2011.0016

Beech, N. (2011). Liminality and the practices of identity reconstruction. *Human Relations, 64*(2), 285–302. https://doi.org/10.1177/0018726710371235

Bennett, T. (2018). Unorthodox rules: The instructive potential of BDSM consent for law. *Journal of Positive Sexuality, 4*(1), 4–11.

Braidotti, R. (2018). Affirmative ethics, posthuman subjectivity, and intimate scholarship: A conversation with Rosi Braidotti. In K. Strom, T. Mills, & A. Ovens (Eds.), *Decentering the researcher in intimate scholarship: Critical posthuman methodological perspectives in education* (pp. 179–188). Emerald Publishing Ltd.

Brown, A. M. (2019). *Pleasure activism*. AK Press.

Brown, B. (2017). *Braving the wilderness: The quest for true belonging and the courage to stand alone* (1st edn.). Random House.

Cairo, A. (2021). *Holding space: A storytelling approach to trampling diversity and inclusion*. Aminata Cairo Consultancy.

Caretta, M. A., & Riaño, Y. (2016). Feminist participatory methodologies in geography: Creating spaces of inclusion. *Qualitative Research, 16*(3), 258–266. https://doi.org/10.1177/1468794116629575

Dewhurst, M. (2010). An inevitable question: Exploring the defining features of social justice art education. *Art Education, 63*(5), 6–13.

Edmondson, A. (2018). *The fearless organization: Creating psychological safety in the workplace for learning, innovation, and growth*. Harvard University Press.

Elferink, J., & Emmen, M. (2017). *LHBT-asielzoeker of –vluchteling zijn in Nederland. Ondersteuning en acceptatie van LHBT-asielzoekers – en vluchtelingen: behoeften, ervaringen en good practises*. Kennisplatform Integratie & Samenleving.

Fanghanel, A. (2020). Asking for it: BDSM sexual practice and the trouble of consent. *Sexualities, 23*(3), 269–286.

Ghorashi, H. (2018). Decolonizing the Islamic other: The changed conditions of critical thinking. In M. F. Weiner & A. Carmona Báez (Eds.), *Smash the pillars: Decoloniality and the imaginary of color in the Dutch kingdom* (pp. 185–197). Lexington Books.

Ghorashi, H. (2020). Failed promise of equality: Iranian women's integration in the Netherlands. *International Migration, 59*(4), 88–104. https://doi.org/10.1111/imig.12774

GlobalGRACE. (2018). *Global gender and cultures of equality*. https://www.globalgrace

Halberstam, J. (2011). The queer art of failure. In *The queer art of failure*. Duke University Press.

Hamilton, M. L., & Pinnegar, S. (2014). Intimate scholarship in research: An example from self-study of teaching and teacher education practices methodology. *LEARNing Landscapes, 8*(1), 153–171. https://doi.org/10.36510/learnland.v8i1.680

Holle, F., Rast, M. C., & Ghorashi, H. (2021). Exilic (Art) narratives of queer refugees challenging dominant hegemonies. *Frontiers in Sociology, 6*, article number 641630.

Jagose, A. (1996). *Queer theory: An introduction*. NYU Press.

Jeanes, E. L. (2006). Resisting creativity, creating the new: A Deleuzian perspective on creativity. *Creativity and Innovation Management, 15*(2), 127–134. https://doi.org/10.1111/j.1467-8691.2006.00379.x

Karimi, A. (2020). Limits of social capital for refugee integration: The case of gay Iranian male refugees' integration in Canada. *International Migration, 58*(5), 87–102. https://doi.org/10.1111/imig.12691

Manjikian, L. (2014). *Refugee narratives in Montreal: Negotiating everyday social exclusion and inclusion* (Doctoral dissertation). McGill University.

Medina, J. (2013). *The epistemology of resistance: Gender and racial oppression, epistemic injustice, and resistant imaginations*. Oxford University Press.

Meekosha, H., Shuttleworth, R., & Soldatic, K. (2013). Disability and critical sociology: Expanding the boundaries of critical social inquiry. *Critical Sociology, 39*(3), 319–323. https://doi.org/10.1177/0896920512471220

Oliveira, E. (2019). The personal is political: A feminist reflection on a journey into participatory arts-based research with sex worker migrants in South Africa. *Gender & Development, 27*, 523–540. https://doi.org/10.1080/13552074.2019.1664047

O'Neill, M., Giddens, S., Breatnach, P., Bagley, C., Bourne, D., & Judge, T. (2002). Renewed methodologies for social research: Ethno-mimesis as performative praxis. *The Sociological Review, 50*(1), 69–88. https://doi.org/10.1111/1467-954X.00355

Plett, H. (2020). *The art of holding space: A practice of love, liberation, and leadership*. Page Two Books.

Renkens, J. W. M. (2018). 'Kunt u dat moment eens beschrijven?': Een analyse van de interviewmethode van de IND in lhbti-zaken.

Said, E. W. (1994). *Representations of the intellectual: The 1993 Reith Lectures*. Vintage.

Shortt, H. (2015). Liminality, space and the importance of 'transitory dwelling places' at work. *Human Relations, 68*(4), 633–658. https://doi.org/10.1177/0018726714536938

Spijkerboer, T. (Ed.). (2013). *Fleeing homophobia: Sexual orientation, gender identity and asylum*. Routledge.

Tracy, S. J. (2013). *Qualitative research methods: Collecting evidence, crafting analysis, communicating impact*. Wiley.

Turner, V. (1969). Liminality and communitas. *The Ritual Process: Structure and Anti-Structure, 94*(113), 30–125.

Turner, V. (1979). Frame, flow and reflection: Ritual and drama as public liminality. *Japanese Journal of Religious Studies*, 465–499. https://www.jstor.org/stable/30233219

Van Driel, N. (2020). *Interview. De Zingevers. De Correspondent*. https://decorrespondent.nl/10876/luisteren-dit-zijn-de-professionals-van-de-zingeving/3247793904256-62a27346

Van Gennep, A. [1909] (2019). *The rites of passage*. University of Chicago Press.

Young, I. M. (2001). Activist challenges to deliberative democracy. *Political Theory, 29*(5), 670–690. https://doi.org/10.1177/0090591701029005004

Young, I. M. (2007). Structural injustice and the politics of difference. In K. A. Appiah, S. Benhabib et al. (Eds), *Justice, governance, cosmopolitanism, and the politics of difference: reconfigurations in a transnational world*. Distinguished W.E.B. Du Bois Lectures 2004/2005 (pp. 79–116). Humboldt University.

Zuiderent-Jerak, T. (2015). *Situated intervention: Sociological experiments in health care*. MIT Press.

CHAPTER 12

Dealing with the Pitfalls of Inclusion and Diversity: How to Involve Citizens Experiencing Distance from and Distrust of Science and Governance

Esther de Weger, Aafke Fraaije, Jaron Harambam, and Willemine Willems

12.1 Introduction

Complex and global emergencies such as the COVID-19 pandemic and the climate crisis have made scientific knowledge more relevant and at the same time more contested. Such contestation is amplified by widespread

E. de Weger · W. Willems
Vrije Universiteit Amsterdam, Amsterdam, The Netherlands
e-mail: w.l.willems@vu.nl

E. de Weger (✉)
Verwey Jonker Instituut, Utrecht, Netherlands
e-mail: edeweger@verwey-jonker.nl

A. Fraaije
Technical University Delft, Delft, The Netherlands
e-mail: a.fraaije@tudelft.nl

B. J. Regeer et al. (eds.), *Transdisciplinarity for Transformation*,
https://doi.org/10.1007/978-3-031-60974-9_12

misinformation and a growing scepticism among certain groups in society towards science and policies that seem to be based solely on scientific research (Davies, 2022).

Many have argued that the complexity of such emergencies and the contestation regarding scientifically developed solutions and policies require transdisciplinary research (TDR). With such an approach diverse perspectives and forms of knowledge from a range of different stakeholders are taken into account and integrated, including those of citizens and communities (Kok et al., 2021). The rationale for such participation often includes a substantive, instrumental and normative component, meaning that such engagement is thought to improve the quality and desirability of science and technology, increase the likelihood of the public acceptance of future technologies, and to meet citizens' democratic right to be involved in matters of public concern. Furthermore, (co)researching is thought to have broader methodological benefits—such as greater validity and more generalizable findings (Evans et al., 2014; Fung, 2015; Hueske et al., 2023; Turbe et al., 2019).

In this chapter we present three major pitfalls and potential remedies to help TDR involve citizens (and other stakeholders) in a successful and meaningful way. The examples are based on three studies that investigated the ways in which citizens' involvement, and the concepts of inclusivity and diversity, were approached in science, technology and policy development. We aim to provide insights for transdisciplinary researchers who seek to make their research, and the transitions being investigated and supported, more inclusive by developing spaces and approaches that allow for a wide variety of views and lived experiences to be heard, shared, understood and ultimately acted upon.

Whom to include and how to do so in an inclusive manner in processes of knowledge co-creation is a pressing question in TDR. The search for such answers has a long history in the fields of science communication and citizens' engagement in science, technology and policy, where practitioners and researchers have reflected on and published about successful and meaningful ways to involve a wide range of citizens as a means to

J. Harambam
University of Amsterdam, Amsterdam, The Netherlands
e-mail: j.harambam@uva.nl

improve research and policymaking. When communication and engagement are not sufficiently inclusive or diverse or cannot make good on their substantive, instrumental and normative promises, it damages the very trust or acceptance it seeks to build (de Weger et al., 2023; Lewis, 2014; Luluquisen & Petttis, 2014). Many studies over the years have pointed out that in practice it remains difficult to fulfil these promises. For example, some of the literature has consistently observed that there is an overrepresentation of white, middle-class and better-educated citizens in research and policymaking while people from less advantaged backgrounds are often underrepresented (Cyril et al., 2015; de Weger, 2022; Fraaije, 2023). Some authors have sought to explain this, in part, through the focus of (western) governments on improving efficiency and effectiveness. This literature suggests that this has made it difficult for organizations to alleviate social exclusion, which results in over-including and overproviding to those who are easiest to reach and assist and under-providing to more disenfranchised, underrepresented citizens who may be costlier to assist (Beresford, 2019; Cortis, 2012; Fletcher et al., 2016) (see Table 12.1 for a summary of the key theoretical concepts).

Notwithstanding the increasing insight into the benefits of inclusivity and diversity, and greater awareness of the enablers and barriers to achieving them, it remains difficult to attain them. Despite many public engagement initiatives and projects, scholars in the field are critical of what has been achieved to date (Irwin et al., 2013). Earlier studies have shown that inequitable social involvement stems from the fact that research is largely based on the interests, needs and norms of those who design and commission it (Beresford, 2019). Moreover, in research and policymaking there tends to be a focus on efficiency in terms of time and budgets, rather than investment in outreach and empowerment. Placing the emphasis on efficiency rather than on empowerment is detrimental to the influence of citizens, especially those from disadvantaged backgrounds. As a result, when disadvantaged groups are involved, they often report feeling shut out and unable to share their experiences and perceptions, their skills and insights (see, for instance, Cook & Kothari, 2001; de Weger, 2022). Although there is a growing literature about the pitfalls of engagement, it remains a pressing issue to understand how to overcome and remedy these, and to build a reflexive engagement environment that is inclusive to a wide range of citizens' experiences, perspectives and norms and values. First, the case studies provide examples of well-known pitfalls that can arise when trying to involve a wide range of citizens in an

Table 12.1 Key theoretical concepts

Term	*Definition*
Citizen engagement	This refers to the various ways in which citizens are involved in the planning, designing, governance and/or delivery of research, policymaking and technological developments. Such engagement takes many different forms and occurs at different levels—for example, at the consultation level, where citizens share information with organizations; the communication level, where citizens and organizations enter into dialogue; and participation, where citizens have an active role in the planning, design, decision-making and implementation (de Weger, 2022; O'Mara-Eves et al., 2013; Rowe & Frewer, 2005)
Inclusivity	While inclusivity has many different definitions it is broadly centred on the importance of involving a range of different citizens with different backgrounds, perspectives, experiences and needs. Processes underpinning 'inclusivity' can be broadly categorized according to (a) the openness of the process to participation; (b) the representativeness of those involved of the wider communities; (c) the transparency of the processes to everyone involved (Landemore, 2014)
Diversity	This term has been defined and conceptualized in many different ways and is dynamic and influenced by the context. Broadly, it refers to differences between (groups of) people. These distinctions are often based on different criteria or characteristics and usually refer to a 'meaningful characteristic', namely one that influences the person's identity and way of life (Van Ewijk, 2011)
Science communication	Science Communication (SciCom) encompasses a wide range of activities that aim to connect science and society and has various meanings. SciCom is not merely scientists talking about their work (to improve public lack of understanding), nor is it simply an offshoot of the field of 'communications'. Many define SciCom as being synonymous with public awareness of science (PAS), public understanding of science (PUS), or scientific literacy (SL). As such SciCom does not only pertain to the communication of scientific facts, but rather denotes ways to a more informed and open dialogue of the role of science in society at all levels and relating to all actors; and also includes the participation of a wider variety of actors (Bucchi & Trench, 2021; Burns et al., 2003)
Risk communication	Risk communication centres on the real-time exchange of information and community involvement between 'experts' and 'people facing the risk, hazard or threat' to their survival, health or economic or social wellbeing. The purpose is to enable citizens to make informed decisions and to involve them in decision-making processes in emergencies or emergency preparedness. Risk communication concerns the relationship between risk, science, and science communication and therefore informs difficult matters of dialogue, governance and decision-making processes (Irwin, 2021; World Health Organization, 2017)

effort to achieve inclusivity and diversity for research and policymaking. Secondly, they highlight relevant lessons and solutions to overcome these.

12.2 Assumption 1: Demographics Are an Indicator of Whether Diversity and Inclusivity Have Been Achieved

12.2.1 Background

Aiming to involve a diverse range of citizens, many organizations rely on demographics to inform their research and policymaking processes. Such organizations frame the lack of diversity or inclusivity according to standards they can easily set and measure. The idea is that inclusivity can be achieved if X share of each demographic is achieved—enough low-income households, enough LGBTQI+ citizens or enough citizens with a migrant background, for instance (de Weger et al., forthcoming; 2023).

12.2.2 Pitfall

Such thinking about inclusivity and diversity lacks reflexivity and cultural sensitivity both for TDR and for policymaking. It wrongly assumes that demographics are clear-cut indicators of who is included. By taking demographics as the main indicator for people's experiences of distrust of and distance from science and government, this spuriously generalizes about such demographic groups, and misses the more relevant features of diversity: different conceptions about (trust in) science and government (see Box 12.1). After all, these initiatives follow general statistics on 'trust in science' or stereotyped understandings of certain social groups, to argue that these—be they migrants, less-educated or conservative-voting people—need to be included precisely because they have less understanding of and trust in science and government. While this may be statistically true, it also obscures the diversity *within* demographic groups and deems it irrelevant. Is it even possible to regard these groups as uniform or are there significant differences between people from a similar demographic? Is diversity reached when demographic categories are taken as benchmarks, or are we including only the already trusting segments of such demographic groups?

Indeed, as Butter and Knight (2023) showed in studying citizens' responses to the COVID-19 pandemic, opposition towards science and government is by no means the monopoly of any single demographic category, but is present in *all* demographic groups. The anti-vaxx demonstrations across Europe and North America included various social groups, from highly educated urban hipsters influenced by alternative health and New Age spiritualities to less-educated people in rural areas inspired by nationalist ideologies, and many others besides (Harambam & Voss, 2023). Although opposition to the governing authorities may seem to unite rather different groups, their (distrustful) understanding of science and government differ markedly: the former challenge 'unnatural' biomedical interventions such as vaccinations, while the latter are opposed because 'elites' disregard 'common sense' and 'ordinary people'. Context matters too: although voters for populist political parties generally tend to have less trust in science, this increased sharply during the pandemic while other groups rallied around the flag (Bromme et al., 2022). Political scientists and psychologists have shown in various quantitative studies that political and religious beliefs rather than demographic categories determine trust in science (Rutjens et al., 2018), while the ethnographic studies undertaken by sociologists and anthropologists show that cultural worldviews are important predictors of whether a person aligns with science (Harambam, 2020; Sobo & Drazkiewicz, 2021).

In the Netherlands, studies undertaken during the COVID-19 pandemic found increasing distrust as communication and policymaking were focused on technical or scientific knowledge (e.g. Harambam, 2023; Prettner et al., 2023). Studies such as the one described in Box 12.1, and more recent studies about the Dutch vaccination campaign, found that by portraying anyone who poses critical questions as a 'conspiracy theorist' arguably creates a greater gap between citizens, science and government bodies. The point is that demographic categories may sometimes be good proxies for groups distrustful of science and government, but in principle they are not the best way to include the right people. What is relevant, after all, is people's social, cultural and political distance from science—their concrete ideas about and experiences of science, and not their supposed group affiliation. Moreover, using such general demographic categories as proxies and lumping all kinds of people together in supposedly uniform groups create various adverse and unintended consequences.

Box 12.1 Understanding 'Conspiracy Theories'

Conspiracy theories—explanations of social phenomena involving the covert actions of certain (powerful) people are—now found everywhere. Narratives about the real truth behind terrorist attacks (like the September 2001 attacks in the United States [9/11] and the November 2015 Paris attacks) or behind collective vaccination campaigns (e.g. COVID-19, Human Pappilomavirus [HPV]) feature widely in in western societies. For many, conspiracy theories have become a normalized idiom to account for what actually happens and have been popularized in films such as The Matrix, the Da Vinci Code, The X-Files. Although conspiracy theories have moved from the cultural margins to the centre, they remain little understood. Both among academics and beyond, the prevalent assumption is that conspiracy theories are paranoid, delusional, and irrational interpretations of reality—so whose who believe in them must, therefore, be similarly delusional. Academics tend to view conspiracy theories as paranoid politics: they are systemized, delusional fears of conspiracy and deceit, and they cast the world rather unproductively in an apocalyptic battle between absolute good and evil. Scholars warn of the social dangers were conspiracy theories and the related paranoia to proliferate: demonization, scapegoating, cultural conflicts, political extremism, radicalization, violence, terrorism. From this perspective, conspiracy theories threaten the health and functioning of democratic societies.

This pathological framing which dominates academic work on conspiracy theories is problematic for a number of reasons. First, it may be questioned deluded and paranoid conspiracy theories really are when there are so many cases of state-sponsored conspiracies that have taken place, such as Watergate, the CIA mind-control programme, the Iran–Contra Affair, the LIBOR scandal, and WikiLeaks. More importantly, the pathological framing does not help to understand why so many people are drawn to conspiracy theories. It would be better to explore such conspiratorial understandings without disqualifying or comparing them to specific moral or epistemological standards.

The ethnographic study of the Dutch conspiracy milieu aimed to see the world from their perspective. It shows what conspiracy culture empirically looks like—the ideas, motivations, practices, biographies, and products of people who inhabit this subculture—and how these are related to the mainstream. For this study, the author was immersed in the Dutch conspiracy environment for two years and became acquainted with a range of people, attended their social gatherings, built rapport and was recognized by insiders as a trustworthy person. The author read their posts,

articles and books, held in-depth interviews with them in safe settings, visited their performances and political activities, watched their documentaries and stayed in contact through social media. The study highlights that conspiracy theories are not merely ideas formulated in the abstract, but that they spur real-life action and incite both cooperation and conflict among their adherents. An important lesson this study conveys concerns the issue of diversity. In contrast to the dominant stereotype of conspiracy theorists as a petty-minded paranoia-espousing hatred and bigotry, the conspiracy (sub)culture harbours many kinds of people. Through the fieldwork, the author came into contact with young urban do-it-yourselfers, people in their 60s and 70s drawn to Eastern philosophies, technical pundits, libertarian vagabonds, etc. The popularity of conspiracy theories cuts across demographics and ideological conviction. There is no typical conspiracy theorist.
Source Harambam (2017)

12.2.3 Lessons and Solutions

From this understanding of the different social positions towards science and governments follows an important lesson when thinking about inclusion. Rather than using demographic categories as diversity benchmarks, it makes more sense to create substantive categories that are adapted to each case. Ideally, these categories are based on, or informed by, empirical research. So, when talking about distrust of science, it becomes possible to create better quantitative categories such as 'socio-cultural distance from science' or more qualitative categories such as 'different types of scientific distrust' than merely relying on demographic generalizations that may not even yield the intended diversity of viewpoints.

For example, for a current science communication research project on using citizen assemblies to discuss climate research issues (Climate Research in Dialogue—Vrije Universiteit Amsterdam (vu.nl)), we used a simple four-item questionnaire in the selection procedure to yield a diverse group of people based on their socio-cultural distance from science. This was designed to ensure that we included enough different people based on our most relevant criterion. Similarly, when probing anti-vaccination positions, it is possible to invite different demographic groups and hope for diversity, but it is also feasible to invite different groups based on their specific opposition to these biomedical interventions. The important question is: ***what is the most relevant diversity criterion for our project?*** This way of ensuring diversity and inclusion better aligns with the self-understandings and identifications of these social groups. Rather than regarding them as a uniform (and often inaccurate) group, people are acknowledged for their specific opinions, beliefs, worldviews and value systems. This not only mitigates the risk of reifying demographic categories, but also creates more trust as people (and their ideas) are taken seriously. Diversity and inclusion should be just as much about worldviews as demographics.

12.3 Assumption 2: People Can Be Included Equally If the Contexts, Conditions and Opportunities for Their Involvement Are the Same

12.3.1 Background

In an effort to ensure inclusivity and diversity many knowledge institutes, public-sector organizations and tech companies will try to create conditions and opportunities that are equal for all citizens. Their thinking is that if all conditions are the same for all citizens and everyone has the same opportunity to be involved, inclusivity must therefore follow. This allows organizations to base engagement approaches on their own views and priorities rather than on citizens' varying lived experiences, perceptions, norms and values.

12.3.2 Pitfall

While the aim to create equal contexts, conditions and opportunities for all citizens may seem centred on a desire for inclusivity, this approach in fact enables organizations to avoid reflecting on how different people have different interests, priorities and support needs. By developing and implementing 'one-size-fits-all' engagement approaches, organizations do not have to invest more time and resources to create contexts, conditions and opportunities that would enable a wide variety of citizens' lived experiences, interests and support needs to be taken into account. Scholars have shown that engagement approaches are often based on the systematic processes, structures and goals rather than on citizens' lived experiences. Creating equal contexts, conditions and opportunities for all citizens may thus inadvertently limit access to involvement and potentially widen inequalities between different groups of people (de Weger, 2022; Holley, 2016). When the contexts, conditions and opportunities, are still mostly based on organizational structures and goals rather than citizens' experiences, it often means that only a select group of citizens feel interested and empowered to engage, thus inadvertently excluding a wide range of citizens (de Weger, 2022; Holley, 2016).

This effect is clearly visible in one of the case studies described in Box 12.2. The study shows that equal contexts and conditions led to more vulnerable citizens to feel excluded, increasing their sense of not being heard. This case study examined whether, how, why and when low-income citizens wished to be involved with municipalities and health(care) organizations to contribute to improving the organizations' services and policies—in other words under which contexts, conditions and opportunities. It showed that citizens' personal situations and their perceptions and experiences of the services they used influenced their engagement preferences (i.e. the engagement opportunities and support they would have liked).

Low-income citizens wanted to contribute to the improvement of health and care services and all interviewees had shared solid ideas on how to improve services and policies. However, most did not want to participate because of systemic issues, such as bureaucratic and inaccessible processes and structures. Others indicated they would have liked to contribute but could not do so because of a lack of support for their own physical and/or mental health conditions. These negative experiences—such as organizational apathy, bureaucracy, fragmentation and never being asked about their experiences or whether they would like to be involved—show the importance of creating different types of engagement contexts, conditions and opportunities by tackling constraining systemic factors. Furthermore, organizations' lack of involvement and outreach towards low-income citizens mirrors how these same citizens had experienced the services they had used as impersonal and apathetic, bureaucratic and fragmented. Clearly, then, such 'equal contexts, conditions and opportunities' were not enough for low-income citizens to become involved, although they are often sufficient for the 'usual suspects' (often white, middle-class, retired citizens) (de Weger, 2022; Jansen & Denters, 2018).

For those who had expressed a wish to be involved there were, broadly speaking, three different categories: (a) in a practical or voluntary way;

(b) as a buddy; (c) as a lay expert. However, none of the interviewees had been enabled or asked by organizations to get involved in any way, thus underscoring the systemic issues that negatively affected their ability to be involved. This highlights that creating 'equal' engagement contexts, conditions and opportunities are more geared to the 'usual suspects', namely citizens who are already engaged because they can operate within this system-focused form of engagement. In doing so organizations exclude low-income citizens and prevent them from discussing their experiences, needs and the ways in which they want to be involved and the support they need to enable them to do so successfully and sustainably. Furthermore, it means that organizations are missing out on important ideas to improve their services, policies and organization (de Weger et al., 2022). By applying the same conditions and opportunities to all citizens equally, organizations forgo significant untapped potential of citizens who are more vulnerable and/or are more distrustful of knowledge institutes, government bodies and the tech industry.

Previous literature has suggested that the reason for this standardized approach to engagement stems from the fact that (western) governments' health and care policies have been focused on improving efficiency and effectiveness. This has made it more difficult for organizations to deploy resources to address social exclusion (Beresford, 2019; Cortis, 2012; Fletcher et al., 2016). This could be an important reason why they focus on inclusivity rather than diversity—and create 'equal' contexts, conditions and opportunities. It remains easier for them to include those easiest to reach and thus fail to involve more disenfranchised, underrepresented citizens who may be more costly to assist. This policy environment makes it more difficult for organizations to promote more diverse contexts, conditions and opportunities for citizens' engagement.

Box 12.2 Achieving Engagement

Citizens' involvement is seen as pivotal to the improvement of people's health and wellbeing and to the development of citizen-centred and sustainable health and care systems. The expectation is that by involving citizens, the effectiveness of services and policies are improved and that these can be more tailored more closely to citizens' own needs and experiences. The study was conducted over four years, drawing on qualitative and realist methods and integrated multiple perspectives, including those of citizens, citizen representatives, local health and care providers, and municipalities in six regions in the Netherlands. The overarching question was: *how can community engagement be realized in health and care settings taking into account different contexts, aims, experiences and needs.*

By addressing how involvement can be successfully realized in the health and care domains, and examining which (aspects of) involvement approaches work, for whom, and under which circumstances, this study examined several important facets of involvement in the Netherlands. The study firstly showed 10 guiding principles for the successful engagement of citizens: (1) ensure staff provide supportive and facilitative leadership to citizens, based on transparency; (2) foster a safe and trusting environment enabling citizens to provide input; (3) ensure citizens' early involvement; (4) share decision-making and governance control with citizens; (5) acknowledge and address citizens' experiences of power imbalances between them and professionals; (6) invest in citizens who feel they lack the confidence to engage; (7) create quick and tangible 'wins'; (8) take into account both citizens' and organisations' motivations; (9) Develop a shared CE vision with clear roles for professionals and citizens ensuring communities' diversity, interests and needs are reflected within the vision; (10) invest in the engagement environment to create a cultural change and enhance reflexivity. Secondly, the study highlighted that citizens and professionals perceive and experience involvement differently and that they have distinct perspectives and priorities for involvement and health and wellbeing in communities. It suggested that citizens' involvement has the potential to better align services and policies to their lived experiences and to improve the democratic legitimacy of policymaking. However, it also showed that organizations and already engaged citizens are seeking new visions and roles to better fit in with a decentralized system and a 'participation society'. Furthermore, the empirical findings also show that the engagement environment needs to be improved and invested in, in order to change organizational cultures, structures and processes to ensure these: (a) address power imbalances between citizens and professionals; (b) are better suited and more sensitive to different ways in which different citizens want to be involved; (c) are more accessible to a wider range of citizens and communities. Without such further investments, citizens' engagement will remain no more than 'positive additions' to health and care systems rather than being seen as crucial to restoring accountability and person-centredness to those systems.

Source de Weger (2022)

12.3.3 Lessons and Solutions

To avoid the 'one-size-fits-all' pitfall, the obvious answer is for organizations to improve their processes and structures by focusing on citizens' own experiences and perceptions, and to align their outreach and engagement approaches more closely to citizens' varied lived experiences, interests and needs. The four-year study presented in Box 12.2 highlights the importance of reaching citizens on their own terms, but also that organizations need to take the time and make the effort to contact more vulnerable and diverse groups. The study also highlights that doing this may not be as hard as is often assumed. For this study, the authors first prioritized contacting low-income citizens and other 'harder-to-reach' groups and decided to take the necessary time to recruit citizens through those who are in closest contact with them, such as their support workers, local foodbanks, non-profit emergency funds and local churches. The authors also took the time to build relationships with citizens to build trust and in order to understand and observe their needs, such as by holding interviews in their own comfortable and safe spaces, using English or French for refugees, or holding dual interviews for those who wanted their carers or loved ones to accompany them to feel more secure. Without such adjustments (to help make the participants feel more comfortable, appreciated and safe), it is unlikely that the authors could have been able to involve them in the study (de Weger, 2022; de Weger et al., 2022).

Of course, such an upfront investment in the outreach and in establishing relationships with a wide range of (more vulnerable) citizens is seldom built into organizations' (or professionals') capacity. This four-year study therefore highlights the importance of investing in a more inclusive and diverse engagement environment that stimulates the embedding of a wide range of engagement practices (and outreach and relationship-building) with a broad range of citizens and by ensuring that their engagement is a structural and routine part of research, projects and policymaking, as well as by providing citizens and professionals with the time, space and support to develop creative engagement approaches. This requires long-term financial support in citizens, community-led initiatives, and by helping professionals and organizations to develop engagement skills and know-how—for instance, by providing training and guidelines. There is also a need for additional resources to create the reflexive space for organizations to build relationships with communities and to develop, innovate and expand engagement approaches.

Ultimately, for TDR to be more inclusive and diverse in its methodologies and engagement approaches, research commissioners and research institutes should not only make citizens' involvement a prerequisite in all of their research calls, but should also build more funding in these to help researchers to invest more time and resources to build relationships with a wide range of citizens, especially with those who are more vulnerable and distrustful. Commissioners should also consider engaging and/or employing citizens as advisors to help evaluate and improve research calls and to advise research teams to ensure that future studies are more inclusive and representative of citizens' interests, needs and questions. In this way, research commissioners could be role models in how to involve a wide variety of citizens in the various project stages and would help to foster an improvement in the engagement environments and researchers' underlying reflexivity.

12.4 Assumption 3: Inclusivity Is a Matter of Getting the Right People at the Event

12.4.1 Background

So far, this chapter has focused on the importance of diversity for achieving inclusivity and the requirements for attaining such diversity within engagement activities. However, what we haven't discussed yet is what organizations should do to promote inclusivity beyond those engagement events. If an engagement event lacks support from an inclusive work culture, the engagement activity is likely to become less inclusive, and the outcomes of engagement will likely lack impact on the involved organizations. Therefore, in this section, we will explore how organizations can foster inclusivity throughout the entire organization, extending beyond individual engagement events.

12.4.2 Pitfall

The pitfall described in this section is organizing engagement events without supporting those events with the inclusive organizational regulations, tools, infrastructures, ways of doing and ways of thinking.

This is a common pitfall in all sectors, but a particularly illustrative example is offered by the world of smart cities. The term 'smart cities' refers to an innovation system of IT companies and governments that together develop new data technologies for the urban space. These technologies are typically developed based on the compelling vision that collecting vast amounts of citizen data will ultimately lead to cleaner, safer and more efficient cities (Sadowski & Bendor, 2019). There are, however, also ethical concerns regarding smart city technologies, such as the invasion of privacy and the increased influence of corporations in the public space.

Despite efforts from both the public and private sectors to promote citizen engagement in smart city development, these initiatives often fall short of expectations. In a case study that we performed in 'smart city' Amsterdam (Fraaije et al., forthcoming), we investigated what makes citizen engagement so difficult to achieve in practice (see the last case study in Box 12.3). In this case study, we looked at how citizen engagement connects to the institutional environment of the involved organizations; in other words, we looked at the ways in which citizen engagement is both enabled and restrained by the way organizations think and work. In this way, the case study clarifies what organizations can do to support inclusivity throughout their organization and beyond their engagement events.

The case study looked at four 'institutional logics' four different layers of the involved organizations that together determine how inclusivity in the organization may take shape:

- *Materialities:* the physical context of citizen engagement: the buildings and digital platforms for citizen engagement as well as the shape and form of the technologies in question
- *Formal rules and regulations*: the laws and regulations under which actors have to work, even if they don't always adhere to these standards in practice
- *Practices:* the things actors do when they can be observed by others from the same organization; a way for them to demonstrate to each other 'this is how we do things around here'
- *Narratives:* the way actors explain and justify their actions to each other, thereby normalizing what they do and what they find important in this

We found that despite the strong motivations and efforts of the involved organizations to organize citizen engagement, inclusivity was severely obstructed throughout all of these organizational layers. Sometimes this happened in very simple ways: The physical rooms in which the organization could meet citizens, for example, hardly allowed for two-way interaction (and was rather intended for one-way presentations). Furthermore, the formal rules and regulations around data ethics were so complex that they distracted the involved organizations from other concerns that citizens might have had. And finally, the practices and narratives of the involved organizations were ultimately targeted at getting citizens to conform to the organizations' smart city vision, rather than to challenge it. The organization for example preferred to highlight the intended benefits of the smart city technologies, and its compliance with existing regulations, rather than to explore any of the possible controversies with citizens. As a result, the involved organizations were unlikely to learn anything new from interacting with citizens, and therefore they did not become significantly more inclusive.

Unfortunately, these issues are not unique to the case study described. Despite growing efforts to involve citizens, they often feel uninformed about smart city developments, ill-equipped to voice their criticisms, and perceive smart city technologies as irrelevant to their lives (Engelbert et al., 2019; Jameson et al., 2019; Rijshouwer et al., 2022). This illustrates that if citizen engagement is not sufficiently supported by an inclusive institutional environment, then engagement risks becoming an 'empty signifier': a means for the organization to face adversity without having to change any of its critical work processes in response (Cardullo & Kitchin, 2019).

Box 12.3

In recent decades, many municipalities across the globe have sought more efficiency, sustainability, and inclusivity by leveraging data technologies. Although such a 'smart city' approach offers certain benefits, it also brings forth various challenges, particularly in terms of privacy, security, and social inequality. Our aim, therefore, was to help municipalities navigate these challenges by investigating how arts-based citizen engagement could foster more responsible innovation in the 'smart' city Amsterdam.

To this end, three case-studies were conducted. In this first case, 12 interviews were performed with innovators working in the smart mobility sector in Amsterdam to investigate how they see the future of Amsterdam. This study revealed that these innovators had fairly uniform vision of the future: a clean, frictionless city in which designers have ironed out the undesirable side effects of smart technologies. The case-study highlighted the need for more inclusive conversations to develop a more diverse vision for the future of Amsterdam.

In the second case-study, a new citizen engagement approach was developed to support such a diverse vision for the future of Amsterdam. Street theatre, playful group conversations, and a theatrical dialogue were used to enable a diverse range of citizens to join the conversation about smart city developments in Amsterdam. Various challenges arose as well, especially with respect to making sure that those perspectives could impact ongoing innovation projects. Overall, the study underscored the importance and systemic challenges of involving vulnerable citizens in smart city developments.

In the third case-study, these systemic challenges were explored further. Another arts-based citizen engagement approach was developed but then specifically for the Digital Perimeter project (DP). The DP was a smart city innovation team that aimed to improve public safety by investigating controversial technologies like real-time bodycams and facial recognition. Our study revealed that despite the fact the municipality and industry partners were motivated to engage citizens, the institutional environment of the 'smart city' severely restrained real dialogue. Organizing citizen engagement did, however, initiate minimal changes to this institutional environment which paved the way for more inclusive citizen engagement in the future.

Source Fraaije A. Can (sm)art save the city? Lessons from action research on art-based citizen engagement towards responsible innovation in 'smart city' Amsterdam, 2023.

12.4.3 *Lessons*

The lesson we can learn from this analysis is that to achieve inclusivity, it is not sufficient to organize a series of engagement events. Rather, inclusivity should be performed throughout all layers of the involved organizations. The only way to ensure engagement activities is effective is by ensuring the rationale and belief in citizen engagement is felt at all organisational levels. This is because it is the only way the involved organizations will take the outcomes seriously and be ready to make any necessary changes in response.

So, what does an organization that supports inclusivity throughout look like? In our smart city case study (Fraaije et al., forthcoming), we saw that the involved smart city organizations supported inclusivity in several ways. They primarily supported inclusivity through various practices. For example, the organizations tried to reach out to citizens in numerous ways, applied various ethical design approaches and regularly showed empathy when confronted with diverse participants at citizen engagement events. These practices were also partially supported by a few helpful narratives, like wanting their smart city project to be more 'ethical' than other smart city projects and considering citizen perspectives to be crucial to the project's success. In addition, the organization could create spaces for employees to meet with citizens in informal, conversational settings, and they could try to put the existing data regulations into perspective.

As described above, any supportive institutional logics did have to compete with other institutional logics that ultimately restrained inclusivity. Yet, the same case study also showed that the logics may conflict with each other and change over time through sustained collaboration. We observed, for example, that by organizing citizen engagement events together with the involved smart city organizations, they became more appreciative and understanding of citizen engagement. In other words, by changing the practices together, also the narratives started to change. This offers potential for change, because it means that when one institutional logic starts to change, then others may change with it.

Even though organizing citizen engagement events may not be *sufficient* for achieving inclusivity, it can be a significant step *towards* inclusivity, as long as the organizers remain reflective regarding how the various organizational layers shape inclusivity.

12.5 Conclusion

This chapter has made the case that it is imperative to make transdisciplinary research processes inclusive and diverse in order to achieve more equitable, high-quality and desirable technologies, policies and public services. We argued that to do so, inclusivity and diversity are not merely a matter of 'getting the (demographically) right people at the right table, in the right room at the right moment'. We presented three pitfalls in attempts to achieve inclusivity and diversity in transdisciplinary research for technology and policymaking. The first concerns the assumption that inclusivity and diversity are about demographic representation. We highlighted that it is just as relevant to involve a range of people in terms of their level of (dis)trust in science and governance as it is to involve a wide range of people in terms of their demographics (e.g. age, gender,sexuality, socioeconomic status). The second pitfall is the common idea that equality in participation can be achieved when everybody participates under the same conditions. This approach to inclusivity tends to exclude, for example, low-income citizens, while over-including those who are easiest to reach. The third pitfall is to focus merely on the engagement events without considering the institutional logics of the organization where the input is most needed. This applies to policymaking organizations, and also for transdisciplinary research that seeks to have social rather than simply an academic impact. For participation to be inclusive and diverse and to make a difference, the institutional logics of the organization with which participants are asked to engage are critical. Relevant questions concern how people collaborate within the organization, how they talk about citizens, what they expect from them and how the physical spaces are organized. The pitfalls and solutions in each case study provide an important overarching lesson that research and policymaking organizations should reflect on their ways of thinking and working in order to reach and include a wider range of citizens.

As with many other well-intended efforts to make society, science or policy more equal and just, efforts to make participation more inclusive and diverse may in fact increase distrust and distance if they fail to take into account the pitfalls we have presented. Without such reflexivity, inclusivity and diversity run the risk of becoming empty checkbox exercises that make the practices of science, technology and policy more complex and cumbersome, while not making any difference to how concerns and values are taken into account. This is equally true for transdisciplinary research. Achieving participation that is meaningfully inclusive and diverse calls for continuously asking critical and reflexive questions not only about the methods, but also about the underlying assumptions and logics of the research project or research organization.

References

Beresford, P. (2019). Public participation in health and social care: Exploring co-production of knowledge. *Frontiers in Sociology, 3*, 41. https://doi.org/10.3389/fsoc.2018.00041

Bromme, R., Mede, N. G., Thomm, E., Kremer, B., & Ziegler, R. (2022). An anchor in troubled times: trust in science before and within the COVID-19 pandemic. *PLoS One*, *17*(2), e0262823.

Bucchi, M., & Trench, B. (2021). Rethinking science communication as the social conversation around science. *Journal of Science Communication, 20*(3). https://jcom.sissa.it/article/pubid/JCOM_2003_2021_Y01/

Burns, T. W., O'Connor, D. J., & Stocklmayer, S. M. (2003). Science communication: A contemporary definition. *Public Understanding of Science, 12*(2), 183–202.

Butter, M., & Knight, P. (2023). Covid conspiracy theories in global perspective. In M. Butter & P. Knight (Eds.), *Covid conspiracy theories in global perspective* (p. 414). Routledge.

Cardullo, P., & Kitchin, R. (2019). Being a 'citizen' in the smart city: Up and down the scaffold of smart citizen participation in Dublin, Ireland. *GeoJournal, 84*(1), 1–24.

Cooke, B., & Kothari, U. (Eds.). (2001). *Participation: The new tyranny?* Zed Books.

Cortis, N. (2012). Overlooked and under-served? Promoting service use and engagement among 'hard-to-reach' populations. *International Journal of Social Welfare, 21*(4), 351–360.

Cyril, S., Smith, B. J., Possamai-Inesedy, A., & Renzaho, A.M., (2015). Exploring the role of community engagement in improving the health of disadvantaged populations: A systematic review. *Global Health Action, 18*(8), 29842.

Davies, S. R. (2022). Science communication at a time of crisis: Emergency, democracy, and persuasion. *Sustainability, 14*(9), 5103.

de Weger, E. (2022). *A work in progress: Successfully engaging communities for health and wellbeing. A work in progress: Successfully engaging communities for health and wellbeing*. A realist Evaluation—Tilburg University Research Portal.

de Weger, E., Baan, C. A., Bos, C., Luijkx, K. G., & Drewes, H. W. (2022). 'They need to ask me first'. Community engagement with low-income citizens. A realist qualitative study. *Health Expectations*, *25*(2), 684–696.

de Weger E., Schuring J., Harambam J. et al. (2023). *Beweegredenen van Nederlandse burgers voor COVID-19 vaccinaties*. Vaccinatiedialogen, beweegredenen van Nederlandse burgers voor COVID-19 vaccinaties (Athena Instituut VU)—Eerste Kamer der Staten-Generaal.

de Weger, E., Schuring, J., Haramban J., Zuiderent-Jerak., & Kupper F. (forthcoming) In dialogue with citizens bridging the gap between citizens and civil servants. *Public Administration Review*. [Submitted 2023].

Engelbert, J., van Zoonen, L., & Hirzalla, F. (2019). Excluding citizens from the European smart city: The discourse practices of pursuing and granting smartness. *Technological Forecasting and Social Change, 142*, 347–353.

Evans, D., Coad, J., Cottrell, K. et al. (2014). Public involvement in research: Assessing impact through a realist evaluation. *Health Services and Delivery Research, 2*(36). https://doi.org/10.3310/hsdr02360

Fletcher, D. R., Flint, J., Batty, E., et al. (2016). Gamers or victims of the system? Welfare reform, cynical manipulation and vulnerability. *Journal of Poverty and Social Justice, 24*(2), 171–185.

Fraaije, A. (2023). *Can (sm)art save the city? Lessons from action research on art-based citizen engagement towards responsible innovation in 'smart city' Amsterdam.* Can (sm)art save the city? Lessons from action research on art-based citizen engagement towards responsible innovation in 'smart city' Amsterdam—Vrije Universiteit Amsterdam (vu.nl).

Fraaije, A., Willems, W., Kupper, F., & Broerse, J. E. W. (forthcoming). *Engaging citizens in smart city Amsterdam. How institutional logics restrain and shape conditions for dialogue.* Athena Institute, Vrije Universiteit Amsterdam.

Fung, A. (2015). Putting the public back into governance: The challenges of citizen participation and its future. *Public Administration Review, 75*(4), 513–522.

Harambam. J. (2017). The truth is out there: Conspiracy culture in an age of epistemic instability. RePub, Erasmus University Repository: "The Truth Is Out There": Conspiracy culture in an age of epistemic instability (eur.nl).

Harambam, J. (2020). *Contemporary conspiracy culture: Truth and knowledge in an era of epistemic instability*. Routledge.

Harambam, J. (2023). Distrusting consensus: How a uniform corona pandemic narrative fostered suspicion and conspiracy theories. *Journal of Digital Social Research, 5*(3), 109–139.

Harambam, J., & Voss, E. (2023). The corona truth wars: Epistemic disputes and societal conflicts around a pandemic—An introduction to the special issue. *Minerva, 61*, 1–15.

Holley, K. (2016). *Equitable and inclusive civic engagement: A transformative guide*. https://kirwaninstitute.osu.edu/research/civic-engagement-transformative-guide

Hueske, A., Willems, W., & Hockerts, K. (2023). *Why and how to engage beneficiaries as co-(social) entrepreneurs? Considering hardware, software and orgware for citizen engagement*. Free University Amsterdam.

Irwin, A., Jensen, T. E., & Jones, K. E. (2013). The good, the bad and the perfect: Criticizing engagement practice. *Social Studies of Science, 43*(1), 118–135.

Irwin, A. (2021). Risk, science and public communication: Third-order thinking about scientific culture. In M. Bucchi & B. Trench (Eds.), *Routledge handbook of public communication of science and technology* (pp. 147–162). Routledge.

Jameson, S., Richter, C., & Taylor, L. (2019). People's strategies for perceived surveillance in Amsterdam Smart City. *Urban Geography, 40*(10), 1467–1484.

Jansen G., & Denters B. (2018). Democratie dichterbij. Lokaal Kiezersonderzoek 2018. lokaal-kiezersonderzoek-2018-democratie-dichterbij.pdf (dpes.nl)

Kok, K. P., Gjefsen, M. D., Regeer, B. J., & Broerse, J. E. (2021).Unraveling the politics of 'doing inclusion' in transdisciplinary for sustainable transformation. *Sustainability Science, 16*, 1811–1826.

Landemore, H. (2014). Inclusive constitution-making: The Icelandic experiment. *Journal of Political Philosophy, 23*(2), 166–191.

Lewis, L. (2014). User involvement in mental health services: A case of power over discourse. *Sociological Research Online*, *19*(1). https://doi.org/10.5153/sro.3265

Luluquisen, M., & Petttis, L. (2014). Community engagement for policy and systems change. *Community Development, 45*(3), 223–231.

Prettner, R., te Molder, H., Hajer, M., & Vliegenthart, R. (2023). Light at the end of the tunnel? The staging of expertise during the COVID-19 vaccination campaign. *Journal of Digital Social Research, 5*(3), 140–170.

O'Mara-Eves A., Brunton G., & McDaid D. (2013). Community engagement to reduce inequalities in health: A systematic review, meta-analysis and economic analysis. *Public Health Research*, 4. https://doi.org/10.3310/phr01040

Rijshouwer, E. A., Leclercq, E. M., & van Zoonen, L. (2022). Public views of the smart city: Towards the construction of a social problem. *Big Data & Society, 9*(1), 1–12.

Rowe, G., & Frewer, L. J. (2005). A typology of public engagement mechanisms. *Science, Technology & Human Values, 30*(2), 251–290.

Rutjens, B. T., Sutton, R. M., & van der Lee, R. (2018). Not all skepticism is equal: Exploring the ideological antecedents of science acceptance and rejection. *Personality and Social Psychology Bulletin, 44*(3), 384–405.

Sadowski, J., & Bendor, R. (2019). Selling smartness: Corporate narratives and the smart city as a sociotechnical imaginary. *Science Technology and Human Values, 44*(3), 540–563.

Sobo, E., & Drazkiewicz, E. (2021). Rights, responsibilities and revelations: COVID-19 conspiracy theories and the state. In L. Manderson, N. J. Burke & A. Walberg (Eds.), *Viral loads: Anthropologies of urgency in the time of COVID-19* (pp. 67–88). UCL Press.

Turbe A., Barba, J., Pelacho, M., Mugdal, S., Robinson, L. D., Serrano-Sanz, F., Sanz, F., Tsinaraki, C., Rubio, J. M., & Schade, S. (2019). Understanding the citizen science landscape for European environmental policy: An assessment and recommendations. *Citizen Science: Theory and Practice*, *4*(1), article number 34.

Van Ewijk, A. R. (2011). Diversity and diversity policy: Diving into fundamental differences. *Journal of Organizational Change Management*, *24*(5), 680–694.

World Health Organization. (2017). *Communicating risk in public health emergencies. A WHO guideline for emergency risk communication (ERC) policy and practice*. https://iris.who.int/bitstream/handle/10665/259807/9789241550208-eng.pdf?sequence=2

PART III

Roles and Competencies

CHAPTER 13

Roles and Competencies in Transdisciplinary Practices

Jacqueline E. W. Broerse and Marjolein B. M. Zweekhorst

13.1 Introduction

This part of the edited volume explores how conceptualizations of transdisciplinary knowledge development processes, guided by transformative visions, materialize in engaged research practices. In this context, researchers grapple with navigating diverse roles and reflecting on their positionality. In addition, a reflexive, situated research practice demands a careful assembly and application of a wide variety of competencies to effectively implement and navigate these roles. By connecting theory-based approaches with empirical examples drawn from various transdisciplinary projects, this part of the book aims to cultivate a nuanced understanding of how roles and competencies contribute to shaping the emerging profile of the 'transformative transdisciplinary researcher'. It will address the following questions:

J. E. W. Broerse (✉) · M. B. M. Zweekhorst
Vrije Universiteit Amsterdam, Amsterdam, The Netherlands
e-mail: j.e.w.broerse@vu.nl

M. B. M. Zweekhorst
e-mail: m.b.m.zweekhorst@vu.nl

B. J. Regeer et al. (eds.), *Transdisciplinarity for Transformation*,
https://doi.org/10.1007/978-3-031-60974-9_13

- What different roles may a transformative transdisciplinary researcher assume and how do these roles relate? (Chapter 14)
- What are different ways for transformative transdisciplinary researchers to foster role awareness and role reflexivity? (Chapter 15)
- What competencies are key for transformative transdisciplinary researchers and how may these be obtained? (Chapters 15–17)
- How to train and foster transformation in universities? (Chapters 16 and 17)

The chapter commences by delineating the role landscape of transdisciplinary researchers with a transformative orientation, and acknowledges the complexities involved in navigating these roles. Although we recognize the inconsistencies and ambiguities in the terminology and descriptions of actors' roles in the scientific literature, we will attempt to present a comprehensive role landscape. Subsequently, we delve into the competencies needed to conduct transdisciplinarity for transformation, the acquisition of these competencies, and the role of universities in training and fostering these.

13.2 Navigating Different Roles

To understand the roles that transdisciplinary researchers may adopt in transformative settings, we distinguish three modes of research that can be considered part of transdisciplinarity for transformation (see also Chapter 1, this volume): conventional research, research through participation and transformative research. Each type of research comes with its own distinct conceptualizations of roles for researchers. If we adopt the view that the relation between these three modes is nested rather than distinct, the number of roles expands from conventional to transformative research, reflecting the changing goals and priorities. The evolving nature of these roles underscores the dynamic relationship between research and its interaction with society. We explain the roles in more detail below.

13.2.1 Conventional Research

The primary role attributed to academics is the collection and analysis of data, pattern recognition and the presentation of evidence to explain a certain phenomenon or causal relation, translating individual data points into general statements. 'Good' research is to be conducted at a distance from society, featuring a clear subject-object division and following the ontology, epistemology and methodology of a certain discipline (Fazey et al., 2018). Even if direct intervention in social processes is aimed for, it is to take place in a rather artificial reductionist context to answer knowledge questions considered appropriate from a disciplinary perspective, and/or to reduce the number of variables so as to be able to draw causal inferences—thereby trying to maintain objectivity and minimizing biases with the aim of uncovering generic patterns about the natural and social world. This role is often called the '*traditional researcher*' (e.g. Bulten et al., 2021).

When the traditional researcher assumes a normative stance, aiming to achieve direct social impact, the role of '*engaged academic*' is adopted (Bauer & Kastenhofer, 2018). The role of an engaged academic differs from that of the traditional researcher, as the former actively engages in advocating and lobbying for a specific normative direction, while the latter keeps a more detached stance, claiming a neutral position.

In both roles, scientists communicate knowledge *to* social actors who may or may not act upon it. In society, academics adopting these roles typically assume the position of 'expert'—elucidating issues, scrutinizing existing solution pathways and offering advice. This guidance is grounded in the perceived superiority of scientific knowledge over other forms of knowledge, such as practical, experiential or indigenous knowledge. However, in practice this type of knowledge and knowledge transfer has limited actionability for developing and implementing solution pathways to address the highly complex problems societies are facing, such as climate change and non-communicable diseases, which is often referred to as the 'implementation gap' (e.g. Damschroder et al., 2009).

Initial attempts to counter this lack of impact by better communicating scientific knowledge to society—an approach called the 'deficit model' (Irwin & Wynne, 1996)—have not been successful. A growing number of evaluation studies showed that an important cause of the implementation gap is that knowledge and innovations that science produces are often not sufficiently aligned with the problem context and complexities related to realizing transformative change. Social actors require not just more information, but other, more actionable, knowledge and innovations (e.g. Broerse & van de Sande, 1995; van der Wilt & Reuzel, 2012).

13.2.2 Research Through Participation

In response to the observed implementation gap, research in the field of science and technology studies began to focus more on the research and innovation process itself. Within the positivist and empiricist paradigm, innovation processes had mainly been perceived as developing in a linear fashion; processes that run in a chronological order from fundamental and monodisciplinary research via applied research to product development, production and use (Godin, 2006; Sismondo, 2011). Another perspective on research and innovation processes was developed, which recognized that these processes are comprised of very complex social activities of variation and selection of innovations, in which different factors play a role and various actors interact with each other (Rip et al., 1995). Concomitantly, research that better provides actionable solutions to complex social challenges was understood to actively involve a wider variety of scientific disciplines and engage the people who are affected by the research—the so-called stakeholders. It was argued that the active involvement of stakeholders can result in a win-win situation for both science and society. The introduction of Part II of this volume provides details on the four main arguments for including a wide variety of actors in research and innovation processes. In addition, this shift to multi-stakeholder participation signals not only the transgressing of boundaries between different types of knowledge but also their reordering, where different types of knowledge are integrated and co-created (Regeer & Bunders-Aelen, 2003).

In this mode of research, researchers fulfil different types of roles. The participatory researcher needs to be a '*process facilitator*' and '*knowledge broker*'. As process facilitator, researchers select participants and locations, initiate and facilitate (short-term) actions, and design social engagement processes (Fazey et al., 2018), '*based on respect, openness and deliberation*' and '*oriented towards a common understanding of situations and collective action, as part of a learning process*' (Pohl et al., 2010, p. 277). The process facilitator maintains participants' attention on the designated task for the engagement process while guarding the quality of the process, thereby ensuring that every participant is afforded an equal chance to articulate their knowledge and express their opinions and ideas (Bauer & Kastenhofer, 2018). As knowledge broker, the participatory researcher mediates between the different knowledge and perspectives that, as process facilitator, they elicit, by giving voice to the wide variety of stakeholders, and '*provides space for critical reflection*', while enabling participants to learn from one another, thereby realizing knowledge integration (Wittmayer & Schäpke, 2014, p. 488).[1]

However, as also highlighted in Chapter 1 (this volume), research through participation has a tendency to emphasize process over purpose. This is due to the attention to procedural criteria for 'good' participatory research in relation to collaboration, co-creation, social learning, reflexivity, and related political and power dimensions (Turnhout et al., 2020). This emphasis can inadvertently transform the exploration of these processes into being an end in itself, rather than a means to an end, thereby diluting the transformative essence of this type of research regarding its contribution to addressing complex social challenges.

13.2.3 Transformative Research

The pressing imperative to make a meaningful impact amidst escalating health and sustainability challenges necessitates a strong focus not only on process but also on realizing radical change ('transformation')—i.e. fundamentally different ways of thinking, organizing and doing. Specifically, this entails the imperative to engaging in systemic experimentations, not merely with isolated interventions, but rather with innovative practices along with their corresponding cultures and structures. It is important to (1) monitor the experiments closely and regularly reflect on the results, (2) identify systemic barriers, synergies and trade-offs, and (3) strategize and make adjustments accordingly. This creates so-called action-learning spirals (Kemmis & McTaggart, 2007, p. 278). The role of the transformative researcher in these learning-by-doing processes is that of a '*reflexive facilitator*' (Fazey et al., 2018) or '*reflexive monitor*' (Van Mierlo et al., 2010). The reflexive facilitator/monitor enhances reflexive practices of others by '*using new knowledge from research as it emerges and by asking critical and challenging questions to keep ambitions for transformative change high*' (Fazey et al., 2018, p. 64, see also Chapter 3).

In addition, to further support transformative change, researchers need to ensure that new practices are anchored in new cultures (e.g. mental models, paradigms) and structures (e.g. regulation, procedures and incentives) by taking on the role of '*capacity builder*' (Sarkki et al., 2013). This involves actively networking to keep participants engaged in the change

[1] Schuijer et al. (2021) have combined these two roles into the role of '*deliberative practitioner*'.

process and expanding the network by attracting additional stakeholders, thus cultivating and strengthening social capital that will result in the emergence of a 'critical mass' driving the transformation process. It also involves enhancing the participants' competencies, encompassing professionals, citizens and researchers, in the principles and methodologies of transdisciplinarity for transformation. This includes fostering an understanding of systemic change processes and cultivating the skills of reflexive learning.

Shifting the emphasis of transdisciplinary researchers towards a more (en)activist stance, they adopt the role of a '*change agent*' (Schuijer et al., 2021). Change agents prioritize more radical transformations and are willing to actively participate in political or policy processes to exert influence not only through the enrichment of knowledge but also through lobbying, campaigning or applying pressure in policy and political spheres—all aimed at steering transformative change in alignment with the normative direction supported by the transformative researcher.

Scholars also identify the role of '*project worker*' (Schuijer et al., 2021) or '*project manager*' (Fazey et al., 2018). In the current knowledge economy—rooted in bureaucratic, short-term logic—research and innovation projects are increasingly expected to demonstrate 'measurable impact' according to predetermined goals, which leads to a phenomenon called 'projectification' (Felt, 2009; Godenhjelm et al., 2015). Project workers/managers are tasked with establishing effective relationships with other project partners, ensuring timely achievement of project milestones and deliverables, reporting on project outcomes, and demonstrating responsible and effective use of funding. Project workers/managers, however, tend to prioritize short-term project goals over long-term ambitions.

13.2.4 Eight Ideal-Typical Roles in a Role Landscape

Above, we outlined in total eight ideal-typical roles that transdisciplinary researchers may concurrently adopt or switch between over the course of a transdisciplinary project aiming at transformation: traditional researcher, engaged academic, process facilitator, knowledge broker, reflexive monitor, capacity builder, change agent and project worker. In practice, however, researchers experience the boundaries between the roles as blurry, and there are moments when a researcher may opt to emphasize one role more prominently than others depending on the situation. Schuijer et al., (2021, p. 174) refer to the latter as '*dynamic positioning in a role landscape*'. Drawing inspiration from the work of Schuijer et al. (2021), we have positioned these roles within a landscape across two axes: socio-political orientation within the normative intervention context and the level of transformative change sought (see Fig. 13.1).

The horizontal axis pertains to the kind of contribution associated with a distinct role. At one end of the spectrum, roles are primarily concerned with contributing to academic knowledge production. At the opposite end, roles ambitiously strive to actively influence policymaking, and seek political involvement, while the contribution to the scientific body of knowledge receives comparatively less emphasis. The vertical axis concerns the nature of the change sought—whether it is more incremental or more radical. The incremental end of the change axis involves a focus on pragmatic changes leading towards transformation, occurring gradually and

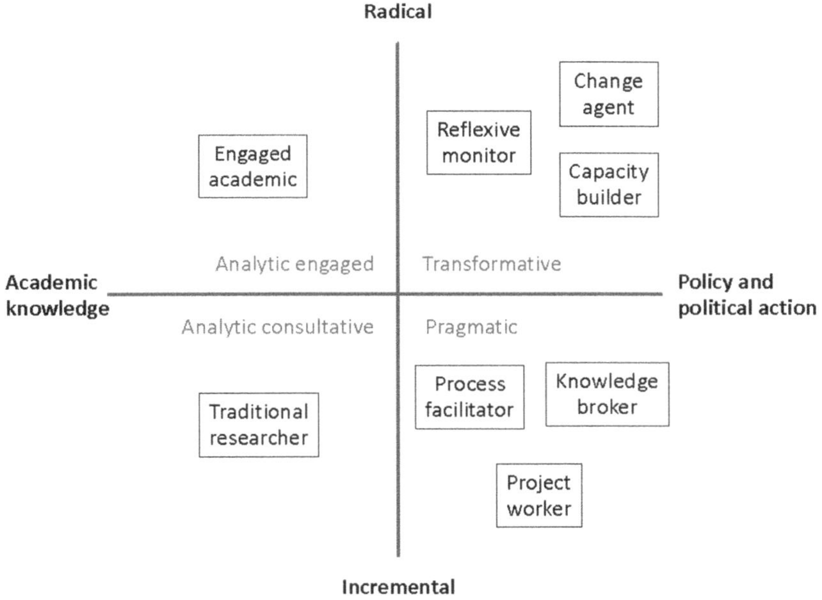

Fig. 13.1 Ideal-typical role landscape of the transdisciplinary researcher (adapted from Schuijer et al., 2021)

with a relatively short-term perspective. At the opposite end of the spectrum, a more radical and ambitious approach is indicated; this entails a pronounced emphasis on challenging and reimagining the entire existing social system. Together, the two axes result in four different dimensions in the transdisciplinary research role landscape: (1) the analytic-engaged dimension, (2) the analytic-consultative dimension, (3) the transformative dimension and (4) the pragmatic dimension (Schuijer et al., 2021).

It is important to acknowledge that the distinctions between roles are not precisely defined, and the boundaries between dimensions are fluid—each influencing the other. Furthermore, researchers are not completely 'free' in selecting a role. Roles are inherently relational and personal; one may be attributed a role in interaction with stakeholders or fellow researchers and some roles 'fit' a researcher's personality and skills better than others (Vinke-de Kruijf et al., 2022).

How to position oneself at what moment in the development of any (transdisciplinary) research project is far from straightforward. It requires a proper understanding of what is needed when in complex and continuously evolving contexts, and self-reflection that does not shy away from engaging with one's own personality and normative stance. Many of the learning questions of early-career transformative transdisciplinary researchers revolve around this complexity (see Chapter 1; this volume). Furthermore, when adopting multiple roles, some of the roles are synergistic, while others may give rise to conflict. Schuijer et al. (2021) analysed the dynamics of synergies and conflicts among roles within the context of their public engagement-related endeavour. Perhaps predictably, roles positioned in close proximity within the role landscape were observed to exhibit less conflict (such as the knowledge broker and the capacity builder, or the capacity builder and the change agent), and to often work synergistically. Other roles introduced tension (for instance, the process facilitator and the change agent, or the project worker and the engaged academic), because of their different normative orientation. However, Schuijer et al. (2021, p. 184) warn that

> the precise benefits and trade-offs of role integration cannot be determined in the abstract but should be assessed and accounted for within the specific context in which the [researcher] operates (…). This requires role awareness and reflexivity.

Chapter 14 (Den Boer, this volume) aims to offer additional insights into the role synergies and conflicts that may arise for a transdisciplinary researcher when integrating multiple roles within a specific context focused on transformative change towards a sustainable food system. In this chapter, Alanya den Boer reflects on her experiential learning as the coordinator of the so-called *City Lab Amsterdam*, one of the eight Labs in the EU-funded FIT4FOOD2030 project (https://fit4food2030.eu/). She subsequently delves into the roles she assumed and the corresponding synergies and tensions she encountered during the project. Rather than anticipating or reflecting on role adoption along the way, she analysed it afterwards. Consistent with Schuijer et al. (2021), den Boer advises transdisciplinary researchers, who are likely to assume various roles, to cultivate role awareness and role reflexivity from the outset. This approach will enable them to navigate role dilemmas more effectively, rather than simply muddling through.

Callum Gunn and his co-authors (Chapter 15, this volume) reflect on different ways to foster such role awareness and role reflexivity in relation to the positionality of the transformative transdisciplinary researcher. Although ideal-typical frameworks are valuable for comprehending the overall role landscape, they argue that these frameworks fall short in addressing the actual process of negotiating, shaping and reinventing these roles in practical scenarios. After all, in a research process already characterized by openness and emergence, can one adequately prepare for knowing how and when to adopt diverse researcher roles? In specific situations, one might even choose to abstain from adopting a particular role or opt to distribute conflicting roles among multiple individuals within a project consortium or transdisciplinary team (as outlined by Wittmayer & Schäpke, 2014; Vinke-de Kruijf et al., 2022). These complexities pose particularly challenging dilemmas for early-career researchers, given the absence—and perhaps impossibility—of formal training or substantial experience in transdisciplinary research practices. However, navigating the multitude of roles that may be assumed, adopted, resisted or otherwise dealt with appears to be inherent in transdisciplinary research.

The contribution of Gunn et al. (this volume) aims to highlight how different roles, positions and identities manifest in the intricate role landscape of transdisciplinary research practice. They offer reflective guidance for researchers grappling with the complexities of positionality in transdisciplinary settings, concluding that it is acceptable '*to experiment and not know exactly what you are doing*,' provided there is a '*safe and reflexive space*' (p. 419). This type of environment empowers transdisciplinary researchers to manage the consequences arising from the tendency of transdisciplinary practices to challenge established identities and roles, and make room for both learning and unlearning.

13.3 Acquiring Competencies

In this section, we highlight the diverse array of competencies essential for transformative transdisciplinary researchers and explore ways for acquiring these competencies. A competency can be defined as an interlinked set of knowledge, abilities, skills, experiences and behaviours that enable effective performance and problem solving (Spady, 1994). Concerning the various roles outlined earlier, it becomes evident that proficiency in one role, such as being a good 'engaged academic', does not necessarily translate into effectiveness in another role, such as that of a 'knowledge broker'. Excelling as an engaged academic involves mastering competencies related to data collection, analysis, pattern recognition, etc. However, the role of a knowledge broker comes with an additional set of tasks, such as orchestrating and enhancing dialogical processes among diverse stakeholders to foster mutual social learning and respectfully explore conflicting positions. These tasks require competencies that include an understanding of knowledge-integration processes and proficiency in organizational communication and mediation, which are unfortunately less frequently addressed in academic curricula compared to conventional disciplinary competencies (Escobar et al., 2014). It is crucial to recognize that new competencies are not merely required in relation to performing tasks associated with specific roles. Equally important is the ability to discern which role to assume when, how to seamlessly integrate various roles, and effectively manage tensions that may arise between them (Levin, 2012). Vinke-de Kruijf et al. (2022) that 'being reflective and self-reflexive is a key competence' for any researchers who is involved in transdisciplinary projects (p. 403).

Scholars have observed that most education on transdisciplinarity and transformation closely adheres to the existing status quo (e.g. Barrett et al., 2019; Redman & Wiek, 2021), predominantly emphasizing the roles of engaged academic, process facilitator and knowledge broker, while largely overlooking more transformative roles. Consequently, graduates from these programmes are primarily equipped to make incremental improvements rather than serving as capacity builders and change agents capable of driving substantial transformations (Gordon et al., 2019). To enhance the effectiveness of transdisciplinary researchers in transformational settings, it is crucial to gain comprehensive insights into the diverse competencies required, particularly those related to the transformative aspect of transdisciplinarity, and how these competencies can be acquired through both formal education and informal education.

The literature on competencies for transdisciplinary researchers in transformative settings has expanded significantly over the past two decades, with a predominant focus on contexts related to sustainability transformations. This often takes the form of long lists of competences related to certain tasks and roles. Here we will refrain from such an approach, because it lacks coherence and tends to overlook meta-competencies. In this part of the volume, we build on the valuable work of Redman and Wiek (2021), who developed a comprehensive competency framework by systematically reviewing 272 relevant publications on sustainability learning objectives spanning 1997–2020. Their framework outlines eight key competencies deemed crucial for graduate students in sustainability science: systems thinking, futures thinking, values thinking, strategies thinking, as well as implementation, interpersonal, intrapersonal and integration competences (see Box 13.1 for definitions). They also identified general competencies such as critical thinking and creativity, along with professional competencies like responsive project management. Although Redman and Wiek's framework primarily centres around 'sustainability science', it has garnered significant interest from designers of transdisciplinary courses who have incorporated (some of) these competencies as intended outcomes.

Box 13.1 Eight sustainability-specific key competences and their definitions (quoted from Redman & Wiek, 2021: Table 1)

.

- *Systems-Thinking Competence*: Ability to apply modelling and complex analytical approaches: (1) to analyse complex systems and sustainability problems across different domains (environmental, social, economic) and across different scales (local to global), including cascading effects, inertia, feedback loops and other system dynamics; (2) to analyse the impacts of sustainability action plans (strategies) and interventions (how they change systems and problems).

- *Futures-Thinking Competence*: Ability to carry out or construct simulations, forecasts, scenarios and visions: (1) to anticipate future states and dynamics of complex systems and sustainability problems; (2) to anticipate how sustainability action plans (strategies) might play out in the future (if implemented).
- *Values-Thinking Competence*: Ability to identify, map, specify, negotiate and apply sustainability values, principles and goals: (1) to assess the sustainability of current and/or future states of complex systems; (2) to construct sustainability visions for these systems; and (3) to assess the sustainability of action plans (strategies) and interventions.
- *Strategies-Thinking Competence*: Ability to construct and test viable strategies (action plans) for interventions, transitions and transformations towards sustainability.
- *Implementation Competence*: Ability to put sustainability strategies (action plans) into action, including implementation, adaptation, transfer and scaling, in effective and efficient ways.
- *Interpersonal Competence*: Ability (1) to collaborate successfully in inter-disciplinary and inter-professional teams and (2) to involve diverse stakeholders, in meaningful and effective ways, in advancing sustainability transformations.
- *Intrapersonal Competence*: Ability to avoid personal health challenges and burnout in advancing sustainability transformations through resilience-oriented self-care (awareness and self-regulation).
- *Integration Competence*: Ability to apply collective problem-solving procedures to complex sustainability problems: (1) to develop viable sustainability strategies (action plans) and (2) successfully implement them, in collaborative and self-caring ways.

In this part of the volume, Wolfgang Stark and Hussain Zeidan and colleagues explore some of these competencies and their interconnection in greater depth, suggest additional competences and discuss implications for positionality and role shifting/integration. They also reflect on the role of universities in developing and fostering these competences.

In Chapter 16, Wolfgang Stark underscores the importance of intertwining 'head, hand, and heart' to effectively address and transform real-world problems (Scharmer, 2009). He emphasizes the need to blend rational knowledge, experiential understanding, and creative thinking and action in innovative ways. Beginning with the essential 'System-Thinking Competence', Stark introduces the additional general competency of 'Artistic Thinking'. A robust systems approach necessitates fostering agile relationships among diverse actors, worldviews and disciplines, leveraging creativity, intuition and the art of improvisation. The 'Artistic-Thinking Competence' pertains to the ability to apply creativity and intuition to 'pattern recognition' within complex systems, combining identified patterns into a cohesive 'pattern language'. Stark draws parallels with jazz as a relevant art form, highlighting its continuous re-designing and re-arranging of implicit and explicit procedural patterns based on experiential (implicit) knowledge. This approach is not only vital for comprehending system dynamics but also for exploring innovative avenues to navigate uncertainty and ambiguity in complex transformative settings. To creatively redesign patterns and structures, Stark underscores the significance of 'improvisation', which he characterizes as '*a technique which allows us to integrate serendipity as a learning process and involves proactive learning*' (this volume: p. 441).

Stark explicitly does not exclude rational analysis; on the contrary, the performative aspect of learning takes centre stage. Improvisation heavily relies on experiential knowledge—learning by doing—and may result in the development of 'practical wisdom' (Schwartz & Sharpe, 2010), 'deep smarts' (Leonard & Swap, 2005) or 'phronesis' (Loeber, 2007). The awareness of others' and one's own practical wisdom is a crucial prerequisite for innovation processes in transformative change, according to Stark. Rather than providing a conceptual or theoretical overview of innovation processes and associated tools, success factors and strategies are deconstructed into individual patterns of action. Unlike a linear and rigid guideline, improvisational actions can be flexibly selected, combined and applied based on the specific transformational setting.

In Chapter 17, Hussain Zeidan Sarju Raj and Marjolein Zweekhorst delve into competencies and their development through the dual perspectives of (1) positionality, which explores the interplay between an individual and the external environment, and (2) the internal synergies and tensions arising from shifting roles within transformative settings. Using the framework of Redman and Wiek (2021; see also Box 12.1), they try to link the different competencies to the various roles that have been identified in relation to transdisciplinary research for transformation, specifically referring to the personal reflections provided in Chapters 14 and 15 (this volume). Based on this analysis, they identify that a specific competence is lacking. Navigating the role landscape and dynamically balancing various commitments, attitudes, dilemmas and tensions necessitate what Zeidan et al. (this volume) term 'Navigation Competence'. This implies that transdisciplinary researchers need skills to effectively express, debate, convey and communicate the perspectives of different stakeholders as well as their own perspective, coupled with a mindset of 'epistemic humility' and 'self-reflexivity' to reflect on own positionality and normativity while navigating a highly dynamic environment and varied role landscape to facilitate transformative change.

Both Chapters 16 and 17 reflect on educational approaches and formats of transformative and lifelong learning, which are considered essential for nurturing the development of transdisciplinary research competencies. Traditional linear models of knowledge transmission, where teachers impart knowledge to students, are viewed as inadequate. Instead, approaches focusing on experiential learning or learning by doing are considered more pertinent. These novel methods immerse students in real-life complex problems and establish physical or virtual spaces for articulating and integrating various types of knowledge and experiences, such as community service learning (CSL) and challenge-based learning. Annemarie Horn and Marjoleine Van der Meij (Chapter 18, this volume) provide an inspiring practical example of a tool they developed to enhance reflexivity in Master's students during an inter- and transdisciplinary course—called Frame Reflection Lab (FRL). The FRL tool facilitates the cultivation of awareness regarding both one's own perspectives on science and those of others and hence stimulates reflection on one's own and other's academic identity (Horn et al., 2022). The tool enhances discussions about diverse viewpoints on science by incorporating 'identity-first language', thereby fostering a more personal and less cognitively oriented discourse. Following the viewing of video portraits featuring four distinct types of researchers addressing climate change, students are carefully led through interactive workshops designed to be both playful and safe. Noteworthy advantages observed include heightened awareness of one's positionality and enhanced reflexivity on academic identities, beliefs and roles within the realm of inter- and transdisciplinary research.

Nevertheless, the multitude of competencies can be overwhelming for any student embarking on a transdisciplinary journey, especially for those aspiring to master transdisciplinarity for transformation. Zeidan et al., therefore, assert that educators should transcend the design of isolated courses purporting to instil transdisciplinary and transformational competencies, as this is inherently unattainable. Instead, educators ought to perceive their courses as integral components within a broader chain or scaffold of diverse courses. Such an approach leverages students' normative learning and competency development while fostering a lifelong learning attitude. Although this seems to be an 'open door' in relation to learning 'disciplinary' competencies, such a scaffolding is surprisingly rare in the case of learning relevant competencies for 'transdisciplinary research for transformation'.

From an institutional standpoint, this evolution means that universities will have to transform into what Stark (this volume) terms 'Resonance Spaces', bridging research and learning, civil society, policy and business. Universities should establish an ecosystem wherein innovative ideas and improvisational patterns, identified in academia, civil society, policy or business, resonate and persist, fostering the requisite reflection for sustainable innovations addressing social challenges. Through such a transformative approach, universities play a pivotal role in shaping a future where research, learning and social impact seamlessly intertwine. In doing so, universities can become dynamic agents of positive change, addressing complex social issues. At the same time, we should not forget that, as Zeidan et al. argue, universities should first and foremost equip students to create their own identity and self-learning capacity in order to develop their own path to becoming an experienced transformative transdisciplinary researcher, and to develop their own 'compass of a sense of purpose' (this volume, p. 469).

References

Barrett, M. J., Alphonsus, K. B., Harmin, M., Epp, T., Hoessler, C., McIntyre, D., Reeder, B., & Singh, B. (2019). Learning for transdisciplinary leadership: Why skilled scholars coming together is not enough. *BioScience, 69*(9), 736–745. https://doi.org/10.1093/biosci/biz072

Bauer, A., & Kastenhofer, K. (2018). Policy advice in technology assessment: Shifting roles, principles and boundaries. *Technical Forecasting and Social Change, 139*(c), 32–41. https://doi.org/10.1016/j.techfore.2018.06.23

Broerse, J. E. W., & van de Sande, T. (1995). Technology transfer or alternative technology: Biotechnology and low-external-input-agriculture. In B. Mepham, G. Tucker, & J. Wiseman (Eds.), *Issues in agricultural bioethics*. Easter School Series in Agricultural Sciences No. 55 (pp. 361–386). Nottingham University Press.

Bulten, E., Hessels, L. K., Hordijk, M., & Segrave, A. J. (2021). Conflicting roles of researchers in sustainability transitions: Balancing action and reflection. *Sustainability Science, 16*, 1269–1283. https://doi.org/10.1007/s11625-021-00938-7

Damschroder, L. J., Aron, D. C., Keith, R. E., Kirsh, S. R., Alexander, J. A., & Lowery, J. C. (2009). Fostering implementation of health services research findings into practice: A consolidated framework for advancing implementation science. *Implementation Science, 4*(1), 1–15. https://doi.org/10.1186/1748-5908-4-50

Escobar, O., Faulkner, W., & Rea, H. J. (2014). Building capacity for dialogue facilitation in public engagement around research. *Journal of Dialogue Studies, 2*, 87–111.

Fazey, I., Schäpke, N., Caniglia, G., Patterson, J., Hultman, J., van Mierlo, B., Säwe, F., Wiek, A., Wittmayer, J., Aldunce, P., Al Waer, H., Battacharya, N., Bradbury, H., Carmen, E., Colvin, J., Cvitanovic, C., D'Souza, M., Gopel, M., Goldstein, B., …, Wyborn, C. (2018). Ten essentials for action-oriented and second order energy transitions, transformations and climate change research. *Energy Research & Social Science, 40*, 54–70. https://doi.org/10.1016/j.erss.2017.11.26

Felt, U. (Ed.). (2009). *Knowing and living in academic research: Convergence and heterogeneity in research cultures in the European context*. Academy of Sciences of the Czech Republic.

Godenhjelm, S., Lundin, R. A., & Sjöblom, S. (2015). Projectification in the public sector—The case of the European Union. *International Journal of Managing Projects in Business, 8*(2), 324–348. https://doi.org/10.1108/IJMPB-05-2014-0049

Godin, B. (2006). The linear model of innovation: The historical construction of an analytical framework. *Science, Technology, & Human Values, 31*(6), 639–667.

Gordon, I. J., Bawa, K., Bammer, G., Boone, C., Dunne, J., Hart, D., et al. (2019). Forging future organizational leaders for sustainability science. *Nature. Sustainability, 2*, 647–649. https://doi.org/10.1038/s41893-019-0357-4

Horn, A., van der Meij, M. G., Willems, W., Kupper, F., & Zweekhorst, M. B. M. (2022). Developing interdisciplinary consciousness for sustainability: Using playful frame reflection to challenge disciplinary bias. *Sustainability: Science, Practice & Policy, 18*(1), 515–530. https://doi.org/10.1080/15487733.2022.2095780

Irwin, A., & Wynne, B. (Eds.). (1996). *Misunderstanding science? The public reconstruction of science and technology*. Cambridge University Press.

Kemmis, S., & McTaggart, R. (2007). Communicative action and the public sphere. *The Sage handbook of qualitative research* (pp. 559–603). Sage.

Leonard, D. & Swap, W. (2005). *Deep smarts: How to cultivate and transfer enduring business wisdom*. Harvard Business School Press.

Levin, M. (2012). Academic integrity in action research. *Action Research, 10*, 133–149. https://doi.org/10.1177/1476750312445034

Loeber, A. (2007). Designing for *phronèsis*: Experiences with transformative learning on sustainable development. *Critical Policy Analysis [Critical Policy Studies], 1*(4), 389–414.

Pohl, C., Rist, S., Zimmerman, A., Fry, P., Gurung, G. S., Schneider, F., Speranza, C. I., Kiteme, B., Boillat, S., Serrano, E., Hadorn, G. H., & Wiesmann, U. (2010). Researchers' roles in knowledge co-production: Experience from sustainability research in Kenya, Switzerland, Bolivia and Nepal. *Science and Public Policy, 37*(4), 267–281. https://doi.org/10.3152/030234210X496628

Redman, A., & Wiek, A. (2021). Competencies for advancing transformations towards sustainability. *Frontiers in Education, 6*, 484. https://doi.org/10.3389/FEDUC.2021.785163/BIBTEX

Regeer, B. J., & Bunders-Aelen, J. F. G. (2003). The epistemology of transdisciplinary research: From knowledge integration to communities of practice. *Interdisciplinary Environmental Review, 5*(2), 98–118.

Rip, A., Misa, T. J., & Schot, J. (Eds.). (1995). *Managing technology in society*. Pinter Publishers.

Sarkki, S., Heikkinen, H. I., & Karjalainen, T. P. (2013). Sensitivity in transdisciplinary projects: A case of a reindeer management in Finland. *Land Use Policy, 34*, 183–192. https://doi.org/10.1016/j.landusepol.2013.03.004

Scharmer, O. (2009). *Theory U—Learning from the future as it emerges*. Berett-Koehler Publishers.

Schuijer, J. W., Broerse, J., & Kupper, F. (2021). Juggling roles, experiencing dilemmas: The challenges of SSH scholars in public engagement. *NanoEthics, 15*, 169–189. https://doi.org/10.1007/s11569-021-00394-8

Schwartz, B., & Sharpe, K. (2010). *Practical wisdom: The right way to do the right things*. Penguin.

Sismondo, S. (2011). *An introduction to science and technology studies*. Wiley.

Spady, W. G. (1994). *Outcome-based education: Critical issues and answers*. American Association of School Administrators.

Turnhout, E., Metze, T., Wyborn, C., Klenk, N., & Louder, E. (2020). The politics of co-production: Participation, power, and transformation. *Current Opinion in Environmental Sustainability, 42*, 15–21. https://doi.org/10.1016/j.cosust.2019.11.009

Van der Wilt, G. J., & Reuzel, R. P. B. (2012). A transdisciplinary approach to the evaluation of medical technology: The case of cochlear implants for prelingually deaf children. In J. E. W. Broerse & J. F. G. Bunders (Eds.),

Transitions in health systems: Dealing with persistent problems (pp. 115–128). VU University Press.
Van Mierlo, B. C., Regeer, B., van Amstel, M., Arkesteijn, M. C. M., Beekman, V., Bunders, J. F. G., de Cock Buning, T., Elzen, B., Hoes, A. C., & Leeuwis, C. (2010). *Reflexive monitoring in action. A guide for monitoring system innovation projects*. Communication and Innovation Studies, WUR & Athena Institute, VU.
Vinke-de Kruijf, J., Verbrugge, L., Schröter, B., den Haan, R. J., Cortes Arevalo, J., Fliervoet, J., Henze, J., & Albert, C. (2022). Knowledge co-production and researcher roles in transdisciplinary environmental management projects. *Sustainable Development, 30*(2), 393–405. https://doi.org/10.1002/sd.2281
Wittmayer, J. M., & Schäpke, N. (2014). Action, research and participation: Roles of researchers in sustainability transitions. *Sustainability Science, 9*, 483–496. https://doi.org/10.1007/s11625-014-0258-4

CHAPTER 14

Running a Real-World Lab to Stimulate Urban Food System Transformation: Navigating Between Different Actor Roles as a Transdisciplinary Researcher

Alanya C. L. den Boer

14.1 Introduction

Real-world Labs are increasingly used to catalyse systemic change by creating experimental spaces in which knowledge co-creation is stimulated among actors from the domains of research, business, policy, and civil society, known as the quadruple helix (Schäpke et al., 2018, p. 86). However, relatively little attention is given to the challenge of facilitating such transformative Labs as a transdisciplinary researcher. The normative orientation of these Labs and the use of participatory and creative methodologies as part of their transdisciplinary research approach mean that these researchers need to go beyond the boundaries of scientific disciplines and become part of the transformation process by adopting different actor roles in addition to their traditional role as scientist

A. C. L. den Boer (✉)
Wageningen University & Research (WUR), Wageningen, The Netherlands
e-mail: alanya.denboer@wur.nl

B. J. Regeer et al. (eds.), *Transdisciplinarity for Transformation*,
https://doi.org/10.1007/978-3-031-60974-9_14

(Loorbach et al., 2011; Turnhout et al., 2013; Wittmayer & Schäpke, 2014).

In the literature, the terms and the descriptions of actor roles are not always used consistently and unambiguously. For instance, the terms 'knowledge broker' and 'intermediary' are both used to describe someone who makes diverse perspectives explicit and mediates between them. Because of differences in terminology and overlapping role descriptions, we designed our own role framework for this study (Table 14.1), which includes the roles of *scientist*, *change agent*, *capacity builder*, *process facilitator*, *reflexive facilitator*, *knowledge broker*, and *project worker*. Although some scholars refer to the *self-reflexive* scientist as an additional actor role (e.g. Fazey et al., 2018; Wittmayer & Schäpke, 2014), which could be defined as 'being reflexive about one's positionality and normativity, and seeing oneself as part of the dynamic that one seeks to change' (Wittmayer & Schäpke, 2014, p. 489), in this chapter, (self)reflexivity is seen as an inherent competence of TD researchers rather than as a separate role. Adopting these actor roles means that TD researchers become involved in different types of activities, which requires different sets of competences.

For Lab-facilitating transdisciplinary researchers, adopting roles can lead to tensions or even conflicts between them (Bulten et al., 2021; Fazey et al., 2018; Hilger et al., 2021; Wittmayer & Schäpke, 2014). For example, tensions might arise because a researcher is expected to take a descriptive–analytic, neutral, and objective position, consistent with the role of scientist, but the same researcher may wish to adopt a role as a reflexive facilitator or change agent, which would mean taking on a normative or even activist position (Bulten et al., 2021; Wittmayer & Schäpke, 2014). Especially for early-career researchers, adopting and mediating between these roles is challenging (Sellberg et al., 2021). Currently, little is known about the tensions and potential trade-offs that transdisciplinary researchers face when adopting various roles in practice (Bulten et al., 2021). This chapter therefore presents a reflexive study in order to gain a better understanding of the challenges that transdisciplinary researchers face when adopting and mediating between different roles and how they navigate these in relation to their ambitions to contribute to transformative change. In doing so, this study seeks to acquire a better understanding of learning in the context of transformation processes. Learning is considered essential for stimulating transformative change, but is also poorly understood (e.g. Van Poeck

Table 14.1 Potential role framework for transdisciplinary researchers

Actor role	*Description*
Scientist	• '*Reflecting on how collecting, analyzing, interpreting and reporting data from an observer point of view can be improved in accordance to the quality criteria of their disciplines and in relation to the reliability of findings*' (Fazey et al., 2018, p. 64, referred to as '*reflective scientist*') • Similar to '*engaged academic*' by Schuijer et al. (2021) and '*traditional scientist*' by Bulten et al. (2021)
Change agent	• '*Advocat[ing] for radical system transformations, create[ing] a collective sense of importance around the desired change, empower[ing] actors to take part in the transformation*'. It is about '*organizing people around an idea, voicing a transformative message, lobbying/campaigning*' (Schuijer et al., 2021, p. 176) • '*Explicitly participating in the learning processes or short-term action with the aim to address real-world problems or motivating and empowering participants*' (Fazey et al., 2018, p. 64) • Similar to '*transition participant*' by Bulten et al. (2021)
Capacity builder	• '*Capable of increasing know-how to enable the participants to implement the co-production processes independently. Translate the practices and rationales behind the co-production paradigm and should focus on explicit training of the sensitiveness needed in knowledge co-production in order to avoid shortcomings (e.g., regarding the scope of participants) in future co-production efforts*' (Sarkki et al., 2013, pp. 186, 191) • Similar to what has been referred to as '*expert in learning*' by Fazey et al. (2018): '*Assisting practitioners or citizen scientists to become better learners and researchers, such as helping them design processes and methods of data collection and analysis, including reflexive practices*' (p. 64) • Similar to '*dialogue capacity builder*' by Schuijer et al. (2021): '*Networking, guiding, and training organizations in public dialogue, professionalization*' (p. 176)

(continued)

Table 14.1 (continued)

Actor role	*Description*
Process facilitator	• '*Facilitating the learning process including initiating the process; selecting participants, locations; initiating and facilitating (short-term) actions, designing the social engagement*' (Fazey et al., 2018, p. 64) • '*Capable of enhancing communicative processes between thought collectives, based on respect, openness and deliberation. Promotes joint reflection oriented towards a common understanding of situations and collective action, as part of a learning process*' (Pohl et al., 2010, p. 277) • Similar to '*deliberative practitioner*' by Schuijer et al. (2021): '*Orginiz[ing] and facilitat[ing] dialogue events, creat[ing] an inclusive dialogue atmosphere in which different perspectives can be made visible, introduce[ing] missing perspectives to dialogical exchanges, develop[ing]new/optimized engagement formats*' (p. 176)
Reflexive facilitator	• Aims to '*encourage reflexive practices of others*' (Fazey et al., 2018, p. 64). '*The capacity for researchers to encourage reflexivity emerges both from using new knowledge from research as it emerges and by asking critical and challenging questions to keep ambitions for transformative change high*' (Fazey et al., 2018, p. 64) • Similar to '*reflexive monitor*' when using Reflexive Monitoring in Action (Van Mierlo et al., 2010) as the monitoring approach (see Fazey et al., 2018). This highlights the longitudinal character of the role of reflexive facilitator, which distinguishes it from process facilitator
Knowledge broker	• '*Mediat[ing] between different perspectives, provide[ing] space for critical reflection and engag[ing] in making sustainability relevant and tangible in different contexts*' (Wittmayer & Schäpke, 2014, p. 488) • Similar to '*intermediary*' by Pohl et al. (2010): '*Able to make different thought styles visible and to link them around common interests*' (p. 277)
Project worker	• Aims to '*set up and carry out a collaborative project, reach milestones, and accomplish deliverables*' (Schuijer et al., 2021, p.176) • Similar to '*project manager*' by Fazey et al. (2018): '*[…] coordinate[ng] and steering of projects to achieve desirable outcomes of a project*' (p. 64)

et al., 2020). It is important to better understand learning processes to guide and support actors who aim to stimulate transformative change more effectively, and to further advance support mechanisms, especially for early-career and future transdisciplinary researchers (Jaeger-Erben et al., 2018; Sellberg et al., 2021).

In this chapter, I reflect on my learning journey as a PhD researcher who conducted transdisciplinary research in the field of food system transformation. As a transdisciplinary researcher, I coordinated the so-called City Lab Amsterdam, which was one of the Labs of the EU-project FIT4FOOD2030. In the next section, I describe the project and its City Labs and go on to outline in brief the methodological approach adopted. Subsequently, I present my personal learning journey before elaborating on the roles I adopted and the role synergies and conflicts that I faced over the course of the project. The chapter ends with a critical discussion and a set of recommendations for the design and architecture of future projects that aim to stimulate system transformation via transdisciplinary Real-world Lab approaches.

14.2 The European FIT4FOOD2030 Project and Its City Labs

The FIT4FOOD2030 project was a three-year Coordination and Support Action (CSA) that was funded by the European Union's Horizon 2020 research and innovation (R&I) programme (EC, 2021), and was led by the Vrije Universiteit Amsterdam. The project aimed to stimulate transformation towards 'future-proof' food systems (i.e. sustainable, resilient, responsible, diverse, competitive, and inclusive food systems) through R&I. Its main objective was to establish the FOOD 2030 platform: a sustainable and multi-actor platform aiming to (1) strengthen R&I policy coherence and alignment; (2) build R&I competences among current and future food system professionals; and (3) raise awareness of the need for transformation of the food system (EC, 2021; Kok et al., 2019). These objectives were to be realized through three interlinked structures—an EU Think Tank, 11 Policy Labs, seven City Labs and seven Food Labs[1]

[1] Mid-way through the project, additional local/regional Labs were selected via an open call. These Labs were called Food Labs rather than City Labs to emphasize that their focus went beyond the city level. Because of the short timeline, the Food Labs did not co-create educational modules but focused on implementing multi-stakeholder workshops (for more information, see EC, 2021).

(see EC, 2021). Since this chapter focuses on my experiences as a transdisciplinary researcher who coordinated one of the City Labs (CLs), I elaborate upon the specific activities of the CLs below.

The formal mandate of the CLs was to stimulate the development of competences for Responsible Research and Innovation (RRI) among (future) food system professionals and to stimulate awareness of the need for transformation of the food system. The CLs' work was based on the need to use a systems approach to this and the need to strengthen RRI. Within the proposed RRI framework, the purpose of R&I is to actively contribute to solving real-world problems in responsible ways, which includes ethical reflection and stakeholder involvement (Stilgoe et al., 2013). Building on the notion of the need for a systems approach to R&I to achieve food system transformation, the CLs were expected to engage a diverse set of actors and to stimulate knowledge co-creation processes during the project's four phases (FIT4FOOD2030, 2019):

- *Phase 1 (started November 2017)—Actor identification and mobilization, visioning, and system understanding*: CLs identified and engaged with food system stakeholders to build or connect to a diverse multi-stakeholder network. The CLs organized multi-stakeholder workshops to co-develop an understanding of the local food system and a vision of a 'future-proof' food system, including the role of R&I in achieving it.
- *Phase 2 (started August 2018)—Developing roadmaps:* Built on the visions arising from Phase 1, roadmaps were co-developed. These roadmaps comprised identification of the competences required for contributing to that vision, audiences that would then need educational modules, the design of educational modules, and—optionally—the development of local–regional food-related R&I and/or policy agendas. The educational modules could be based on either a 'light' or a 'deep' learning approach (Fenollosa & Paca, 2018), the latter referring to transdisciplinary approaches.
- *Phase 3 (started February 2019)—Action planning and experimentation:* CLs prototyped and piloted their educational modules and developed generic educational tools from these for others to use.
- *Phase 4 (started December 2019)—Scaling-up and continuity:* CLs developed strategies to further scale up, embed, and/or translate their efforts so as to make these 'sustainable'.

The Lab coordinators' learning processes were stimulated via four two-day training sessions and a set of webinars. In addition, Lab coordinators were encouraged to formulate specific learning questions concerning major challenges that they had faced in order to create a dynamic learning agenda (DLA) (Regeer et al., 2009). Finally, Lab coordinators were supported through tailored tools that had been developed by the project's consortium partners (EC, 2021; Kok et al., 2019).

Four CLs were embedded in science museums, and the other three, including the CL Amsterdam, which is the focus in this chapter, were based in research institutes. More specifically, the CL Amsterdam was embedded in the 'Science Shop' of the Vrije Universiteit Amsterdam, which meant that it was particularly suitable for the development of 'deep-learning' modules.

14.3 Methodological Approach

The study used a self-reflective approach that was inspired by *autoethnography*,[2] which is a combination of autobiography and ethnography (Ellis & Bochner, 2000; Ellis et al., 2011). Other (early-career) transdisciplinary researchers have also used such self-reflective autoethnographic approaches (see, for example, Patterson et al., 2013; Sellberg et al., 2021). Part of autoethnographic approaches involves retrospectively analysing personal experiences and making use of storytelling techniques. This enables them to provide readers with opportunities for vicarious experiences and learning (Ellis & Bochner, 2000). In order to follow a structured approach to reflection, I used the reflective journal that I kept during the three years of the project as a basis for developing my personal learning journey (next section) as well as four narratives. To further enrich and structure these narratives, a number of in-depth reflection sessions were held after the project had ended with the 'sounding board', which comprised three experienced transdisciplinary researchers who were all engaged in the project. Over the course of the project, they adopted the role of 'critical friend' with regard to the CL Amsterdam. During reflection sessions, the sounding board asked me critical questions that stimulated me to be introspective and to make explicit the emotions and

[2] *Auto* refers to self, *ethnos* to culture, and *graphy* to the research process (Reed-Danahay, 1997, as cited in Ellis & Bochner, 2000, p. 740).

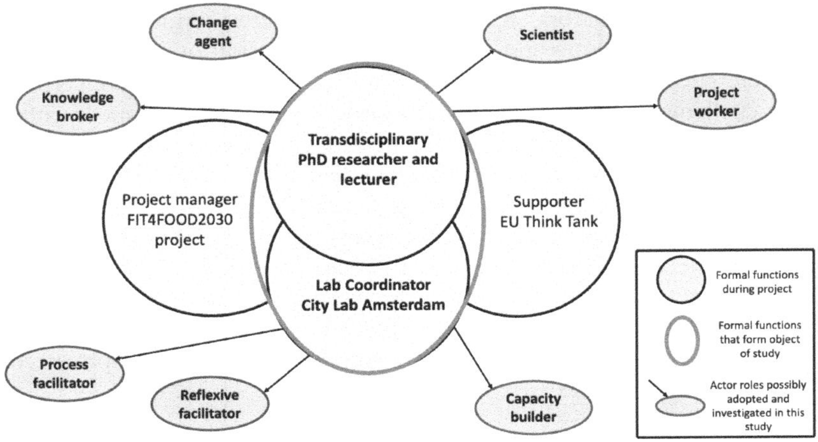

Fig. 14.1 Overview of the four formal functions carried out during the project and the focus of this study

tensions that I had felt as a transdisciplinary researcher and Lab coordinator. After the narratives were co-constructed, I analysed them to identify which roles I had adopted and which synergies and tensions had arisen. The quotes included in Sect. 14.6 come from the narratives.

It is important to note that I differentiate between functions and roles. Besides my formal function as a transdisciplinary researcher and lecturer and my formal function as CL coordinator, I had two other formal functions, including that of FIT4FOOD2030 project manager during the project's first eighteen months and a supporter of the project's EU Think Tank. The functions of transdisciplinary researcher/lecturer and Lab coordinator called for different roles (see Table 14.1). These functions and associated roles form the main focus of the study (Fig. 14.1).

14.4 Personal Learning Journey: Sketching the Starting Position

The personal learning journey set out below is a brief story that highlights my starting position and significant moments of learning. It gives the relevant background information for the rest of the chapter.

When I started working within the FIT4FOOD2030 project, I had just graduated in Applied Communication Sciences, with a specialization in Health and Society (Wageningen University and Research). Because of this background, I knew that it is important to adopt a holistic perspective in the context of health, and had a basic understanding of the importance of interdisciplinary research. During the first 18 months, I was rather absorbed by the functions of project manager, supporter of the EU Think Tank, and lecturer. I had no experience of project management and teaching. Fortunately, mid-way through the project an experienced project manager was taken on, which I was incredibly happy about.

In parallel to the project management and education-related activities, I tried to think about what we meant when we said that the project was going to support '*the urgently needed transformation of R&I on food and nutrition security (FNS)*'. I was aware of disciplinary knowledge gaps, but frequently thought about questions such as 'What do we actually mean by "the R&I system"?' and 'What do we mean by "a food systems approach"?' In the autumn of 2018, I wrote down: '*Make crystal clear to yourself why "the R&I system" needs to be transformed. This is what the EC is saying, but why exactly?*'. Initially, I thought the problem with R&I was 'simple': R&I funds were just unevenly distributed, meaning that certain topics and disciplines received less funds than others. But why did we design interactive workshops with Post-its and other colourful materials? Was that part of our idea about '*the role of R&I in stimulating food system transformation*'? Or was it just our way of working in the project and the institute?

As part of my role as Lab coordinator, I had to follow four training sessions with the other Lab coordinators. This training was helpful in that it provided the space in which to learn about and practise using participatory methodologies. However, I also felt that this training and the DLA sessions were often not applicable to my situation and ambition. For instance, several other CLs were based at science museums, which meant that their context, objectives, and challenges were different from mine.

Although the training with other coordinators did prepare me to coordinate my Lab, it did not answer my questions about why R&I has an important role to play in system transformation. Because of my function as a PhD researcher and my involvement in the entire project, I gradually started to learn about concepts, terms, and/or fields of research, such as *sustainability transitions research, sustainability science, and transdisciplinarity*, and the importance of related participative methodologies. I taught myself what was meant by terms such as *knowledge co-creation*. I remember coming across some documents from a food-related research project that was not connected to our project, which showed pictures of researchers working with Post-its and other materials, just like we were doing! This was an eye-opener, since that was the moment when I realized that this way of working was an integral part of research.

Although my undergraduate and Master's degrees were a good basis for my function as a researcher and Lab coordinator, I was not trained in transdisciplinarity nor in systems thinking. However, at a certain point, I clearly understood that this type of research—transformative transdisciplinary research—could be seen as an intervention in itself. Moreover, I started to realize that not only are there challenges associated with 'doing' this type of research, but also with creating the space to facilitate it, which is strongly linked to how R&I is funded.

14.5 Transdisciplinary Researcher and Lab Coordinator: Actor Roles Adopted

During the course of the project, I adopted all of the roles as introduced, except for that of reflexive facilitator. In this section, I give examples, using an overview of the main activities of the CL Amsterdam (Fig. 14.2) and quotes from the four co-constructed narratives.

As *scientist*, I performed a stakeholder analysis, which quickly showed that there were already many food initiatives in the Metropolitan Region Amsterdam (MRA). Moreover, the analysis showed that two food policy networks (FPNs) were emerging—the Food Council MRA and Food Connects ('Voedsel Verbindt')—both aiming to work at the metropolitan level. Food Connects was co-initiated by the two provinces of the MRA and was still in its pre-formation stage when the CL started. As scientist, I also performed academic work which taught me, for instance, about underlying theories and methodologies with regard to stimulating system transformation.

As *change agent*, I aimed to link to these emerging FPNs in the region (upper part of Fig. 14.2) by developing a transformative food-related R&I agenda, as illustrated below:

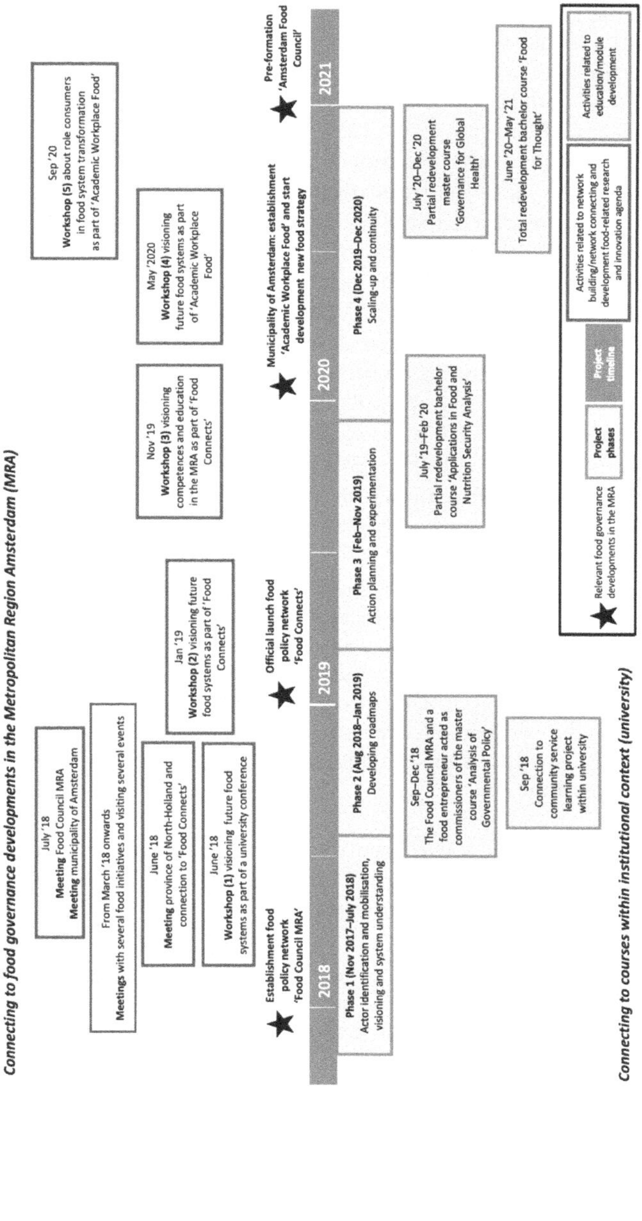

Fig. 14.2 Overview of the main activities of the City Lab Amsterdam as part of the EU FIT4FOOD2030 project

> Whereas the development of educational modules was obligatory, the development of an R&I agenda was not. However, I thought that such an agenda would offer most opportunities to create mutual benefits. [...] I hoped the CL could stimulate the adoption of a so-called food systems approach in the MRA, both by the process of developing such an R&I agenda as well as with the R&I agenda itself. [...].

Moreover, as change agent, I aimed to strengthen the connection between those FPNs and the educational system at the Vrije Universiteit, for which I considered the R&I agenda to be a major link (lower part of Fig. 14.2). I reasoned that the development of an extracurricular deep-learning module would be possible, but it would probably not be sustainable in the long term given the already overloaded curricula and the opportunities within existing courses. Therefore, I aimed to transform university courses by integrating community service learning (CSL). CSL is a pedagogical approach that stimulates experiential learning in which students '*can achieve real-life experiences in and with the surrounding community*' (Tijsma et al., 2020, p. 391). One of the activities that was used to connect one of the main actors of the FPNs to the university is described below:

> Through my connection to FPN 'Food Connects' as well as to a large CSL project within the institute, we came up with the idea of asking the programme manager of Food Connects to record a video in which the need for food system transformation and the ambitions of Food Connects would be explained. I supported the programme manager with writing the script. We showed this video to lecturers [...] to motivate them to integrate food-related CSL in their courses. One person [...] became enthusiastic [...] and contacted me.

As *capacity builder*, I was active regarding both the upper part and the lower part of Fig. 14.2. For instance, I connected individual stakeholders in the region where possible (upper part):

> Through our conversations with the municipality and some citizen food initiatives, I was able to inform one of the food initiatives about the possibility to pitch their ideas to the alderperson. This stakeholder was happy to learn from us about this opportunity and contacted me a few weeks later to tell me that they had been invited by the municipality and were working towards partnership.

As *capacity builder*, I also supported other actors in designing their own multi-stakeholder workshops (upper part of Fig. 14.2), as illustrated here:

> 'Even before the multi-stakeholder workshop took place, the relevant official of the provincial government we had spoken with earlier asked me whether I was willing to support other working group leaders of Food Connects as well.

Moreover, I adopted the role of capacity builder in the educational context (lower part of Fig. 14.2), as shown in the following extract:

> I noticed that the lecturers were not yet familiar with the concept [of CSL], and, therefore, not necessarily enthusiastic. After some time, they asked me to give them a 'lecture' about CSL. I was happy [with this question], because that meant they were willing and open to learn more about the concept. [...] Because of this experience, I realized that the entire process of redeveloping a university course was interesting in itself.

As process facilitator, I designed and facilitated several workshops (upper part of Fig. 14.2). For instance, I had the opportunity to not only facilitate but also to design a workshop for Food Connects (workshop 2):

> The programme manager said that they aimed to develop an 'action agenda' within Food Connects. I suggested enriching this agenda with relevant underlying R&I questions. She was fine with that. Moreover, I suggested co-designing a visioning workshop (workshop 2) for the first meeting of the working group leaders of Food Connects. The programme manager was happy with [this] suggestion [...] and let me design it.

Also, when we organized a multi-stakeholder workshop in collaboration with the municipality of Amsterdam (workshop 4), I adopted the role of process facilitator, as did some of my colleagues:

> Early in 2020, my promotor came into contact with the chief science officer of the city of Amsterdam. This officer [...] was planning to initiate the so-called Academic Workplace Food (AWF) [...]. Because of this contact, the CL got the opportunity [...] to co-design and facilitate the first (online) multi-stakeholder visioning workshop (workshop 4) of the AWF [...]. We were happy [...]: finally, there was momentum to work further on the aspiration to develop a transformative food-related R&I

> agenda with the municipality [...]. We developed an interactive workshop script that we revised and refined with the members of the AWF [...]. During the workshop, we, as process facilitators, tried to make sure everyone had the opportunity to speak [...]. We also asked participants whether they thought any aspects were missing in the vision of the future food system and/or in the R&I agenda that was co-developed.

An important moment occurred when I adopted the role of *knowledge broker* and tried to ensure that the outcomes of workshop 4 (upper part of Fig. 14.2) became 'actionable':

> It was unclear how our workshop was going to influence the municipal food strategy [...] it seemed to me as if the municipality was just ticking the box regarding stakeholder involvement. [...] However, I soon realized I had a role to play. I tried to stay in close contact with the relevant municipal officer who was responsible for writing the food strategy. I made detailed suggestions about where I thought the strategy could be enriched based on our workshop. I presented this to him and he reacted enthusiastically. In that moment it really felt that I had made an impact.

Moreover, also with regard to connecting to university courses (lower part of Fig. 14.2), I adopted the role of knowledge broker:

> I contacted the relevant municipal officer whom I had already collaborated with. I explained the idea concerning [this] course. Initially, the officer was [...] not interested in becoming engaged. However, as soon as I pointed to the opportunity to link some of the research topics of the municipal food strategy and/or the AWF to this course, the officer became very enthusiastic [...]. The municipal officer and I formulated an assignment around healthy and sustainable food consumption.

Finally, as *project worker*, I wrote and submitted reports on the multistakeholder workshops that we facilitated, and wrote the educational modules that had to be submitted.

As the above shows, I did not adopt the role of *reflexive facilitator*, which would have required me to be intensively involved over a longer period of time in food networks such as Food Connects, Food Council MRA, and/or the AWF. It would have required me to monitor these networks and their processes and activities closely and to stimulate learning and reflexivity among actors over a period of time. Although this is an essential role in the context of transformation, an important reason why I did not adopt this role is that I had a mandate to develop educational modules. In addition, these networks were in their early stages. Therefore, it seemed to me that they were not necessarily open (yet) to external persons becoming involved over a longer period of time. Moreover, it felt as if I would need to be more 'senior' to have the authority to create the space for adopting the role of reflexive facilitator.

14.6 Synergies and Tensions Between Actor Roles

In this section, I present synergies and tensions that I experienced when adopting and mediating between the roles. Figure 14.3 gives an overview, highlighting the tensions between the need to fulfil academic requirements as a scientist, the need to fulfil project requirements as project worker, and the need and ambition to contribute to transformation by adopting transformative action-oriented roles, namely the roles of change agent, capacity builder, process facilitator, and knowledge broker.

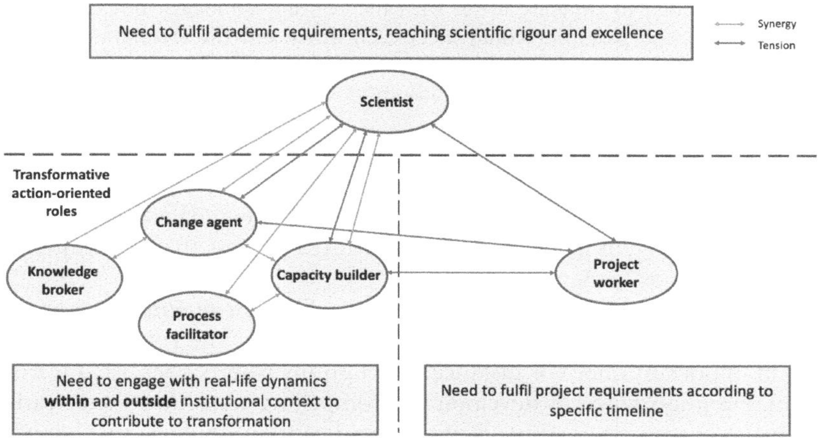

Fig. 14.3 Overview of identified synergies and tensions between actor roles adopted as TD researcher and Lab coordinator of the City Lab Amsterdam of the EU FIT4FOOD2030 project

14.6.1 Synergies

First, I experienced an important synergy between the roles of *change agent* and *capacity builder*. As capacity builder, I supported actors in designing their multi-stakeholder workshops. This felt valuable from a change-agent perspective.

Second, I experienced synergy between the roles of *knowledge broker* and *scientist*. It was necessary to use my analytical and reflection skills as a scientist when making a comparison between the multi-stakeholder workshop results (workshop 4) and the draft municipal food strategy in order to make the workshop outcomes 'actionable' for the municipal officer in my role as a knowledge broker. Moreover, as a scientist, I was aware of the importance of adopting a holistic perspective, for instance with regard to the importance of looking at both health and environmental sustainability aspects of food consumption. As knowledge broker, I brought different perspectives together and created space for critical reflection:

> To my surprise, it was clear that the lecturers for the course did not think that the link between health and environmental sustainability was important. I decided to organise a call with the relevant municipal officer and the lecturers so that they could meet each other and the lecturers could learn about the importance not only of focusing on healthy but also on environmentally sustainable food consumption.

Third, I experienced synergy between the ambitions of a *change agent* and those of a *scientist*, since connecting to emerging FPNs in the region and developing a transformative food-related R&I agenda was interesting from a change-agent perspective, but was also a process that I was aiming to write about as a scientist.

Fourth, I felt synergy between the roles of *process facilitator*, *capacity builder*, and *scientist*. Since I was a PhD researcher, the role of scientist was an important one. For instance, through my role as scientist, I learnt about the importance of developing a compelling transition vision early on in the process. As process facilitator and capacity builder, this knowledge was essential to design and support the multi-stakeholder visioning workshops as part of the FPNs in the region:

> The working group leader [of Food Connects] was happy with my suggestion for collaboration and organizing a workshop (workshop 3), and she let me design it. It felt as if they (again) saw me as someone who had

experience with designing such multi-stakeholder workshops based on participatory methodologies and tools.

14.6.2 Tensions

I also experienced tensions between roles. First, between the roles of *project worker* and *change agent*, because of different logics and timelines. As a project worker, I had to comply with the project's deadlines, which was important to be able to show the funder what was happening and to reach our impact objectives. However, as change agent, I was willing to immediately anchor the multi-stakeholder workshops to emerging FPNs in the region, which meant the timeline of the project did not match the real-world dynamics:

> To comply with the project requirements, we organised and facilitated our first multi-stakeholder visioning workshop in June 2018 (workshop 1). Although this was an interesting workshop from a project perspective, the workshop was not connected to any of the food governance developments in the MRA. [...] it just felt like ticking a box, and I wondered if it had any additional value.

This was also the case with regard to the development of educational modules, for which there was a specific timeline. As change agent, I aimed to transform courses, since the development of a 'separate' course would be difficult to integrate into or add to the existing curriculum without support from a higher institutional level. Moreover, I aimed to transform some of the courses in which I was involved as a lecturer, since doing so required me to first 'know' the course. However, the process of trying to start a collaboration with other lecturers was more time-consuming than I expected. It also meant that I had to deal with the timeline of the teaching system. One of the lecturers from a course on which I taught was open to the suggestion of integrating CSL provided I was able to formulate student assignments and could arrange and take care of the stakeholders who were going to commission the assignments. This resulted in an enthusiastic commissioner and an assignment. However, I soon experienced resistance, which hindered me as project worker who had to develop and submit an educational module before a rather tight deadline:

> According to the lecturers, there was almost no flexibility to make adjustments to the course planning and set-up. However, without the integration of assignments that stimulate reflexivity, I was afraid it would just turn into a superficial consultancy project [...]. It showed me that the ability to influence [...] the lecturers around me was essential for me as TD researcher [...] Although this process was interesting from a scientist's perspective, it did not help me to fulfil my requirements as project worker with regard to producing educational modules that were suitable to be piloted in other contexts.

When I started collaborating with other lecturers at the university who were more interested and willing to change one of the courses, the project was near its end, which meant it was interesting from the perspectives of a change agent and capacity builder, but not from that of a project worker:

> Together with three [lecturers ...] I worked on the redevelopment of a bachelor's course. This was the result of the video that we had recorded earlier. It was an enlightening experience that they were fully open to integrating elements of CSL and RRI and were aware of the links between health and environmental sustainability. [...] However, the FIT4FOOD2030 project was in its last phase [...], so as project worker I did not have the mandate or incentive to put substantial time into [this]. Moreover, although [they] were enthusiastic, they did ask me to actually do the work.

Second, the roles of *project worker* and *scientist* resulted in tensions for the same reasons. Also, from a scientist's perspective, connecting to ongoing developments in food governance and making use of opportunities (such as the initiation of the AWF by the municipality of Amsterdam) was more interesting than following the strict project timeline as project worker.

Third, although my ambitions as *change agent* and *scientist* were aligned in relation to connecting to ongoing developments, I also experienced tensions between those roles. For example, as change agent, I aimed to follow up on how—if at all—the municipal officer was going to use my detailed analysis (as knowledge broker), which was given to him after workshop 4 (see above). However, I refrained from doing so from a scientist's perspective. As a scientist, I was supposed to write academic papers, for which there was little available time and which was possible regardless of whether the municipal officer was really going to do anything with the workshop results. In addition, I felt that the municipal officer might not appreciate my following up on this issue for a longer period of time (as change agent). Moreover, as scientist I had several ideas about the aspects about which I wanted to write academic papers. However, as change agent, I first had to start a collaboration with significant FPNs in the region, which took a substantial amount of time. For example, the first multi-stakeholder visioning workshop as I envisaged it (workshop 4) took place in May 2020. Because of these real-life dynamics, I had to follow a different timeline as a scientist. On several occasions, I changed the focus of the academic papers. Although this is part of transdisciplinary research, it felt tense since I was constantly thinking about how I would be able to maintain the proper scientific rigour of these papers.

Fourth, there was a major tension between the roles of *capacity builder* versus *scientist* and *project worker*. From a change-agent perspective, supporting the FPNs in designing their multi-stakeholder workshops as *capacity builder* was relevant (see synergies above). However, as *scientist* and *project worker* this was not desirable:

> I was contacted by several actors who participated in our workshops. They were so enthusiastic that they asked me to support their workshops as well. However, my roles of scientist and project worker were already demanding and this was not something I was going to write about as scientist or deliver products on as project worker. [...]

14.7 Discussion and Conclusion

This reflexive study sought to gain a better understanding of the challenges that transdisciplinary researchers face when adopting and mediating between roles and how they navigate those challenges in terms of their ambitions to contribute to transformative change. This study showed that I adopted all roles presented in Table 14.1 except that of reflexive facilitator. It is important to note that my interpretation of the role of change agent explicitly includes the efforts to anchor solutions in the institutional context. This anchoring aspect is not made explicit in the transdisciplinary literature on actor roles as shown in Table 14.1, although it has been explicitly highlighted by De Haan and Rotmans (2018), who point to the importance of 'transformative change agents' for stimulating transformation. One of these transformative change agents has been referred to as the 'connector', which they describe as a person who '(1) connect[s] solutions to systems—be they emerging or incumbent—by embedding or anchoring them in the institutional context [...] and (2) connect[s] actors with other actors' (p. 279). This description includes elements of the roles of change agent and capacity builder, and highlights the activity of trying to anchor solutions to systems. The analysis of my role as transdisciplinary researcher and Lab coordinator showed that I had to continually combine and switch between the different roles, which resulted in synergies but also tensions.

The tensions experienced reflect the broader observations of early-career transdisciplinary researchers of the trade-offs between aspirations and requirements with regard to *Science* (scientific rigour and excellence), *Society* (societal impact and engagement), and *Self* (self-care and reflexivity in terms of a researcher's own role and position) (Sellberg et al., 2021), or the challenges of simultaneously having to deal with the logic of the science system and that of the political-administrative system (Jaeger-Erben et al., 2018). Strikingly, three of the five tensions were related to the role of project worker, which shows that navigating between Science, Society, and Self became especially complicated because of problems associated with a dominant project logic, which refers to the project's time-bound character with its specific timeline requirements. Although the project was the basis for conducting transdisciplinary research since it provided time, resources, and 'status', its logic also substantially hindered me in terms of adopting the more transformative action-oriented roles, those of change agent and capacity builder in particular. The project required all CLs to organize a number of multi-stakeholder workshops according to a specific timeline and to develop and pilot two educational modules, without substantially taking into account the Labs' specific institutional contexts and the regional dynamics. In my case, the Lab was hosted by a university, which provided different opportunities than those hosted by science centres and museums. For instance, the municipality of Amsterdam was eager to work with the CL and its scientists in relation to their ambition to develop a food strategy (workshop 4). As noted by Bansard et al. (2019), scientists do indeed have important roles to play in supporting local governments to develop sustainability plans. However, timelines and requirements were the same for all CLs and left little room for manoeuvre based on an assessment of what is most relevant in the local situation. It became clear to me that the project logic did not value how important it was for me as a change agent and capacity builder to first connect to emerging regional FPNs to develop a food-related R&I

agenda and to start motivating university lecturers around me to integrate pedagogies that are characterized by experiential learning in order to sustainably stimulate competence development (the ultimate objective of the CLs). As a consequence, the project logic also hindered me as a scientist who was aiming to write about this process. Moreover, the need to adhere to the project logic as project manager during the first 18 months further hindered me both as a scientist and as a change agent, giving me the feeling that I had moved from the 'ivory tower' of science to the '*iron cage of the project rationality*' (Maylor et al., 2006, as cited in Felt, 2021, p. 5). The tensions between scientist, project worker, and change agent might have been made worse because the project was not a research project but a *Coordination and Support Action*, which is characterized by its focus on, for example, mobilization, communication, and dissemination. This is in line with previous findings of Schuijer et al. (2021), who highlight the challenge of integrating research activities within a project that is primarily focused on action.

Based on this reflexive study, I would make three key recommendations in relation to the realization of an enabling environment for future transdisciplinary researchers who aim to engage in challenging transformative transdisciplinary practices.

1. **Flexibility:** To reduce tensions caused by a dominant project logic, it is important that future projects (and funders) allow for a certain degree of flexibility with regard to the timeline and type of 'products' that Labs have to deliver. This would allow Labs to have the space to try to align their activities with the institutional context and relevant regional dynamics. Such flexibility is in line with the recommendation of Torrens and Von Wirth (2021) regarding paying more attention to the qualitative evaluation of experiments rather than only focusing on outputs that are easy to measure. As it is desirable that the roles of change agent and capacity builder are also adopted by non-academics, such as policymakers, which was indeed the ambition within the FIT4FOOD2030 project, such flexibility is also relevant for projects that work only with Labs that are hosted by non-scientific institutes.

2. Training: This study highlighted the intensive learning trajectory of an early-career transdisciplinary researcher and pointed to the importance of effectively guiding researchers who have no prior experience in transdisciplinarity or systems thinking through training programmes. According to Schneider et al. (2019), learning and competence building within transdisciplinary Communities of Practice (CoP) is considered an important mechanism for generating impact, since this can stimulate the development of '*reflective leadership*' among the participants. However, to activate this impact mechanism, researchers themselves need to develop competences so that they are able to adopt the necessary roles to stimulate social learning and competence-building processes. In line with Loorbach et al. (2011), who argue that training of researchers should be explicitly integrated into transition research processes, and with the experiences outlined in this chapter, I recommend that training programmes are integrated into the architecture of future transformation projects and that these training programmes are tailored as much as possible. Although a training programme was integrated into the FIT4FOOD2030 project, an empirical analysis of learning among Lab coordinators within the project (Svare et al., 2023) highlighted the importance of allowing and supporting such coordinators to develop their own learning paths in order to meet their individual learning needs that change over time and are highly context-dependent.

Moreover, I recommend that such training programmes are explicit about the different roles that one might adopt in practice, and the possible tensions between them, which may give rise to feelings of uncertainty among researchers (Bulten et al., 2021). Moreover, such training should stimulate the development of '*navigation skills*' (Schuijer et al., 2021, p. 186). I agree with Redman and Wiek (2021) that such training programmes should pay attention to intellectual and emotional development. Explicit emphasis on the development of *intrapersonal competency*, which refers to the ability '*to avoid personal health challenges and burnout in advancing sustainability transformations through resilience-oriented self-care (awareness and self-regulation)*' (Redman & Wiek, 2021, p. 6), and emphasis on the development of reflexivity with regard to a researcher's own role and position, both reflected in the *Self* (Sellberg et al., 2021), is of crucial importance as both aspects affect the ability to fulfil the different types of requirements described above. In addition, it is important to facilitate joint learning spaces for early-career and more experienced transdisciplinary researchers who work in the same or similar institutional environments.

3. Teamwork: Although competences and personality traits are important for individual researchers' ability to combine and switch between roles (Bulten et al., 2021), I agree with Raven et al. (2010) that 'probably a team of transition practitioners is practically more realistic than a single "superman" transition practitioner with all necessary competences' (p. 13). Moreover, a team approach is important in order to divide conflicting roles between different persons (Bulten et al., 2021) and to allow for '*role flexibility*' (Sol, 2018, p. 116), which requires critical reflection. I would argue that such a team should also include more experienced transdisciplinary researchers who have already gained a certain degree of credibility to guide and support young researchers and to be present at important points, and thus to actively be part of the team. This is especially important given the time-consuming and challenging tasks of network building and at the same time seeking to align the objectives and strategy of a Lab and the objectives and input of actors working on change processes in the region. This important and challenging phase is what Horcea-Milcu et al. (2022) have referred to as *Phase 0* of a transdisciplinary process, that is, the phase before the actual transdisciplinary collaboration starts. This phase requires leadership (Horcea-Milcu et al., 2022), and, therefore, the engagement of more experienced transdisciplinary researchers. Moreover, as highlighted by Schneidewind et al. (2018), the reputation and credibility of scientists and their institutes are of great importance when starting to engage with power structures, which is the case during Phase 0. The findings of this study confirm this, as I found that I was involved in Phase 0 for a substantial amount of time. I faced several challenges during this phase. My promotor's active engagement with municipal food developments immediately gave the Lab a prominent role in designing and facilitating a visioning workshop that was meant to provide input to the municipal food strategy, and this gave me the opportunity to continue with the development of a transformative food-related R&I agenda.

As most early-career scientists are highly committed to contributing to transformative change—but also face significant trade-offs—it is important to act on these recommendations as a matter of urgency. Otherwise, we might end up with academics writing papers about how they were hindered while trying to adopt transformative action-oriented roles rather than about how the actual adoption of these roles supported transdisciplinary practices in contributing to much-needed transformation. Given the importance of thinking about the transferability of findings, and, most importantly, to support and empower early-career researchers, I end by sharing my most important *lessons learnt*. Although these are aimed at junior researchers, some of them clearly highlight the importance of the active engagement of senior transdisciplinary researchers:

1. Take care of yourself, because only then you will be able to contribute to transformation and make academic impact.
2. Make use of the authoritative power in your environment at important points, especially when you start to connect to established governance networks and/or powerful actors in the region ('Phase 0').
3. Regularly reflect with experienced transdisciplinary researchers on the entire process, especially when you are involved in a large, ambitious, and complex transformative project, and be explicit about the tensions that you experience so that you can determine a strategy together.
4. Be aware of, and keep in mind, that transformation processes are slow and may meet with resistance from actors in your environment so that you do not become too stressed or frustrated when the process takes longer than you expected or hoped. Moreover, remember that these dynamics might be interesting in themselves when viewed from a scientist's perspective.
5. Collect data, but try to take it easy. Make (extensive) reflexive notes after meetings, calls, workshops, etc., but do not try to capture everything, because that is exhausting. Engaging in transdisciplinary research also means that you need to develop another perspective on the questions of 'what is data?' and 'what is scientific rigour?' Keep notes about your experiences and actions so as to stimulate awareness about your own role and position (reflexivity).
6. Keep in mind that your research objective or question might evolve because of unexpected contextual developments. That is part of research, but even more so in transdisciplinary research. However, make sure to frequently reflect upon the process with senior transdisciplinary researchers.

References

Bansard, J. S., Hickmann, T., & Kern, K. (2019). Pathways to urban sustainability: How science can contribute to sustainable development in cities. *GAIA-Ecological Perspectives for Science and Society, 28*(2), 112–118. https://doi.org/10.14512/gaia.28.2.9

Bulten, E., Hessels, L. K., Hordijk, M., & Segrave, A. J. (2021). Conflicting roles of researchers in sustainability transitions: Balancing action and reflection. *Sustainability Science, 16*(4), 1269–1283. https://doi.org/10.1007/s11625-021-00938-7

De Haan, F. J., & Rotmans, J. (2018). A proposed theoretical framework for actors in transformative change. *Technological Forecasting and Social Change, 128*, 275–286. https://doi.org/10.1016/j.techfore.2017.12.017

Ellis, C., Adams, T. E., & Bochner, A. P. (2011). Autoethnography: An overview. *Historical Social Research/Historische sozialforschung, 36*(4), 273–290. https://www.jstor.org/stable/23032294

Directorate-General for Research and Innovation (European Commission). (2021). *Research and innovation for accelerating food system transformation—Operationalizing food 2030 through living labs*. Publications Office of the European Union. https://data.europa.eu/

Ellis, C., & Bochner, A. P. (2000). Autoethnography, personal narrative, reflexivity. Researcher as subject. In N. K. Denzin & Y. S. Lincoln (Eds.), *Handbook of qualitative research* (pp. 733–768). Sage.

Fazey, I., Schäpke, N., Caniglia, G., Patterson, J., Hultman, J., van Mierlo, B., Säwe, F., Wiek, A., Wittmayer, J., Aldunce, P., Al Waer, H., Battacharya, N., Bradbury, H., Carmen, E., Colvin, J., Cvitanovic, C., D'Souza, M., Gopel, M., Goldstein, B., … Wyborn, C. (2018). Ten essentials for action-oriented and second order energy transitions, transformations and climate change research. *Energy Research & Social Science, 40*, 54–70. https://doi.org/10.1016/j.erss.2017.11.026

Felt, U. (2021). Making and taking time: Work, funding and assessment infrastructures in inter- and transdisciplinary research. In B. Vienni-Baptista & J. Thompson Klein (Eds.), *Institutionalizing interdisciplinarity and transdisciplinarity: Collaboration across cultures and communities*. Routledge. https://doi.org/10.4324/9781003129424

Fenollosa, C., & Paca, C. (2018). *Deliverable 6.1. Catalogue on analysis of contents, formats and needs for trainings*. https://fit4food2030.eu/reports-publications/

FIT4FOOD2030. (2019). *Midterm review report; FIT4FOOD2030*. Amsterdam. https://fit4food2030.eu/wp-content/uploads/2019/06/FIT4FOOD2030-Mid-term-summary-FINAL.pdf

Hilger, A., Rose, M., & Kell, A. (2021). Beyond practitioner and researcher: 15 roles adopted by actors in transdisciplinary and transformative research processes. *Sustainability Science, 16*, 2049–2068. https://doi.org/10.1007/s11625-021-01028-4

Horcea-Milcu, A. I., Leventon, J., & Lang, D. J. (2022). Making transdisciplinarity happen: Phase 0, or before the beginning. *Environmental Science & Policy, 136*, 187–197. https://doi.org/10.1016/j.envsci.2022.05.019

Jaeger-Erben, M., Kramm, J., Sonnberger, M., Völker, C., Albert, C., Graf, A., Hermans, K., Lange, S., Santarius, T., Schröter, B., Sievers-Glotzbach, S., & Winzer, J. (2018). Building capacities for transdisciplinary research: Challenges and recommendations for early-career researchers. *GAIA, 27*(4), 379–386. https://doi.org/10.14512/gaia.27.4.10

Kok, K. P. W., den Boer, A. C. L., Cesuroglu, T., van der Meij, M., de Wildt-Liesveld, R., Regeer, B. J., & Broerse, J. E. W. (2019). Transforming research and innovation for sustainable food systems—A coupled-systems perspective. *Sustainability, 11*(24), 7176. https://doi.org/10.3390/su11247176

Loorbach, D., Frantzeskaki, N., & Thissen, W. (2011). A transition research perspective on governance for sustainability. In C. Jaeger, J. Tàbara, & J. Jaeger (Eds.), *European research on sustainable development*. Springer. https://doi.org/10.1007/978-3-642-19202-9_7

Patterson, J. J., Lukasiewicz, A., Wallis, P. J., Rubenstein, N., Coffey, B., Gachenga, E., & Lynch, A. J. J. (2013). Tapping fresh currents: Fostering early-career researchers in transdisciplinary water governance research. *Water Alternatives, 6*(2), 293–312. http://hdl.handle.net/11071/3778

Pohl, C., Rist, S., Zimmerman, A., Fry, P., Gurung, G. S., Schneider, F., Spearnza, C. F., Kiteme, B., Bolliat, S., Serrano, E., Hadorn, F., & Wiesmann, U. (2010). Researchers' roles in knowledge co-production: Experience from sustainability research in Kenya, Switzerland, Bolivia and Nepal. *Science and Public Policy, 37*(4), 267–281. https://doi.org/10.3152/030234210X496628

Raven, R., Van den Bosch, S., & Weterings, R. (2010). Transitions and strategic niche management: Towards a competence kit for practitioners. *International Journal of Technology Development, 51*(1), 57–74. https://doi.org/10.1504/IJTM.2010.033128

Redman, A., & Wiek, A. (2021). Competencies for advancing transformations towards sustainability. *Frontiers in Education, 6*, 785163. https://doi.org/10.3389/feduc.2021.785163

Regeer, B. J., Hoes, A., van Amstel-van Saane, M., Caron-Flinterman, C., & Bunders, J. F. G. (2009). Six guiding principles for evaluating mode-2 strategies for sustainable development. *American Journal of Evaluation, 30*(4), 515–537. https://doi.org/10.1177/1098214009344618

Sarkki, S., Heikkinen, H. I., & Karjalainen, T. P. (2013). Sensitivity in transdisciplinary projects: A case of a reindeer management in Finland. *Land Use Policy, 34*, 183–192. https://doi.org/10.1016/j.landusepol.2013.03.004

Schäpke, N., Stelzer, F., Caniglia, G., Bergmann, M., Wanner, M., Singer-Brodowski, M., Loorbach, D., Olsson, P., Baedeker, C., & Lang, D. J. (2018). Jointly experimenting for transformation? Shaping real-world laboratories by comparing them. *GAIA-Ecological Perspectives for Science and Society, 27*, 85–96. https://doi.org/10.14512/gaia.27.S1.16

Schneider, F., Giger, M., Harari, N., Moser, S., Oberlack, C., Providoli, I., & Zimmerman, A. (2019). Transdisciplinary co-production of knowledge and sustainability transformations: Three generic mechanisms of impact generation. *Environmental Science & Policy, 102*, 26–35. https://doi.org/10.1016/j.envsci.2019.08.017

Schneidewind, U., Augenstein, K., Stelzer, F., & Wanner, M. (2018). Structure matters: Real-world laboratories as a new type of large-scale research infrastructure. A framework inspired by Giddens' structuration theory. *GAIA-Ecological Perspectives for Science and Society, 27*(1), 12–17. https://doi.org/10.14512/gaia.27.S1.5

Schuijer, J., Broerse, J. E. W., & Kupper, F. (2021). Juggling roles, experiencing dilemma's: The challenges of SSH scholars in public engagement. *NanoEthics, 15*(2), 169–189. https://doi.org/10.1007/s11569-021-00394-8

Sellberg, M. M., Cockburn, J., Holden, P. B., & Lam, D. P. M. (2021). Towards a caring transdisciplinary research practice: Navigating science, society and self. *Ecosystems and People, 17*(1), 292–305. https://doi.org/10.1080/26395916.2021.1931452

Sol, A. J. (2018). *Reflexively stumbling towards sustainability. Understanding social learning in regional governance networks.* PhD thesis, Wageningen University, The Netherlands. https://doi.org/10.18174/448662

Stilgoe, J., Owen, R., & Macnaghten, P. (2013). Developing a framework for responsible innovation. *Research Policy, 42*, 1568–1580. https://doi.org/10.1016/j.respol.2013.05.008

Svare, H., Gjefsen, M. D., Den Boer, A. C. L., & Kok, K. P. W (2023). Learning systems and learning paths in sustainability transitions. *Ecology and Society, 28*(1), 22. https://doi.org/10.5751/ES-13868-280122

Tijsma, G., Hilverda, F., Scheffelaar, A., Alders, S., Schoonmade, L., Blignaut, N., & Zweekhorst, M. (2020). Becoming productive 21st century citizens: A systematic review uncovering design principles for integrating community service learning into higher education courses. *Educational Research, 62*(4), 390–413. https://doi.org/10.1080/00131881.2020.1836987

Torrens, J., & von Wirth, T. (2021). Experimentation or projectification of urban change? A critical appraisal and three steps forward. *Urban Transformations, 3*(1), 1–17. https://doi.org/10.1186/s42854-021-00025-1

Turnhout, E., Stuiver, M., Klostermann, J., Harms, B., & Leeuwis, C. (2013). New roles of science in society: Different repertoires of knowledge brokering.

Science and Public Policy, 40(3), 354–365. https://doi.org/10.1093/scipol/scs114

Van Mierlo, B., Regeer, B., van Amstel, M., Arkesteijn, M., Beekman, V., Bunders, J., De Cock Buning, T., Elzen, B., Hoes, A. C., & Leeuwis, C. (2010). Reflexive monitoring in action. A guide for monitoring system innovation projects. *Communication and innovation studies*, WUR; Athena Institute, VU. https://edepot.wur.nl/149471

Van Poeck, K., Östman, L., & Block, T. (2020). Opening up the black box of learning-by-doing in sustainability transitions. *Environmental Innovations and Societal Transitions, 34*, 298–310. https://doi.org/10.1016/j.eist.2018.12.006

Wittmayer, J. M., & Schäpke, N. (2014). Action, research and participation: Roles of researchers in sustainability transitions. *Sustainability Science, 9*(4), 483–496. https://doi.org/10.1007/s11625-014-0258-4

CHAPTER 15

Reassembling the Scholar: A Conversation About Positionality in Transdisciplinary Processes

Callum Gunn, Sabine Hoffmann, Morten Sager, Julia Wittmayer, and Teun Zuiderent-Jerak

15.1 Introduction

Transdisciplinary research (TDR) is a framing of scientific practice in which collaboration on problems or issues of common interest is located outside disciplinary approaches, and is geared towards a transformation of the current situation towards something more desirable, albeit without necessarily knowing what that is. Any form of 'transformation' invariably has a normative component, and the same goes for positionality. A TDR context thus has much to do with both. The issue of positionality specifically in inter- and transdisciplinary research settings has become a more

C. Gunn (✉)
University Medical Centre Utrecht, Utrecht, The Netherlands
e-mail: c.j.gunn@umcutrecht.nl

S. Hoffmann
Swiss Federal Institute of Aquatic Science and Technology, Dübendorf, Switzerland
e-mail: sabine.hoffmann@eawag.ch

B. J. Regeer et al. (eds.), *Transdisciplinarity for Transformation*,
https://doi.org/10.1007/978-3-031-60974-9_15

prominent object of study within these scholarly traditions (e.g. Freeth & Vilsmaier, 2020).

Our contribution takes the positionality of the researcher as the departure point. In much the same way that is impossible to conduct TDR in a way that is detached from the practices it seeks to change, we cannot remove our academic identities from the change process. We are 'alive to the world too'. Essentially, we are talking about taking the effort to relate to the worlds we are trying to transform in one way or another (Law & Singleton, 2013). Furthermore, in the context of transformative research (Fazey et al., 2020, p. 6), there is a call for more 'reflexive second order science' which 'shifts focus away from studying a system as if looking in from the outside to conducting research as if from within. This includes reflexively examining one's own role in the way a system is reproduced. This opens space for inclusion of more diverse forms of knowledge and knowing'. For Bartels and Wittmayer (2018, p. 6), knowing 'is thus not a monological process of "discovering" an external, static reality in which researchers can abstain from any responsibility for it; rather, it is a dialogical process of intervening in actual situations with immediate consequences for who is and who is not affected, included, and empowered'. We work dialogically to enact this—to couple our contributions more carefully with our own identities. The first author (CG), an early career researcher, speaks with the co-authors (SH, MS, JW, TZJ), who are more experienced researchers from various fields of science and society working in inter- and transdisciplinary settings.[1] We introduce the different themes of the conversation in connection with relevant literature

M. Sager
University of Gothenburg, Gothenburg, Sweden
e-mail: morten.sager@theorysc.gu.se

J. Wittmayer
Erasmus University Rotterdam, Rotterdam, The Netherlands
e-mail: j.m.wittmayer@drift.eur.nl

T. Zuiderent-Jerak
Vrije Universiteit Amsterdam, Amsterdam, The Netherlands
e-mail: teun.zuiderent-jerak@vu.nl

[1] As group leader of inter- and transdisciplinary research, Sabine Hoffmann's research focus includes management and leadership of integration processes in TDR. Morten Sager leads a group that combines the production of systematic review evidence for use in health

situate insights from this dialogue to scholarly discussions throughout the chapter.

The following activities were integral to the crafting of this conversation. The first author initially undertook an analysis of a dataset of 'learning questions' from a postgraduate training programme on TDR at Vrije Universiteit Amsterdam (2015–present). During the programme, participants develop and address personal learning questions in relation to setting up, conducting and evaluating TDR. These questions reflect on what researchers find important (often at the beginning of their career) in making sense of their own efforts in implementing this form of research practice. In an analysis of over 500 learning questions, 53 were coded as the researcher's roles and competences in transdisciplinary research settings. The analysis then fed into a workshop, co-designed by the authors, convened at the International Transdisciplinarity Conference in September 2021. Fourteen workshop participants discussed the challenges and opportunities of the roles of researchers in transdisciplinary processes.

To an extent, the dialogue here aims to express a broader perspective and experience than our own, while acknowledging the difficulties in generalizing the nature of transdisciplinary work. With that caveat, we frame TDR (and thus our discussion on roles within it) with three salient features. First, the centrality of learning processes as part and parcel of research practice that seeks to contribute to transformative processes in some way (Hirsch Hadorn et al., 2008). Second, we emphasize the pertinence of experimentation in such processes. Third, since the research aims to contribute to transformation related to some form of a real-world challenge or problem, the research setting tends to be particularly normatively charged. The examples we discuss below—from our experience in transdisciplinary initiatives—illustrate these features in action, and their implications for the roles and positionalities of the researcher.

care, social work, and education organizations with epistemological analyses of evidence-based practices. Julia Wittmayer leads various initiatives on transformative social innovation and studies the role of research and transdisciplinary engagements in (the governance of) sustainability transitions. Teun Zuiderent-Jerak leads projects on transdisciplinary social studies of health care and medicine and co-founded the Making & Doing programme at the Society for Social Studies of Science (4S) that seeks to support transformative scholarship in this field.

15.2 Breaching Researcher Identities in Transdisciplinary Work

15.2.1 Breaching Experiments: Taking Apart Researchers' Identities in Transdisciplinary Research

Understanding the role of the researcher should not only be a case of learning by doing since there are many valuable resources to help us prepare for transdisciplinary research. These include guidance on the many roles of the researcher, based on previous empirical work (Wittmayer et al., 2021, p. 11). These can be learned *up to a point*. But we should emphasize that roles are developed and crafted in specific contexts. In relation to *acting with* others, with different worldviews, interests and framings, our roles and identities are *breached* as we navigate through making sense of shared issues of concern.

CG: A very common question postgraduate students pose concerns how they are supposed to know what roles to assume in their project at what time. Some are more comfortable with being an expert scientist, others with being a facilitator or change agent.

JW: The thing about researchers' roles in TDR is that they are made in practice. They aren't something you are or have which you can just apply in the research setting. We have to position ourselves again and again.

Box 15.1. Using roles as resources in urban sustainability research
This research took place in a Rotterdam neighbourhood that had been labelled as a deprived area, with various problems from a policy perspective. Low levels of income, lower levels of education in comparison with the rest of the city—issues that have become considered to be problematic. The work involved trying to understand how sustainability governance could work on the local level: how could problems be addressed in the neighbourhood and by whom? The idea was that a systems analysis—including problem structuring, visioning and experimentation, that we would conduct with about 20 people from the area, together—would elicit insights and actions for the neighbourhood to become more sustainable.

JW: And we met with some resistance initially. Specifically connected to the idea that researchers would come in and 'extract' knowledge from the neighbourhood without any form of reciprocity. To address these concerns and to clarify the type of transdisciplinary action-oriented research we were aiming to do, we convened a meeting to contextualize our methodologies. This meeting included negotiating our roles as action researchers. Eventually, they related our research to some of the participants' experiences in 'engaged research' initiatives in Rotterdam as part of urban renewal programmes in the 1980s—as *activating researchers*. That's where we really found each other in terms of the language we were using, and the positions we were taking.

CG: Yes, finding a *shared language* is important if you want to move forward. But what makes this so applicable to transdisciplinary research? Isn't this relevant for all forms of research? Someone working in a physiology lab surely has to adapt in some way in relation to the surprises thrown up by their experiments. Social studies of science have explored this kind of thing before (see e.g. Star, 1985). Isn't that the same as having to deal with resistance in your research?

JW: To some degree, but there is something else, something different. Being able to talk to one another is surely necessary in transdisciplinary research, where 'the public' is not some vague idea 'out there', but a *collaborating party* in the research. I tried to address the tension resulting from different understandings of what research is and what researchers do by referring to roles as a resource—that these are negotiated through dialogue. I think that this tension comes from the dilemma of doing 'research' and 'engagement'. While you are critical, you also know the system, including the actors involved, and take a critical stance towards injustice and so on. But you also need to be relational—finding ways to relate to the people you are working with in these shared endeavours. One way of doing so is by *negotiating roles* through dialogue (see Bartels and Wittmayer [2018], who discuss the critical–relational position in the context of transformative action research).

MS: All research processes have uncertainty, but this is amplified in TDR settings.

CG: Which means we cannot anticipate the kinds of roles researchers need to take in advance—this has to emerge in practice in some way.

SH: And that uncertainty means we may have to depart from the comfort zones of pre-defined roles as researchers.[2] Assuming other roles than that of the 'traditional scientist' involves leaving your comfort zone, it has a lot to do with vulnerability as well as uncertainty.

CG: Meaning that we could expect to find ourselves in positions that feel uncomfortable?

TZJ: And we also need to consider the ways in which certain roles are ascribed to us.

MS: The example below [Box 15.2] refers to these—it comes from my work with several Swedish health and welfare agencies, where we have been involved in producing systematic reviews as a consequence of the evidence-based Master's programme that I coordinate at my university.

Box 15.2. Lending legitimacy in evidence-based practices
Three cases of these collaborative evidence-based initiatives have been conducted for and with health and welfare organizations.[3] The first was with a public health agency and concerned systematic reviews on interventions to prevent suicide. The second includes two systematic reviews, one on the prevention of violent extremism and another on prevention of antisemitism in schools. In the third case, we collaborated with a provider of social services. As in the other two cases, we developed reviews, this time for disability care. In this case, however, we simultaneously developed the review format, drawing on Science and Technology Studies (STS) scholarship on standardization and informal expertise.

MS: The common denominator in all of the projects in which we were invited to participate was on the basis of our expertise in systematic reviews, i.e. a technique for producing summaries of all available knowledge in a certain area. This type of expertise is highly valuable in cases of

[2] Freeth and Caniglia (2020) elaborate on the interlinked *comfort* zone, *discomfort* zone and *learning* zones in inter- and transdisciplinary work.

[3] This tight relationship with social partner organizations to produce systematic reviews can be seen as the trans-disciplinary aspect of the practice, in contrast to a more straightforward commissioning relationship between a funding body and an academic institute that designs and carries out a project in line with pre-defined requirements.

high scientific, professional or political uncertainty. In all of the cases, I would claim that this expertise had the potential to lend legitimacy to the commissioning organizations. What differed, however, were the commissioners' interest in and openness to our STS-related analytical competencies. In short, all of them were interested in legitimacy, but their interest in reflexive scholarship varied...

CG: So your role became ascribed through your position as an academic researcher?

MS: Yes and no. My role was ascribed through a particular set of research competencies that met the needs of the commissioning body. Other potential academic roles—such as my reflexive abilities trained in STS—that were less familiar, understood or desired by the commissioning body were sometimes *not* ascribed to me, at least not formally. This does not have to become a problem, but it *may* do so in at least two ways, both related to methodological norms. One problem concerns different actors' requirements and another concerns academic requirements. In most situations, I suspect that my reflexive abilities were picked up implicitly by actors as a certain relaxed attitude to the legitimating methodologies that had prompted the commission. The STS motto 'it could have been different', I think, may create spaces for actors who sometimes are burdened by constrained methodological requirements, not least in areas influenced by the evidence-based policy movement. Several explicit interactions in all of the three cases lend support to this conclusion; our broad methodological approach to systematic reviews was interesting and to some extent liberating. We came in with a legitimating method in combination with a broader-than-usual methodological attitude.

CG: Both you and the commissioning bodies benefited from your dual position! But you suggest that this dual position can become a problem, too?

MS: I believe so. In one particular case our methodological breadth became a problem. While our team's work had previously been appreciated in one agency, its internal processes at one point deliberately moved to redefine and constrain the methods we had at our disposal. This started a series of negotiations between me and the commissioning party that ended in divorce. The ascribed role, you could say, became too one-sided, too detached from our analytical interest and conscience. We had to ask for a painful, premature ending of the contract. Even if, initially, there had been an informal acceptance of our methodological breadth, the agency's evolving internal processes clearly trumped this tacit agreement. At some point the institution declared quite clearly: 'it cannot be different'.

Another problem of roles concerns the issue of building up a career. Since I have tenure, I have considerable space for deciding how to use my time and what interests to cater to. Clearly, this is more perilous in a more junior position! As a PhD student, disciplinary detours can threaten a burgeoning academic identity. But also for my situation: in order to be promoted and for collegiate recognition, I needed to adjust to the methodological criteria of my own discipline, where I had to publish in specific journals, using particular theoretical resources. It is not enough to use STS to widen collaborating actors' methodological spaces. This is understandable. We have to conform to the methodologies of our disciplines. But, in another sense it surely isn't understandable. My university boasts of its aim to engage collaboratively with the surrounding society, and many colleagues are curious about my group's collaborations, but for promotion disciplinary-bounded publications are more or less everything. Thus, in my view, transdisciplinary work is the sometimes perilous, and more often fruitful, navigation between more or less implicit methodological norms.

CG: The other cases were more fruitful than perilous?

MS: Fortunately, yes! The second case resembled the first one, but without the evolving institutional constraints on all the available methods. However, the commissions clearly did not involve any fully fledged reflexive processes, typical of STS analyses. The boundary conditions of the deliverables were pre-set, but the path to producing these allowed for some methodological breadth. In the third case, there was far greater scope. We are developing new ways of doing knowledge support and reviews. There are some pre-set rules we are using, which we are trying to expand and work with reconstructively. As such we are changing our ways of doing things, adjusting to certain demands to deliver what is needed, while our collaborators are also changing in this process. We are both changing and learning—it is a two-way thing.

CG: Would it be correct to say that the third case combines the best of two worlds: both legitimacy and reflexivity, both concrete deliverables for the actors and analytical deliverables for academia? If so, what are the preconditions of such win-win collaboration?

MS: A very relevant question. In my view, the interaction between epistemology and institutional arrangements is key here. In my three cases *institutional arrangements* have been so important. In the first case there had been epistemological affinities for several years, but they did not hold under changing institutional pressure. In the second case, a common epistemological awareness facilitated collaborations, but the institutional frames did not promote a more fully fledged exploration of the methodological formats that would have been analytically interesting for our group. The institutional frames of the third case—through funding and ownership—were designed, from the outset, to combine legitimating delivery and analytical explorations and even reconstructions. It is very interesting. What sort of institutional spaces or frameworks enable all actors involved to adapt, learn and transform? And of course, you are not always in a position to decide on your own.

CG: And that may be a good thing, may it not?

MS: I think you are right. Too much control may not be compatible with the nature of transdisciplinary work. It is true that the evolving boundaries on available methods in the first case were detrimental—to us, perhaps not to the agency—but you cannot avoid running into boundaries. What is crucial to me is the possibility to experiment with these boundaries. Haven't we all experienced 'breaching experiments': places, and times, when our ambitions, competencies or roles were breached, because of the nature of transdisciplinary research? It's that opening up that we've all felt in some way: we had all come into these situations thinking 'this is what I want to do?' and then realizing that the actors, the collaborators, want to do something else. And they know things—*things I didn't know that I didn't know*. We have all had these experiences of the world (our worldviews? framings?) being 'broken apart' to some extent. These experiences tell us that our roles as researchers are not fixed in TDR—as part of the co-learning process, our positions, relations to others and so on, must be reassembled in the practice of research.

15.2.2 *The Consequences of Breaching: Academic Quality and Qualifying*

Quality and qualifications are crucial when we refer to positioning in transdisciplinary research. Early career researchers are well aware of the dilemma posed by doing TDR in the academic environment. They want to know how they can navigate between taking different positions, in particular in becoming an expert in 'their field' while conducting research work beyond disciplinary boundaries. This raises the question: to what kinds of quality criteria can and should we subscribe? For instance, many early career researchers in the global health field want to address how they can situate themselves more as activists while at the same time remaining part of the status quo, being part of the university infrastructure in academic research. Although efforts to qualify good scholarly work should not be downplayed, others have argued that conventional comparisons between disciplinary and transdisciplinary work places them on the same plane, which might not necessarily be very useful (see Fochler & Rijcke, 2017; Rafols et al., 2012). Following this line of thought, the point is that transformative research settings have a different set of implications.

CG: A common question or concern is that it appears difficult to know what good scholarly inquiry looks like in transdisciplinary settings. It seems far more elusive than in disciplinary spaces, because of the highly context-specific nature of transdisciplinary research products and findings. Especially if, as we are saying, the craft of the research process demands flexibility and creativity,[4] we are always situated in a specific time and space.

SH: We started by conducting a systematic review, trying to develop indicators for TDR 'success' with a focus on integration. We struggled! For four years now, I have had this huge database with a wide range of indicators. Although it's not worth poring over the details, there is something to be said about assuming and balancing a dual role in TDR settings, that is, what we called a '*creative science role*' and, at the same time, a '*supportive service role*'. I should emphasize the distinction between the two roles here. The supportive service role implies, for instance, becoming familiar with a particular method or tool to be applied in a specific TDR setting and facilitating the subsequent TDR process in that setting.[5] The creative science role means digging into different disciplinary perspectives, identifying relevant gaps and critical connections,

[4] See Cerwonka and Malkki (2007) on the improvisation of fieldwork practice. Law (2004) frames it as the mess of research methods.

[5] Deutsch et al. (2021) go into more depth on the distinction. Similarly, Guggenheim (2006) refers to the procedural aspects of transdisciplinary research. 'The procedural' is

and linking them to the broader literature. Assuming and balancing these two roles in TDR settings is challenging, but also allow you to develop expertise in facilitation and integration—so they are really key elements of quality transdisciplinary work. A prerequisite to a transdisciplinary expertise, even.

CG: Is it a case of rethinking what 'the scientific role' consists of?

SH: It's also about visibility. Being acknowledged for this kind of important work of creating the bigger picture by combining previously unconnected perspectives; this work of establishing critical connections represents a very important intellectual contribution of researchers working at the interface between different disciplines, but also between research, policy and practice. Unfortunately, that work is rarely made visible, and is therefore rarely acknowledged and recognized.

CG: So in making room for different perspectives on an issue, fostering their articulation and so on, the researcher facilitating that process needs, in some way, to do away with their own specialist or expert framing on the matter, while simultaneously being able to make their own 'intellectual contribution'.

SH: Yes! This work led to us developing the notion of 'integration experts'—researchers who lead, administrate, manage, monitor, assess, accompany, and/or advise others on inter- and transdisciplinary integration.[6] This requires taking on and balancing many different roles, which is a role *in itself*.

CG: In that it demands a certain type of 'expertise' to be able to navigate through these positions?

SH: Yes, definitely! I think you need different kinds of expertise; you need a sort of *interactional expertise* (following Collins & Evans, 2002). That means the expertise to speak the language of a discipline or field without necessarily being able to contribute to that discipline or field in depth. You also need *contributory expertise*, namely the expertise to

different to 'science' because it does not travel; it cannot be frozen into a mechanism of scientific method; procedures are by definition the things that cannot travel.

[6] Hoffmann et al. (2022) discuss the institutional implications of a growing focus on the 'integration expert' in inter- and transdisciplinary research settings.

contribute to research in a discipline or field. But such expertise, I would say, is not enough; when we conducted a workshop on integration experts at the ITD Conference in 2019, participants claimed that critical personal qualities were at least as important as such different kinds of expertise. These qualities included, for instance, openness, curiosity, creativity, sociability, but also persistence and patience as well as degrees of reflexivity and humility.[7] I think, if we really believe that TDR can contribute to solving pressing societal problems of our time, we need to make room for these kinds of experts, this sort of expertise—which I consider crucial for realizing the integrative potential of TDR.

CG: Yes, but I suppose the point here comes back to the struggle you have faced in classifying 'successful' TDR. It is difficult to do because TDR (and the dimensions of expertise needed to enable it) is very different, or perhaps needs to be so, in different places and spaces. So it is probably useful to think about the *qualities of TDR work* to get a sense of in what ways we can qualify good work.

MS: I have a deep scepticism of and suspicions about theory—generalizations in particular. So I see it as a mark of quality that I'm not *just* doing 'academic work' when I'm engaging with health agencies, social care providers and so on. In doing so I am being challenged by their worldviews, their needs, having to translate what I think I know into their languages—that's really a driver for me. It's so easy to get stuck in thinking 'because I wrote a paper about it, accepted by this journal' equates to having an authoritative understanding of the matter. That to me is not quite good enough.

JW: For me, it's better—more intellectually stimulating—to be making sense of things together with others in the room. I 'grew up', academically speaking, in a very 'engaged-activist' kind of institute—I eventually stepped out, towards this 'normal', disciplined way. I also had to gain my credits there. So I had research endeavours that were less transdisciplinary. I missed that embodied learning with people in the room. And I think better theory comes out of it.

CG: In what sense can 'better' theory come out of it?

[7] See Hoffmann et al. (2022) for more on these qualities.

TZJ: If we're not doing it removed from practices that we're trying to be involved in changing. It's trying to live those as not being separate. That's where I think we can really do more interesting theorizing!

SH: If we're trying to bridge the perceived divide between the theory and practice of transdisciplinary research and more strongly articulate our conceptual and empirical insights, this is something that is currently lacking in the TDR literature. It's so crucial, though, to see these not as two separate spheres but a hybrid one.

15.2.3 Summary: Breaching Research Identities

Taking roles and positions is less a matter of responsibility, a kind of social contract, and more about the kinds of *positioning work* (Felt et al., 2013) within transdisciplinary processes that we researchers find ourselves 'doing'. Such work involves leaving the comfort zone of pre-defined research roles and positions, negotiating new ones that transcend prescriptions of the researcher's identity and synchronizing multiple roles/ positions that may be adopted, assumed, ascribed or resisted. As a consequence of all our *breaching experiments*—exposure to and interaction with different worldviews and logics, different problem definitions and so on in transdisciplinary work—our own identities are breached. Although certain qualities are useful for handling this, we cannot always know what and when these are going to be useful. Then, carving out different spaces (physical, institutional, intellectual, etc.) is integral to supporting different sorts of positioning work in order to deal with the breaching of our identities in transdisciplinary scholarship.

15.3 Carving Out Spaces: Anchoring Transdisciplinary Scholarship

Graduate students often express the dilemma posed by 'growing up' in transdisciplinary academic environments. At the same time, the lack of a disciplinary home to tap into makes one's intellectual identity and career trajectory somewhat ambiguous (Haider et al., 2018). Felt et al. (2013) interviewed candidates in a transdisciplinary PhD programme at the University of Vienna and elicited these themes. Felt develops the term *epistemic living space* to describe the 'entanglements of institutional rationales, epistemic work, life course decisions, and wider research and teaching politics' (Felt, 2022, p. 207) that contribute to how (early career) researchers make sense of manoeuvrability in TDR. While others may 'happily retreat to their own specific fields' when they like, young scholars in TDR might be faced with an unclear present, and an uncertain future. There is a lack of clear incentives for transdisciplinary scholarship as the mainstream academic reward structures tend to promote disciplinary work. Transdisciplinary research work tends to be overlooked or misrepresented in 'classic' approaches to the evaluation of academic performance (e.g. Fochler & de Rijcke, 2017; Rafols et al., 2012) while traditional incentives for university scholarship tends to marginalize transdisciplinary work (Müller & Kaltenbrunner, 2019). PhD and post-doc contracts are often short, leaving little space to balance disciplinary academic development with the uncertain nature of transdisciplinary work. Transdisciplinarity has thus been viewed often as an 'add-on' rather than integral to an individual's research practice (Schmidt & Pröpper, 2017). With more early career researchers working on explicitly transdisciplinary projects, and growing up in transdisciplinary environments, this dynamic is probably shifting (Felt et al., 2016).

Many questions we have discussed relate to the idea of *socialization:* what we are growing into, to what can we anchor, to what sort of intellectual communities can we subscribe and contribute. Many postgraduates note that growing up in academia doing TDR or positioning themselves as transdisciplinary scholars—might be detrimental to their research career as opposed to working and progressing within the confines of a specific discipline, which offers a clearer trajectory and fewer risks. Conducting transdisciplinary processes during one's academic 'training' means less time is spent on learning and applying specific disciplinary methodologies and competencies.

15.3.1 Making Space for Experimenting: At Ease with the Unease of Transdisciplinary Research

How can the early career researcher deal with the friction of working in transdisciplinary settings when academic recognition is oriented towards disciplinary outputs? Perhaps we should be thinking differently about this issue of socialization. We know that current academic structures should be more inclusive of inter- and transdisciplinary research. And it is clear that different sorts of spaces need to be made to accommodate those doing transdisciplinary work.

TZJ: The struggle from my experience has been how to get transdisciplinary STS to count as scholarship.

CG: Less of a focus on papers as *the* academic product?

TZJ: But more actually doing and being part of the research. Recognizing the intricacies of doing non-linear scholarship and knowledge production.

CG: Right. How can that sort of recognition become instilled in an academic setting?

TZJ: There are initiatives that seek to achieve just that. I have been part of a programme called 'Making and Doing' in the STS field where others agreed or felt this struggle. It became more of a question of how you get that other kind of work to count. We have attempted to infrastructure something that could do that with the Making and Doing programme (Downey & Zuiderent-Jerak, 2021). So, for instance, during making and doing programmes held at conferences, people get a 2 × 1 metre table where they can show what kind of things they've been involved in.

CG: Any more instructions?

TZJ: That's pretty much it. And that's the point—there is room to experiment with this set up. Contributors can show their experiments with other forms of knowledge production, expression and travel. The format allows them to express things differently, and it is interesting to consider what happens as a consequence: how have your assumptions been challenged, how has your theorizing changed? It speaks to people in that it doesn't force them to make a choice between academic and other versions of the self. And this is especially appealing to doctoral researchers, as it means they don't have to choose between commitments as scholars and commitments in their other 'roles', or other positions, other sides to oneself.

CG: So why is getting more or different things 'to count as scholarship' important for understanding researchers' positions?

SH: I think it's about making other kinds of important work visible. This is similar to the dual role we often assume in leading inter- and transdisciplinary processes. We have tried to establish more visibility for the many sorts of roles researchers adopt in these in-between positions. For instance, at my own institute we created a Community of Practice (CoP) in 2015, which includes current programme, platform managers, coordinators and leaders who share a common interest in leading integration at the interface between science and practice (see Hoffmann et al., 2017). The CoP meets three or four times a year. Creating such a community has been important: it provides a 'safe' space to jointly reflect on our shared experiences in working at the interface, the challenges and opportunities it implies; it also provides a home ground for researchers to anchor, thus countering the potential feeling of intellectual homelessness (Lyall, 2011, p. 80). I think it is really important to feel at home somewhere.

CG: That's important. Especially as a researcher still finding your feet, if you don't have a clear peer community with whom you can make sense of your experiences, you can easily feel as if you are stuck in between legitimate academic spaces.

SH: Yes, it easily produces a sense of in-betweenness; a community can counterbalance this sense, while giving greater visibility—in our case—to programme/platform managers/coordinators/leaders.

CG: And as a collective, is this community recognized by others?

SH: Yes, it is recognized within our institute; we interacted, for instance, with our director, who asks the CoP to discuss certain topics and issues and provide particular inputs. Conversely, we asked our director, for example, to review two scientific papers that have been written by CoP members.

CG: Meaning you have a sort of collective voice in wider spaces.

SH: Yes, I would say so.

15.3.2 Coupling the Human and Scholarly Self: On Being an Idiot, and Other Commitments

The approaches we discussed above also address the second aspect of visibility, which involves recognizing the human element of the researcher, which is so often rendered invisible in academic work. The advice of Patricia Hill Collins regarding the positioning of an 'outsider within' status in the social sciences seems particularly salient here. Those taking an outsider within position 'learn to trust their own personal and cultural biographies as significant sources of knowledge. In contrast to approaches that require submerging these dimensions of self in the process of becoming an allegedly unbiased, objective social scientist, outsiders within bring these ways of knowing back into the research process' (Collins, 1986, p. 28). Framing TDR as a boundary-crossing practice invites us to reflect more explicitly on our commitments as scholars: what exactly mandates drawing certain boundaries to perform certain empirical approaches using specific concepts and methods. Furthermore, inter- and transdisciplinarity offers a space to think carefully about the relations between the objects and subjects of research, how these are defined at the level of scientific knowledge practice of which we scholars are part. Marres and de Rijcke (2020) call for exploring the object–subject relations further in TDR. What does it mean for us to be alive to the world in the research process?

SH: This is sometimes lacking, I mean, the focus on the human beings in the academic system. Sometimes it feels like I am surrounded by heroes and I then ask myself: where are the human beings, with all their failures and successes?

CG: Like superheroes wearing a disguise?

TZJ: It is important to address the performance of the academic self as a hero. I'd like to ask some idiotic questions, throw in some confusion, rather than necessarily having to solve something, or resolve a tension. And at the same time I'm so chicken when it comes to actually speaking up in a certain setting and thinking they could exclude me. But I want to build relations, and make sure they still take me seriously, although I bring a different perspective. So if you have a space in which you feel somewhat legitimized, that could be your institution, or could be people from whom you take inspiration ... sometimes I'd like to dare to take a little more space. The safe space helps you to be a bit of an idiot sometimes.

CG: Throwing in confusion sounds as though it could be unproductive. Are there certain times when it's better to be an idiot than in others?

TZJ: It's not really about being confused as such, but being comfortable with not knowing where we are going in the process.

MS: It's about abandoning the idea of competency at some level. Accepting incompetence.[8]

CG: Right—because we are not exactly encouraged to be incompetent.

SH: The academic system forces us to be an expert. We are trained not to ask incompetent questions.

CG: So in terms of our positioning work, these spaces should help us take those risks—be more comfortable in more vulnerable positions perhaps—so that they offer a space to reflect and learn.

TZJ: These different initiatives serve as examples of building infrastructure that allows for these kinds of roles and (in)competencies. So, the question is not only about how can someone 'learn it' but more about how can we carve out academic spaces where it is legitimate to experiment—meaning, in a broad sense, *to not know exactly what you are doing.* For that to be a good thing.

CG: Legitimacy is a good term to return to here. I think a lot of postgraduate researchers can feel illegitimate because of the nature of TDR. It feels logical, to me, to assume that the space in which to experiment in very complex environments, where the stakes seem high and so on, is best reserved for those who actually do know what they're doing. But I suppose that's the point, carving out spaces needs to make this kind of work legitimate.

SH: Another issue is physical and emotional exhaustion. Taking on the many roles as an inter- and transdisciplinary researcher often implies a significant workload. Moreover, being forced to navigate and manoeuvre the discrepancies between high-flown conceptualizations of an ideal-typical TDR process, and the lived experiences of numerous challenges in the process itself, only adds to this exhaustion (see Hoffmann et al., 2022). That's where these safe spaces serve to help, to support and learn from each other and nurture new forms of reflexive scholarship.

CG: It's true that the human side of the scholar is rendered invisible across pretty much all disciplinary work. In what sense is it important for transdisciplinary positionality to make the human aspect more visible?

[8] Thanks to Sheila Ramaswamy for articulating this point.

Why should this be a specific comment in a book about transdisciplinary research?

JW: Because TDR is ultimately about being relational: trying to relate to the people and fields you are trying to change. Here it is important to gain experience to begin being comfortable with 'not knowing exactly' about the very specific process, but learning to rely on your accumulated experience, knowledge and trust the process.

15.3.3 Summary: Anchoring Transdisciplinary Scholarship

We have discussed different ways of making room for TDR and the roles and qualities it demands of researchers. A transdisciplinary process breaches our identities and roles, and the carving out of safe and reflexive spaces enables us to handle the consequences; efforts to refigure institutional frameworks and boundaries to create such spaces where it is legitimate to take different positions, to experiment, and to *not know exactly what you are doing*. That does not mean being incompetent, but being comfortable with uncertainty, or at ease with the unease that arises from breaching experiments. There may be ways that (in)competencies can be taught to some extent. But it has as much to do with learning as it has with *unlearning*.

Ultimately, the themes we have discussed here may provide a loose framework relevant to the craft of empiricizing researcher roles—a much-needed element of TDR praxis. This practice is important for maintaining our scholarly commitments that include bridging between the conceptual and empirical spheres, improving our theorizing, challenging our own worldviews by being part of those things we have an interest in changing. Despite these commonalities that have brought us together in this conversation, it is important to stress that transdisciplinary scholarship needs to be heterogeneous—different in places and spaces. Working to develop our understanding of role dynamics—how roles are negotiated, ascribed and so on—provides opportunities for learning more about the fields to which we are trying to relate, as well as our own academic and analytical identities. This is critical for improving our understanding of how the relations between the subjects and objects of research become redefined in transdisciplinary knowledge practices.

References

Bartels, K., & Wittmayer, J. (2018). Introduction: Action research in policy analysis and transition research. In K. P. R. Bartels & J. M. Wittmayer (Eds.), *Action research in policy analysis* (pp. 1–17). Routledge.

Cerwonka, A., & Malkki, L. H. (2007). *Improvising theory: Process and temporality in ethnographic fieldwork*. University of Chicago Press.

Collins, H. M., & Evans, R. (2002). The Third Wave of Science Studies. *Social Studies of Science, 32*(2), 235–296. https://doi.org/10.1177/0306312702032002003

Collins, P. H. (1986). Learning from the outsider within: The sociological significance of Black feminist thought. *Social Problems, 33*(6), S14–S32. https://doi.org/10.2307/800672

Deutsch, L., Belcher, B., Claus, R., & Hoffmann, S. (2021). Leading inter- and transdisciplinary research: Lessons from applying theories of change to a strategic research program. *Environmental Science & Policy, 120*, 29–41. https://doi.org/10.1016/j.envsci.2021.02.009

Downey, G. L., & Zuiderent-Jerak, T. (Eds.). (2021). *Making & doing: Activating STS through knowledge expression and travel*. The MIT Press. https://doi.org/10.7551/mitpress/11310.001.0001

Fazey, I., Schäpke, N., Caniglia, G., Hodgson, A., Kendrick, I., Lyon, C., Page, G., Patterson, J., Riedy, C., Strasser, T., Verveen, S., Adams, D., Goldstein, B., Klaes, M., Leicester, G., Linyard, A., McCurdy, A., Ryan, P., Sharpe, B., … Young, H. R. (2020). Transforming knowledge systems for life on Earth: visions of future systems and how to get there. *Energy Research & Social Science, 70*, 101724. https://doi.org/10.1016/j.erss.2020.101724

Felt, U. (2022). Making and taking time: Work, funding, and assessment infrastructures in inter- and trans-disciplinary research. In B. V. Baptista & J. Thompson Klein (Eds.), *Institutionalizing interdisciplinarity and transdisciplinarity*. Routledge.

Felt, U., Igelsböck, J., Schikowitz, A., & Völker, T. (2013). Growing into what? The (un-)disciplined socialisation of early stage researchers in transdisciplinary research. *Higher Education, 65*(4), 511–524.

Felt, U., Igelsböck, J., Schikowitz, A., & Völker, T. (2016). Transdisciplinary sustainability research in practice: Between imaginaries of collective experimentation and entrenched academic value orders. *Science, Technology, & Human Values, 41*(4), 732–761. https://doi.org/10.1177/0162243915626989

Fochler, M., & de Rijcke, S. (2017). Implicated in the indicator game? An experimental debate. *Engaging Science, Technology, and Society, 3*, 21–40. https://doi.org/10.17351/ests2017.108

Freeth, R., & Caniglia, G. (2020). Learning to collaborate while collaborating: Advancing interdisciplinary sustainability research. *Sustainability Science, 15*(1), 247–261. https://doi.org/10.1007/s11625-019-00701-z

Freeth, R., & Vilsmaier, U. (2020). Researching collaborative interdisciplinary teams: Practices and principles for navigating researcher positionality. *Science & Technology Studies, 33*(3), Article 3. https://doi.org/10.23987/sts.73060

Guggenheim, M. (2006). Undisciplined research: The proceduralisation of quality control in transdisciplinary projects. *Science and Public Policy, 33*(6), 411–421. https://doi.org/10.3152/147154306781778795

Haider, L. J., Hentati-Sundberg, J., Giusti, M., Goodness, J., Hamann, M., Masterson, V. A., Meacham, M., Merrie, A., Ospina, D., Schill, C., & Sinare, H. (2018). The undisciplinary journey: Early-career perspectives in sustainability science. *Sustainability Science, 13*(1), 191–204. https://doi.org/10.1007/s11625-017-0445-1

Hirsch Hadorn, G., Biber-Klemm, S., Grossenbacher-Mansuy, W., Hoffmann-Riem, H., Joye, D., Pohl, C., Wiesmann, U., & Zemp, E. (2008). The emergence of transdisciplinarity as a form of research. In G. Hirsch Hadorn, H. Hoffmann-Riem, S. Biber-Klemm, W. Grossenbacher-Mansuy, D. Joye, C. Pohl, U. Wiesmann, & E. Zemp (Eds.), *Handbook of transdisciplinary research* (pp. 19–39). Springer Netherlands. https://doi.org/10.1007/978-1-4020-6699-3

Hoffmann, S., Deutsch, L., Klein, J. T., & O'Rourke, M. (2022). Integrate the integrators! A call for establishing academic careers for integration experts. *Humanities and Social Sciences Communications, 9*(1), 147. https://doi.org/10.1057/s41599-022-01138-z

Hoffmann, S., Pohl, C., & Hering, J. G. (2017). Exploring transdisciplinary integration within a large research program: Empirical lessons from four thematic synthesis processes. *Research Policy, 46*(3), 678–692. https://doi.org/10.1016/j.respol.2017.01.004

Law, J. (2004). *After method: Mess in social science research*. Routledge.

Law, J., & Singleton, V. (2013). ANT and politics: Working in and on the world. *Qualitative Sociology, 36*(4), 485–502. https://doi.org/10.1007/s11133-013-9263-7

Lyall, C. (Ed.). (2011). *Interdisciplinary research journeys: Practical strategies for capturing creativity*. Bloomsbury Academic.

Marres, N., & de Rijcke, S. (2020). From indicators to indicating interdisciplinarity: A participatory mapping methodology for research communities in-the-making. *Quantitative Science Studies, 1*(3), 1041–1055. https://doi.org/10.1162/qss_a_00062

Müller, R., & Kaltenbrunner, W. (2019). Re-disciplining academic careers? Interdisciplinary practice and career development in a Swedish Environmental

Sciences Research Center. *Minerva, 57*(4), 479–499. https://doi.org/10.1007/s11024-019-09373-6

Rafols, I., Leydesdorff, L., O'Hare, A., Nightingale, P., & Stirling, A. (2012). How journal rankings can suppress interdisciplinary research: A comparison between innovation studies and business & management. *Research Policy, 41*(7), 1262–1282. https://doi.org/10.1016/j.respol.2012.03.015

Schmidt, L., & Pröpper, M. (2017). Transdisciplinarity as a real-world challenge: A case study on a north–south collaboration. *Sustainability Science, 12*(3), 365–379. https://doi.org/10.1007/s11625-017-0430-8

Star, S. L. (1985). Scientific work and uncertainty. *Social Studies of Science, 15*(3), 391–427. https://doi.org/10.1177/030631285015003001

Wittmayer, J. M., Loorbach, D., Bogner, K., Hölscher, K., Hendlin, Y., Lavanga, M., Vasques, A., von Wirth, T., & de Wal, M. (2021). *Transformative research: Knowledge and action for just sustainability transitions* (DIT Working Papers). Design Impact Transition Platform, Erasmus University Rotterdam. https://www.eur.nl/en/media/2021-11-dit-working-paper-1dit-platformerasmus-university-rotterdam2021

CHAPTER 16

Beyond Interdisciplinary Research: Transdisciplinarity and Transformative Literacy Through Artistic Thinking and Research

Wolfgang Stark

16.1 Why Transdisciplinarity and Art?

While there is now more than enough scientific knowledge to make changes towards sustainability, responsible business and a just society, we are failing to change individual, collective and entrepreneurial mindsets. In other words, the dominant ways we organize our world, think and act, obtain knowledge and learn, are still based on rational-technical thinking that is promoting industrialized processes and a myth of everlasting growth and effectiveness.

In order to enhance our individual and collective minds and skills to adopt open mindsets for sustainability and social responsibility, we need to activate the economic, political, sociological and psychological drivers.

W. Stark (✉)
University of Duisburg-Essen, Duisburg, Germany
e-mail: wolfgang.stark@steinbeis-isl.de

Steinbeis Center for Innovation and Sustainable Leadership, Berlin, Germany

B. J. Regeer et al. (eds.), *Transdisciplinarity for Transformation*,
https://doi.org/10.1007/978-3-031-60974-9_16

Change and social transformation cannot be achieved only on the basis of rational choice and planning. In complex systems, there is always a need for transdisciplinary approaches. In particular, art-based approaches to sustainable transformation may have the power not only to combine the four drivers into a truly transdisciplinary and transformative methodology, but also to add the 'tacit dimension' of change to a traditional model of knowledge production. Approaches that take into account the 'tacit dimension' (Polanyi, 1966) and its built-in patterns of implicit knowing (Neuweg, 2004) have been used in science and are frequently used to create art in many ways; this 'tacit dimension' is needed to co-create in innovative organizations and entrepreneurial communities. 'Implicit wisdom' (Dewey, 1938) based on 'intuition' or experience-based 'deep smarts' (Leonard & Swap, 2005), is known to be crucial for successful change, especially in unpredictable and ambiguous settings.

A truly transdisciplinary and 'transformative' science therefore needs to integrate natural science, social science *and* the arts (music, dance, theatre, visual arts) in order to affect individual and collective mindsets and ways of thinking for current and future leaders, decision makers and entrepreneurs. An experience-based and creative 'knowledge and action base' will be needed to reveal and teach the tacit knowing patterns we need to develop, in order to take the next steps towards social sustainability. The example of JR (see Box 16.1) may be inspiring in this sense—as a unique approach to both understanding and transforming communities. He and his team go beyond analysing and documenting situations. They highlight hidden patterns, and by enlarging particular aspects of those patterns, they change—albeit only temporarily—people´s mindsets, their behaviour and their relationships. They give power and voice back to communities in 'fuzzy' situations.

Box 16.1: JR—A case study in transformative research using the arts and street wisdom

Being raised in 'Les Bosquets', the 'ghetto' of Clichy-Montfermeil, one of the typical banlieus in metropolitan Paris, JR started out as a notorious street artist and sprayer. He evolved into a transformative researcher using art and artistic thinking, doing transformative research in 2004, when he started the 'Portrait of a Generation'. He photographed its young inhabitants and pasted enlarged photocopies to the walls.

JR is eager to enlarge both situations of groups and communities all over the world, and the experiential street wisdom which you will find unexpectedly in prisons, slums, among the elderly and children, in border-like segregation (such as Israeli and Palestinian people, or migrants and security policy at the Mexican border).

'Enlarging' for JR is meant in the original sense of the word. After photographing and recording stories of individuals and groups, his main act is to paste giant pictures of individuals on city walls, containers, water reservoirs, border walls—you name it.

JR's unique artistic process creates more than temporary art: his innovative way of doing street-art worldwide ('I have the largest gallery in the world') with different vulnerable groups (elderly people, women, children, prison inmates, slum dwellers) also works as a transformative community building process. Street Art and community building in JR's sense emerge to a transformative process, researching scenarios of conflict and strength in local and global communities.

Of many single projects, the 'Inside Out' project emerged to integrate community members all over the globe into a piece of common art, which JR has called 'Infiltrating art'. During his collage activities, local communities take part in the act of artistic creation, with no stage separating actors from spectators. Now, artistic creation and community building based on 'Inside Out' are taking place in close to 2,500 community projects in 148 countries.

See for examples https://www.insideoutproject.net/en/explore

Thomas Kuhn (1962)—more than 60 years ago—highlighted the need for transdisciplinary approaches to real-world problems at the individual, group or structural level in his seminal work, *The structure of scientific revolutions*. This early insight has since been adopted by many disciplines (e.g. Kahnemann, 2011; Scharmer, 2009; Wahl, 2016; Wilber, 2000). Nevertheless, my own discipline—psychology—is structured in disciplines and lives in scientific silos, like many other sciences in scientific institutions (see also Chapter 2; this volume). This fragmentation slows down collaboration urgently needed to cope collectively with 'wicked problems' (Schuler et al., 2020). Therefore, our common challenge is to unleash disciplinary boundaries in order not only to promote human well-being, but to create a planetary sense of community (Francescato, 2020). Successful solutions to social challenges manage to link the scientific results 'of the few' with the tacit knowledge 'of the many' and to cooperate between various disciplinary and transdisciplinary knowledge domains. Although disciplinary approaches are important in deepening our understanding of the world, only transformative and transdisciplinary approaches are able to promote human well-being. Thus each scientific discipline needs to have both: sound disciplinary-based 'deep knowledge' and 'linking' sub- or meta-disciplines to detect, set up, design and evaluate helpful links to transform and solve real-world problems.

This requires being able to 'connect head, hand and heart', says Otto Scharmer's *Theory U* (2009, 2019). Rational knowledge, experiential knowing, and creativity in doing and thinking need to be mingled in innovative ways. A sound systems approach that nurtures agile relationships between different actors, worldviews and disciplines, the creativity and intuition of artistic thinking and doing, and the art of improvising are basic elements of a transformative science. Following my own transdisciplinary research experiences, working together with artists in fields like 'Improvisation Research', this chapter will go on a journey in which I will explore the how and why of (1) 'crossing borders' between the arts and science, and (2) unchaining research from a purely cognitive exercise by linking 'head, heart and hands' in the research and learning process.

This may lead us towards a land of 'transformative science', where research is all about being able to 'explore, wonder, and transform ourselves in context'. It will be worth starting a joint discourse about the forms of 'Transformative Literacy' (Künkel & Ragnarsdottir, 2022; Scharmer, 2019) and the types of research and teaching we will need to shape our common future.

16.2 Beyond Bounded Rationality

People in social systems, such as scientific communities, economies or societies, can act and create new solutions either by analysis and planning, by intuition or by improvisation. For most of the last century and until today, science is based on a rational cognitive mode: rational planning based on analysis and measurement focused on accountability. This approach has been adopted by society at large and professional communities alike and has 'infected' and limited our everyday way of thinking (Sandel, 2021). Rational thinking is based on one overall assumption: all technical and social challenges can be solved by an objective step-by-step rational approach. This particular way of thinking and organizing also created our world bound with disciplinary specializing. We pretend to know and give answers based on specialized knowledge; we tend to forget to ask questions to open up our mind in unpredictable settings (Berger, 2014).

Yet, most social challenges and scientific settings are governed by unknown situations, subjective personal creativity and implicit knowing and intuition. Although many social scientists, as well as many practitioners, agree that a rational approach captures only a small part of the processes and dynamics existing in both social systems and organizations, it seems to work well for traditional organizations (both profit and non-profit) that are based on the hierarchical model of top-down decision making and planning. However, the more complex a situation and settings become, the more planning and rationality are losing ground. To deal with complexity we need to learn how to use emergent and creative processes based on the tacit knowing of the arts. Modern, network-type social systems need to encourage system-oriented factors like relation-building and open-minded cultures in order to survive in their complex and constantly changing social environment.

The concept of 'Bounded Rationality' (Gigerenzer & Selten, 2002) has already challenged rationality in decision making based on experimental models. The concept proved that rational choice is only one part of human choice; non-rational factors are highly influential in everyday decisions. Nevertheless, the majority of professionals pretend to make rational choices, although relying heavily on 'intuition'. The reason? Complex social systems, such as modern companies, universities and research institutions, but also non-profits, political and informal communities, are very often not determined by clearly defined goals and strategies. Innovative processes in research are based—in addition to rigorous research principles—on the idea of serendipity. That is, they use opportunities that emerge from non-planned networking. Gradually, we (re)discover that many settings in which we live and work are governed by unknown situations and ill-defined factors. The ability to be creative, to design innovative environments and to improvise in an ostensibly rational and structured situation may be key for our survival in a world that is in reality unpredictable and subject to serendipity.

Indeed, the dynamic process of organizing (Weick, 1995)—although still bound to a culture of numbers, results and rationality—displays a complex network of relations and 'tacit knowledge', which is neither seen nor addressed, since practice and perception are both oriented towards attaining goals, maintaining control and setting strategies. More problematically, scientists, entrepreneurs, decision makers and managers—in both traditional and sustainability-driven organizations—typically lack a language to describe their 'tacit knowledge' (Polanyi, 1966) in this land of uncertainty.

16.2.1 Generating Patterns Towards 'A Performative Pattern Language'

Research and organizational cultures are both performative and dynamic, and will develop through action and establishing relationships, not only by setting rules and structures (Kuhn, 1962). Paul Bate (1995)—one of the most prolific researchers on cultural change—compares organizational culture in social systems to a river bed; habits and dynamic patterns of employees as individuals and (formal an informal) groups, and structural patterns of the organization try to find a viable path, which then (temporarily) becomes the riverbed. The basis here is the principle of 'viability' (Glasersfeld, 2002), which is also crucial for the theory of self-organization and entrepreneurship.

Cultural patterns in social systems therefore can be detected, analysed and documented in a sector-, type- or situation-specific manner. The basis for the analysis of cultural patterns is real-life participant observation and interviews; in some cases also document analyses (expressed values and artefacts) or other forms of data collection like digital pattern recognition (Bishop, 2006).[1] Organizational patterns, used in scientific processes and research, can reveal parts of the collective tacit knowledge base in organizations and scientific institutions: patterns will be discovered and documented as part of a shared learning process that incorporates the perspectives of different actors regarding a specific challenge and their viable approaches. In doing so, they also will create open spaces of conducive and successful actions that can be applied differently in diverse situations.

'Pattern theory', as a systems approach (Leitner, 2015), emphasizes both the importance of experiential knowledge and cooperation, and

[1] For example, modern AI—like ChatGPT and more everyday AI-applications to come—is based on 'pattern recognition'. Human brains, like other advanced brains in nature, act on the basis of pattern recognition.

importance of relationships between different parts of a system. The 'performance of patterns', i.e. the dynamics of relationships between patterns of a system, is crucial for the success of the whole social system (Stark et al., 2018). According to Pattern Theory, a number of patterns related to each other create 'a pattern language'[2] (Alexander et al., 1977) to understand the hidden knowledge or wisdom of social systems. Therefore, to identify and recognize successful patterns in organizations, social systems and eco-systems is highly relevant for common challenges such as sustainability, quality, innovation and learning ability. The role of patterns in transdisciplinary research helps to manage open structures, ambiguity, processes that are difficult to plan and knowledge transfer (Stark et al., 2017).

Patterns may help to address challenges of everyday work within open processes:

- Which patterns of action or design can be successfully combined?
- How can different participants communicate with each other about this?
- Which combinations are viable, which need to be changed?
- How is experiential knowledge made transparent and usable for all participants?

If the basic structures of the patterns are known, their dynamic application in practice can be developed—as in jazz improvisation and similar to ecology (Hutchinson, 1953) or modern quantum physics (Alexander, 2016). Central to this creative process is the triad of challenge, context and solution with the forces acting between them (Keidel, 1995). By revealing the underlying principles in the tension between these forces, not only is the procedure itself conveyed, but also the insight behind it. Following Borchers (2001), each pattern can be represented like a

[2] http://groupworksdeck.org is an early outstanding example of a pattern language based on social science which revealed implicit and tacit knowledge of community practitioners, and which is a ready-to-use tool for researching reality from different angles. Already a growing body of knowledge on pattern languages has been developed for different forms of social systems—communities, special groups, business, among others (for a sample please see https://pattern-publishing.de).

functional equation:

$$p = \left(nc, f^1 \ldots f^i, set, t, sol, e^1 \ldots . e^i, con\right)$$

Each pattern (p) is the function of a typical challenge (nc = challenge), different forces acting here in context (f = forces), temporal dynamics (t = time) and spatial circumstances (set = setting), one or more solution variants (sol = solution), different application examples (e = example) and possible consequences (con = consequences). This describes a typical, situation-specific system of relationships in organizations or communities. Documenting patterns in this form helps to reflect and describe their principles, and to extract invariable elements and apply them to different situations.

If one develops a system of patterns for social systems, they will describe the current state of culture formation in organizations and/or communities in a condensed fashion. A system of relationships (comparable to a grammar) between single patterns (Keidel, 1995) makes it possible to combine related and complementary patterns into a 'pattern language' (Alexander et al., 1977; Leitner, 2015; Stark, 2014). A 'pattern language' represents essential elements of the co-creation process to deal with certain challenges in different subjects or disciplines. Pattern languages, although situation-specific and unique, represent general procedures proven to be viable in a certain field of application. The principle of the 'pattern language' is therefore applicable in all social systems; the goal is to discover one's own—often hidden and underused—patterns of success (viable patterns), especially for application in ambiguous situations that are difficult to plan. Patterns provide access to a 'deep understanding', to the 'unnamed' of social systems. Since patterns and pattern languages will open potentials of becoming agile rather than static, they need to initiate creative processes of inventing rather than just finding (Dell, 2002), and should relate to each other in the sense of a 'living (pattern) language'.

Patterns and corresponding pattern languages for a transdisciplinary approach towards a transformative science may be central components of a dynamic data-bank of knowledge, fed by pattern-generating interviews, document analyses, participant observation and the notation of organizational scores (Vossebrecher, 2017). The approach of a pattern language for tacit knowledge in social systems breaks new transdisciplinary ground in this context. Similarly, it develops a language for (1) a scientific (evidence based), (2) a tacit (experience based) and (3) a creative, artistic dimension of our world. That, again, opens up new ways how to deal with uncertainty and ambiguity and to develop a new approach towards collective resilience (Stark, 2021).

16.2.1.1 Tacit Knowing and the Improvisational Field

The culture of societies and its organizational research systems unfold in a texture of co-creation of diverse models of partnership. Co-Creation may be found in different formats:

(a) the small cooperative cell (the informal group, team) within a scientific community, organization or social system;
(b) the socio-dynamic design of a company or non-profit organization as a distinct entity in itself with explicit structures and tacit knowledge;
(c) the strategic alliances between different types of groups, organizations and stakeholders.

In contrast to most organizational and scientific settings, in which processes are determined by 'rational planning' that does not grasp the hidden power and potentials of tacit knowing, there is a common challenge shared by all stakeholders; they are driven inherently by implicit and 'tacit' knowing, and, quite often, emotionally based decision making and processes. This is what we call the 'improvisational field', which appears as a layer beneath planning and acting. It is built upon tacit knowing and experiential wisdom.

To uncover the language of tacit knowing, it is useful to experiment with new sensorial channels; if we could 'sense' the dynamic processes of social systems, the communicative sensorium in the workplace could be expanded to a new and deeper level which would allow us to access both aesthetic and emotional dimensions of processes. Performing arts, such as music, dance and theatre, as well as modern, performative ways of painting, can be helpful in detecting the potentials of tacit knowledge beyond rational planning (Forsythe, 2003). It can be one key to the '...deep level of organizing and innovative processes' (Stark & Dell, 2013, p. 252), which can be used as a reflective tool for both managers and employees but also for people in communities to start a dynamic and creative process of learning for social systems and individuals.

Improvisation and its performative patterns do not replace the rational, cognitive mode; just as the muscle system in the body is needed for a skeleton to move, to balance and to be alert, the improvisational field and performative patterns are needed to balance structures and rules, as well as ambiguities in each situation new to routine and to be alert to innovations and creative opportunities.

In this chapter, I particularly focus on 'music and sound', and use this form of performing art to 'imagine processes in social systems (communities, organizations) as a piece of music' (Stark & Dell, 2013, p. 251). I argue that this will open up social and organizational systems, which are often stuck in strategic plans and workflows, and help them to creatively redesign the system. To detect the dynamics of this hidden (implicit and tacit) system within a visible and well-documented system, a special form of musical production is helpful in order to foster learning processes in complex and constantly changing settings, which call for the ability for continuous sense-making and serendipity (Weick & Westley, 1996). The technology of improvisation has already inspired organizational theory as an analogy (Barrett, 2012; Hatch, 1999; Weick, 1995). Improvisational, performative patterns will be even more important for social systems, if we look at music not only as something that can be received or interpreted, but also as a tool for sense-making. Then, improvisation and 'musical thinking' (Zürn, 2022) will open up the ability not only to cope with unknown potentials and uncertain processes, but also to redesign patterns and minimal structures in a creative way (Dell, 2012; Stark, 2014).

Adaptive organizational cultures work with improvisational processes (Cunha & Cunha, 2006), well known from jazz or contemporary music (Dell, 2002); that is why they need highly qualified employees with a large degree of freedom to recognize innovation potential and to act flexibly, but they do not require complex structures. Patterns of improvisation in innovative, transformative settings, according to Cunha (2005), are intended but unplanned deviations from routines. It is through these deviations that unexpected problem solutions and development opportunities can be identified and exploited. By breaking existing rules, a new 'figure' is achieved. This is a dynamic which we call the 'improvisational field'—useful, if not necessary, for unplanned and ambiguous settings.

Coping with unpredictable processes also is an everyday challenge in contemporary society, in organizations and communities. In addition to codified rational procedures, members of social systems will usually develop a set of tacit procedures which prove to be viable (Glasersfeld,

1992). Similar to improvisation in jazz music, where musicians interact on the basis of well-known explicit and implicit 'jazz patterns' (Coker et al., 1990), this kind of process can be viewed as continuously redesigning and re-arranging implicit and explicit procedural patterns based on experiential (implicit) knowledge; they interact based on already known patterns, and they also refer to other, already existing or traditional patterns, and by redesigning and re-arranging they also create a constant flow of new patterns, which are added to their body of experiential knowledge (Barrett, 2012).

16.2.1.2 The Art of Listening

Researchers and consultants in social systems are often like deaf observers who enter a room in which someone is playing the violin. In addition to seeing someone using the violin as an instrument, they may measure vibrations with calibrated instruments, and sometimes they can even draw conclusions about the pitch and form of the music. However, being deaf, they will never experience the sensory and emotional perception of the sound. They will never learn from what music offers or triggers in terms of experience as heard (Hayek, 2006), because they are oriented to the usual and rather narrow modes of rationality cognition/language and measurability. Social research by and large focuses on the rational part of a given field by referring primarily to directly identifiable parameters and to countable measures. In contrast, organizational cultures 'made audible', sonically or understood musically, will expand the language of visibility and '(ac)countability' in a senso-emotional way. Deep dimensions of social systems, of social structures and organizations, and of transforming innovation processes will be experienced through the channel of music and can be used for understanding and reflection. Since music is complex, it can reflect complex systems, as well as temporal and performative aspects that are lacking in static models. Music can give feedback, both at a structural level and on an emotional level, beyond language codes. Musical feedback stimulates learning and development processes of the organization, especially related to elements such as social interaction, emotion and values (Zürn, 2022).

Self-reflection skills are crucial for innovation, success and sustainable survival. Patterns of innovative processes can be experienced at a new level of reflection. (Self-)Reflection skills are based on basic competences we will need to re-detect in order to develop a transdisciplinary and transformative science. Otto Scharmer (2009), in his seminal *Theory U: Leading from the Future as it Emerges*, highlights how significant the 'art of listening' may be to allow transdisciplinary and transformative processes. Listening may slow down processes in a (self)-reflective mode, if you are able to identify the four 'Levels of Listening' (see Fig. 16.1):

1. *Downloading* will re-confirm habitual judgements towards a person or a situation. You are interested in re-confirming your well-known judgements. Downloading may be typical for political debates, boardrooms or many group discussions. You are talking nicely and politely and re-enacting existing rules.
2. *Factual Listening* is looking for facts which may confirm or disconfirm your data. Switching off your inner judgement also means that you may be open to novel facts. Factual listening is the basic mode of traditional 'good' science: you ask questions and listen carefully to responses and data you may get. You are object-focused and maybe rule-revealing.
3. *Empathic Listening* means you are opening your heart beyond novel data, but to develop an empathic capacity for emotions or

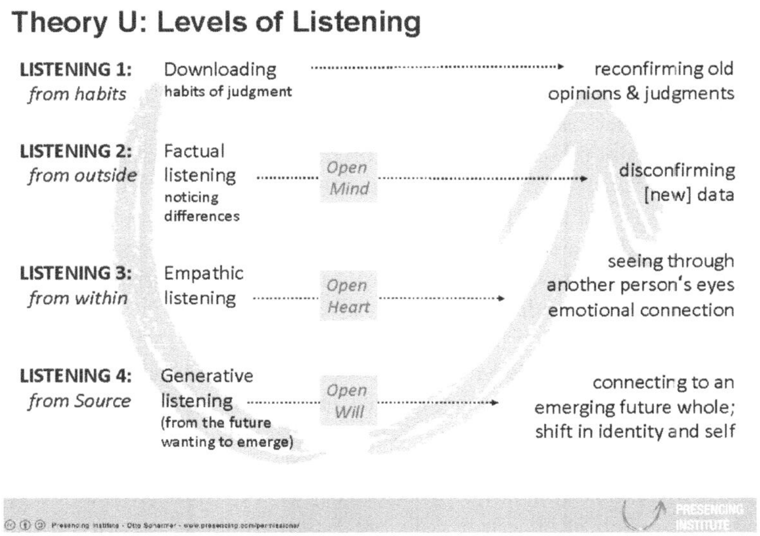

Fig. 16.1 Theory U: Levels of Listening (Scharmer, 2009)

emotional results of a context or situation. In empathic listening you try to connect to another person or situation 'from with-in'. You are reflecting rules and try to reach mutual adjustment. Empathic listening enables dialogue.

4. *Generative Listening* occurs when you realize that a 'third space'—something new, not-yet known—may emerge. Generative Listening may be co-creating in its best sense; it will need a mutual open mind to co-create opportunities and novel potentials.

According to Zenk et al. (2023), a transdisciplinary way of doing research and practice will require 'meta-competences' beyond disciplinary boundaries and methodological technologies. They discovered a number of meta-competences learning from different disciplines and the art of improvisation, among others:

- the capability to learn and act in real time;
- to identify the set of action patterns at hand in different situations; and
- pre-sensing potential links or future possibilities.

Meta-competences are connected to the art of asking questions rather than giving answers or delivering results is another basic competence for transdisciplinarity: To know how to ask open questions may feel like the flipside of the coin of the art of listening. Empathic and generative questioning means to join one others world view and ask 'what if…' (Berger, 2014).

16.3 Artistic Research and Thinking: Improvisation as an Art and Skill

Processes of research can be re-experienced through music in an indirect, non-representational way. In this type of transformative research, questions or results of process analyses and pattern discovery are made available in the form of musical feedback in order to open up spaces for reflection and action that stimulate the system's learning potential. The recognition and representation of patterns serves as a methodological and content-related hinge between organization and music. Patterns are the context-related description of highly typical problem solutions that have proven to be successful (viable); they make implicit action-relevant knowledge explicit. As a form of communication, patterns map a cooperative learning process and convey values. Since patterns exist in improvised (and composed) music as well as in organizations, they have an interconnecting function.

Organizational Scores have been developed and used as a research method of 'musical thinking about organizations' (Vossebrecher, 2017; Zürn, 2022). In general, scores are forms of making sense of music; if we want to understand the structure of a Beethoven symphony, we use a musical score and read along while listening. The scores of New Music (starting in mid-twentieth century), however, no longer completely predict the exact course of the music; they are not representational, but 'diagrammatic'. The performers do not just play or perform the notes, but have to develop their own forms of action from the score. This procedure is valuable for innovation in organizations, because here 'fuzzy' instructions should lead to 'sharp' results: an intended, but unplannable, use of degrees of freedom.

Thus, the medium of music opens up new levels of reflection for the analysis of social system contexts, processes and events. Organizational and musical patterns combine to form a pattern language of organizations that allows access to their deep dimension (Leonard & Swap, 2005). Furthermore, organizational pattern languages can be used to redesign procedures and processes (e.g. for crisis management). In this respect, scores are elements of an instrument portfolio that not only take music as an analogy, but also re-sounds the sensitivity for and enabling of improvisational processes in the sense of learning organizations.

Music, as a performing art (as well as other performing arts like dance or theatre [Fischer-Lichte, 2012; Forsythe, 2003; Johnstone, 1987]), can be a key to the 'deep level of organizational and innovation processes' that can be used as a reflective tool both in professional contexts and in everyday life for people in communities to initiate dynamic and creative learning processes (cf. Zürn, 2022). Imagining research processes and projects as a piece of music opens up social and organizational dynamics that are otherwise often stuck in strategic plans and workflows, helping them to creatively reframe the system.

Capabilities for continuous realignment and serendipity (Weick & Westley, 1996) require systematic procedures—a 'technology'—of improvisation that inspires organizational theory as a metaphor (Barrett, 2012; Dell, 2017; Hatch, 1999; Weick, 1995). If we consider music not only as something that can be received or interpreted, but also as a tool for meaning-making and community building, we can enter and analyse another level of understanding and 'world-making' (Goodman, 1978). Francescato et al. (1992) were early to introduce the idea of multidimensional analysis of communities and social systems incorporating artistic thinking into community psychology. Improvisation as an art and skill (and artistic thinking in general) is now seen by numerous researchers and practitioners as a basic element in processes of community resilience and innovation (co-creation) (e.g. Barrett, 2012; Dell, 2012; Stark, 2014, 2017; Zenk et al., 2023).

In everyday life, we discover the art of improvisation in many sporting activities, such as modern soccer, sailing and skiing. Thus, we can assume that whenever human creativity and playfulness are triggered, the art of improvisation is one of the keys to community and collective self-awareness, in addition to developing skills to deal with ambiguity and uncertainty (Small & Schmutte, 2022). Improvisation then opens up the ability not only to deal with unknown potentials and uncertain processes, but also to creatively reshape patterns and 'minimal structures' (Dell, 2012; Stark, 2014). Thus, the inventive production of improvisation becomes a norm in itself: challenge and possibility.

16.3.1 Learning to Improvise: Navigating the Unexpected

The art of improvisation as a basic competence for transformative research develops practical tools for collaborative resilience and innovation processes in research communities and social systems. We can learn to navigate social systems characterized by abrupt change, uncertainty and insecurity. Research communities, especially those collaborating between disciplines, also may become transformative places choreographed by complex 'rhythms of knowing' in which we simultaneously navigate and act. Complex social systems—modern 'fractal' or 'fluid' organizations or networks (De Vet & Lowette, 2019)—take on the qualities of permanent improvisation. A lifestyle of transition and transformation becomes one of the most important features of daily life, especially in multiple global crises.

Improvisation teaches skills of 'constant readiness' (or, rather, alertness or mindfulness) and ability to act in the moment ('act-in-an-instant'; Dell, 2002, 2012), which may be crucial to act in ambiguity and constantly changing situations and crises. To improvise in situations of ambiguity, vigilance and presence become key features of any organization or social system. Improvisation positions itself as a process that also takes into account commitment and trust, the self-confidence of the actors and their interdependence, and biographical characteristics of the individual in a group process. The knowledge of one's own and organizational success patterns and the ability to combine them in a flexible and creative way opens up new potentials to (re)act on unplannable and unpredictable situations.

Donald Schön (1983) in his well-known description of the 'reflective practitioner' refers to the challenge of jazz musicians to use improvisation to create coherence in unpredictable situations. As musicians collectively attempt to develop a creative and inspiring new sound dynamic, they use metrical, melodic and harmonic patterns with which they are all familiar to shape the melody or sound. Musicians usually just intuitively grasp the idea of where the melody is going based on their performance; they will be able to pick up on the new meaning and direct their individual playing to the new goal. Not only are successful improvisations inspiring examples of 'reflective practice', according to Donald Schön, but collaborative improvisation can also be seen as the foundation of a new practice of organizing complex systems with an innovative character (Johnson, 2011).

Improvisation does not distinguish between thought and action, but intensifies the movement between the systems/components of the organization and the community in the moment. Improvisation therefore acts like a 'regulator' between inter-subjective openness and solipsistic moments of subjectivity. This is when intellectual cognition, social experience and practical-intuitive competence converge—as does the difference between the individual and the collective in social systems and the difference between past and future in time (Scharmer, 2009).

Improvisation works because it asks questions rather than providing answers. It contains difference, gaps, looseness and interstices that are available for the active interpretive work of the actors and thus contribute to the qualification of their experience (Hatch, 1999). In a process of improvisation, actors develop the sensors they need to directly grasp, interpret and harness the ambivalence of a situation. Cunha (2005) simplistically states: 'In the improvisational mode, people act while they learn, and learn while they act' (p. 133).

The art of improvisation thus enables the integration of 'serendipity' as a learning process and promotes proactive learning ('deeper learning')

(Sliwka et al., 2022). Rational analysis is not excluded, on the contrary, but the performative aspect of learning and acting is brought into focus (Stark et al., 2017). Analysis in the context of improvisation focuses on the rearrangement and reinterpretation of material gathered in the improvisational process.

Box 16.2: A Transformative Workshop on Community Resilience
How 'learning' can transform from knowledge transfer to lively exchange with the inspiration of artistic approaches was demonstrated by a social experiment during the 9th International Conference on Community Psychology in Naples 2022. In a one-day workshop, we approached the topic of 'Community Resilience' from three angles: the artistic, the everyday experience and the social science perspective. The workshop has evolved like a dance choreography: different perspectives (represented by the participants) met, approached, moved away, presented themselves or retreated. The participants' movement opened factually and metaphorically the space for 'community resilience'. Three performances examined community resilience from various perspectives and viewpoints:

- An artistic approach used inputs from music, painting and improvisational theatre. These were subsequently reflected upon in small groups with artistic, experiential and social science 'eyes'.
- In the experiential angle, the inputs consisted of 'community stories' representing personal and collective experiences during the pandemic. Again, small group reflection addressed questions like: what is the beauty or artistic value of the stories told; how did they change personal or shared experiences; what social science analyses can be connected to them?
- The social science input (research, findings, concepts) was the most familiar to the participants in the third performance, but here it was already touched by the artistic and experiential perspectives.

A multidimensional new way of looking at 'community resilience' enriched and changed all participants in their respective cultural contexts.

People in social systems learn through analysis, intuition or improvisation, according to Mintzberg and Westley (2001). Analysis is a structured process that may or may not lead to surprising insights. The analytical mode assumes that an ontological basis is externalized from existing situations. The intuitive mode derives its learning outcomes from making connections not previously suggested. The improvisational mode is structured quite differently. Not only do people act to learn, but they also seek to incorporate analytical frameworks into the action, which then itself becomes a learning laboratory for the 'reflective practitioner' (Schön, 1983). Graebner (2004) has shown that an important source of value creation is a kind of sensitivity or sense of feeling ('serendipity') that comes from exposure to different practices. The mode of 'serendipity' (Eco, 2014) underlies improvisational processes. This mode aims to make use of surprise: to trigger a process that continually recomposes existing and identified patterns, thereby opening up new possibilities for solutions that offer different forms of surprise. This means that those who practise improvisation also practise recognizing patterns that others overlook, and using patterns pragmatically and subtly—as a level below rational patterns of planning and design.

16.3.2 A Performative Pattern Language

If we use patterns (and thus the experience and knowledge they embody) as foundational elements for dealing with complexity and ambiguity, we can develop a 'Performative Pattern Language' of performative patterns of action for community resilience in our social systems (Schümmer et al., 2014). The ability to deal creatively with one's own patterns of experience and action and to practise the art of improvisation in a situation that is 'only supposedly' rational and structured can be a key factor for survival in a world of unpredictability and chance. Patterns of creative action must be able to interact with rational patterns in order to release their true potential when combined. Thus, identifying the creative use of patterns of tacit knowledge (performative or improvisational patterns) as they may occur in music is important for understanding and dealing with codified and documented procedures. This is what we call 'improvisational fields' in communities and social systems. It is a level of action where experience and intuition create new structures and movements that parallel rational thought. In scientific communities, performative patterns of action can be used to deal with as yet unknown challenges in creative ways and to find new solutions to given problems. Unlike instruction manuals and user guides, they define the principles of solutions that can be adapted to a variety of environments and situations.

Analysing performative patterns in professional and research context—based on Alexander's (1977) concept of 'pattern language'—develops a transformative use of patterns in social systems (Keidel, 1995; Manns & Rising, 2005; Schuler, 2008). According to this concept, patterns unfold and change within the values and principles of specific professional cultures as flexible forms of problem solving that prove viable and successful in practice. It has much in common with the concept of unfolding wholeness presented in Alexander's more recent work (Alexander, 2004; Alexander et al., 2013).

Collaborative and performative patterns of action may be key to understand the principles of social systems and the deep levels—'of the unknown'—of complex modern civil and organizational cultures. However, community action patterns today need to go beyond the status quo in a research community and promote flexibility in addition to stability. They create and discover new forms of relationships ('serendipity'—Cunha, 2005) between well-known patterns of action, between people and things/spaces (Latour, 2005). They enable an interplay between 'movement' and 'structures': i.e. movements as creative unfoldings of strong centres triggered by perceived tensions and structures as integrative orders that connect the different movements into a coherent whole. The principles of patterns and pattern languages (the 'patterns of patterns') can then meet the dual challenge of providing both continuity and variability found in nonlinear systems (Brockman, 1995). On this basis, our approach can be linked to Arcidiacono's concept of the community psychologist as a 'collaborative-reflective plumber'—a second (meta) level specialist (Arcidiacono, 2017).

16.4 Transformative Literacy

At first sight, improvisation works in a disorderly fashion and seems to be either unprofitable or ineffective. But this first impression also shows that the process works, because it triggers those questions that it wants to trigger. In other words, improvisation works because it contains difference, gaps, looseness and intermediate spaces, which are available for the recipients' active interpretative work, thus helping to qualify their experience (Hatch, 1999). Improvisation thus can be described as a technique which allows us to integrate serendipity as a learning process and involves proactive learning. Rational analysis is not excluded, rather the opposite; the performative aspect of learning is put into focus. Analysis in the context of improvisation concentrates on the rearrangement and reinterpretation of material that is gathered through the improvisational process in such a way that it is connectable to new processes in time. The analytical work then relies on qualified experience and the development of complexity sensors that should lead to a transformation of attitudes and thus enable ecological change. But in order to do this, the improviser needs to develop the abilities needed to recognize change, allow it and help design it.

Unlike an instruction manual or recipe, improvisational patterns describe principles of a solution that can be applied to a particular situation in a situation-specific way. Examples of such patterns include 'develop trust', 'recognize, use and share different skills' or 'use unusual places'. When different patterns are combined in a systematic way, they form a situation-specific pattern language that can be used for problem solving and collaborative innovation. Patterns build on experiential knowledge—on strategies and practices that have stood the test of time.

The tacit knowledge in improvisational processes—collective, experiential and accumulated in social systems over time—is, so to speak, the 'oil in the gears' or the 'muscle on the bones' that gives professional and research processes its character and determines its own dynamics. Schwartz and Sharpe (2010) call it 'practical wisdom' (see also Schonbrun & Schwartz, 2020), which is often only apparent at second glance. It is rarely systematically cultivated, because individuals are usually unaware of a 'collective culture'—as a self-evident part of their everyday lives. Collective improvisational patterns are often the 'building material' from which communities and mutual solidarity emerge and with which professional and research communities become stronger and resilient.

Although the 'practical wisdom' based on collective improvisational patterns often forms the core of innovative communities or teams, it is rarely documented and taken for granted even by experienced practitioners. Identifying experiential wisdom requires intensive conversations and discussions, because practical wisdom is usually 'implicit' and 'tacit', i.e. not directly conscious or 'intuitive' (…). Becoming aware of one's own practical wisdom is an important prerequisite for the process of collaborative innovation and transformative research. Rather than conceptually or theoretically describing innovation processes and their associated tools as a whole, success factors and strategies are broken down into individual patterns of action. In contrast to a linear and rigid guideline, the 'improvisational action patterns' can be flexibly selected, combined and applied depending on the perspective and situation.

Improvisational action patterns function like practical guides, but are flexible in their application. They help to understand the dynamics of processes, and why some communities are innovative and successful while others are not. In most cases, tacit collective knowledge is passed on orally and informally ('This is how we do it', or 'Do's and don'ts') and has an intuitive character ('I have a feeling about this', 'I do this intuitively'). Sometimes, collective documents (minutes, informal newspapers, local history) reveal principles that shape actions. However, when they are used to formulate fixed (behavioural) rules, they often lose their creative and dynamic character. Improvisational patterns of action therefore contain experiential values and principles that can be used to flexibly manage most situations and are constantly evolving.

It is helpful to categorize patterns in a story-like manner (e.g. 'How we work together and cooperate – Community actors and our ecosystem – The real best solution – Community sustainability and responsibility – Time and space'). The goal is to embed the viable improvisational patterns into a structure of different fields of action and their temporal dependency. By combining different patterns, categories and phases, a pattern language is created in which the potentials of the different approaches are condensed. A pattern language is constantly supplemented, developed and improved.

Successful patterns of collaborative action can—like the building blocks of a DNA—be reassembled again and again to trigger new ideas and innovation processes. The technical and methodological know-how, the extensive experience of community members as entrepreneurs, committed citizens, professionals with different backgrounds and their regional (and sometimes international) networks, will bring the success stories behind the patterns to life and hopefully give rise to many more new ones. In order to enhance these new types of transdisciplinary processes, we need to develop supportive infrastructures and 'resonance spaces'.

16.5 Activating Resonance Spaces for Transformative Learning

Researching, teaching and learning have long ceased to be discipline-oriented inventing or the one-dimensional transmission of competencies (knowledge, attitudes and skills). Current and future social, ecological and technical challenges require continuous reflective experience, coordination and negotiation in direct exchange between different actors from art, science, policy, business and civil society.

Universities (and Higher Education in general) will play a central role in creating multidimensional 'Resonance Spaces' to develop formats of transformative and lifelong learning. Resonance Spaces will need to be established between research and learning, and civil society, policy and business. They will create an eco-system in which innovative ideas and improvisational patterns, identified in any of the three spheres, will not simply fade away, but will resonate and enable the necessary reflection for sustainable innovations for social challenges. In this way, socially relevant knowledge about change is generated and kept up to date, and transformative social processes are initiated and accompanied at the regional, supra-regional and global levels. In order to develop the mental attitudes and ways of thinking necessary for this, flexible and customized offers of transformative learning are required.

Future universities therefore may emerge as 'Activating Resonance Spaces' for our society (Rosa, 2016; Stark, 2021). To establish universities as resonance spaces, and to exchange and share implicit and explicit knowledge (Stark, 2017), patterns and skills, we will need to establish an expanded and transparent 'communication and reference framework' for social innovation and improvisational processes (Sailer et al., 2017). We will need to go beyond a mutual understanding of those acting within the academic system. A mutual and collaborative eco-system within the scientific community will still be central, but not sufficient. Rather, through its various formats (teaching, research, transfer) and institutions, universities need to recognize, understand and respond to the demands and challenges of society—in other words, 'relate' and 'resonate'. At the same time, universities as 'resonance spaces' need to be heard and echoed in society, as active members of a social discourse on science-based discoveries, insights and innovations.

Teaching and learning in this context will go beyond a one-dimensional transfer of knowledge (from teacher to student; from university to society). It will be a continuous mutual reflective experience. Learning will take place in coordinated and negotiated ways in a continuous exchange of different actors in physical as well as virtual spaces. A multidimensional and resonating space, which will enable, create and maintain its references for research and learning will not simply fade away once a degree or project has been completed. Mutual knowledge and skills from academia, the arts and experience will resonate with current challenges and enable an urgently needed re-reflection for responsible innovation. In short: 'Activating Resonance Spaces' are needed as innovative enablers for communication between all social actors. Future universities should act as and provide resonance spaces for the future of our societies and the planet. The core of future universities therefore should rather:

- promote transdisciplinary *thinking and acting 'out-of-the-box'*;
- systematically encourage *learning by experimentation and making mistakes*;
- foster *a culture of critical and productive questioning*;
- promote the *development of a learning culture in and between social organizations*; and, last but not least,
- build *the personalities and identities of future generations and leaders* by strengthening social and societal responsibility and a sense of community.

Initial steps and open questions towards transformational teaching and research in 'Universities of the Future' have been started in many places—small 'pockets' of innovative and transformational teaching and research in the universities of the world, in the context of community service learning, in programmes and research projects on sustainability, or as part of other innovative teaching concepts that have been developed at universities in recent years. However, small innovative 'pockets' rarely are connected, so innovation—in a more traditional way—has to be re-invented over and over again. Therefore, common consequences can neither arise from the results and continuous developments can be initiated, nor does the important systematic didactic–methodical exchange between the innovative offers of transformative learning succeed. Yet, at the same time, the vast majority of teaching still is based on the traditional one-way-street.

To break the wave, an interactive, dynamic and adaptive market and information place—for example, an interactive online platform as well as offline elements—might serve. This marketplace makes it possible to match and further develop the different actors with their ideas, competencies, questions, searches and resources as well as existing projects and results in a dynamic process. Future universities will need to add social responsibility and experiential wisdom and practical relevance to academic knowledge—which is how they will contribute to addressing the major future challenges of society.

References

Alexander, C. (1977). *A pattern language: Towns, buildings, construction*. Oxford University Press.

Alexander, C. (2004). *The nature of order* (Vols. 1–4). Center for Environmental Structure.

Alexander, C., Neis, H., & Alexander Moore, M. (2013). *The battle for the life and beauty of the earth*. Oxford University Press.

Alexander, S. (2016). *The jazz of physics*. Basic Books.

Arcidiacono, C. (2017). The community psychologist as a reflective plumber. *Global Journal of Community Psychology Practice, 8*(1), 1–16.

Barrett, F. J. (2012). *Yes to the mess: Surprising leadership lessons from jazz*. Harvard Business School.

Bate, P. S. (1995). *Strategies of cultural change*. Routledge.

Berger, P. L., & Luckmann, T. (1991). *The social construction of reality: A treatise in the sociology of knowledge*. Penguin.

Berger, W. (2014). *A more beautiful question*. Bloomsbury.

Bishop, C. M. (2006). *Pattern recognition and machine learning*. Springer Science.

Borchers, J. (2001). *A pattern approach to interaction design*. Wiley.

Brockman, J. (1995). *The third culture: Beyond the scientific revolution*. Touchstone.

Buttrick, D. (1987). *Homiletic moves and structures*. Augsburg Fortress Publishers.

Coker, J., Casale, J., & Campbell, G. (1990). *Patterns for jazz: A theory text for jazz composition and improvisation*. Alfred Music.

Cunha, M. P. (2005). *Serendipity: Why some organizations are luckier than others*. Nova SBE.

De Vet, A., & Lowette, P. (2019). *The fluid organization*. Utrecht.

Dell, C. (2002). *Prinzip Improvisation* [Principles of improvisation]. Walther König.

Dell, C. (2012). *Die improvisierende Organisation. Management nach dem Ende der Planbarkeit*. Transcript.

Eco, U. (2014). *Serendipities*. Columbia University Press.

Fischer-Lichte, E. (2012). *Performativität. Eine Einführung*. Transcript.

Forsythe, W. (2003). *Improvisation technologies: A tool for the analytical dance eye*. ZKM Karlsruhe.

Francescato, D. (2020). Why we need a planetary sense of community. *Community Psychology in Global Perspective, 6*(2/2).

Glasersfeld, E. v. (2002). *Radical constructivism: A way of knowing and learning*. Routledge.

Graebner, M. (2004). Momentum and serendipity: How acquired leaders create value in the integration of technology firms. *Strategic Management Journal, 25*, 751–777.

Gigerenzer, G., & Selten, R. (2002). *Bounded rationality*. MIT Press.

Halpern, J. Y. (2003). *Reasoning about uncertainty*. MIT Press.

Hatch, M. J. (1999). Exploring the empty spaces of organizing: How improvisational jazz helps redescribe organizational structure. *Organization Studies, 20*, 75–100.

Hayek, F. (2006). *The constitution of liberty*. Routledge.

Hutchinson, G. E. (1953). The concept of pattern in ecology. *Proceedings of the Academic of Natural Science of Philadelphia*, 1–12.

Johnson, S. (2011). *Where good ideas come from: The natural history of innovation*. Penguin.

Johnstone, K. (1987). *Impro: Improvisation and theatre*. Routledge.

Kahnemann, D. (2011). In S. Farrar, & Giroux (Eds.), *Thinking fast and slow*.

Kant, I. (1987). *Critique of judgement*. Hackett Publishing Company.

Keidel, R. W. (1995). *Seeing organizational patterns*. Berrett-Koehler.

Kuhn, T. (1962). *The structure of scientific revolutions*. University of Chicago Press.

Künkel, P., & Ragnarsdottir, K. V. (Eds.). (2022). *Transformation literacy: Pathways to regenerative societies*. Springer.

Latour, B. (2005). *Reassembling the social: An introduction to actor-network-theory*. Oxford University Press.

Leitner, H. (2015). *Pattern theory*. Introduction and Perspectives on the Tracks of Christopher Alexander. Nausner & Nausner.

Leonard, D., & Swap, W. (2005). *Deep smarts: How to cultivate and transfer enduring business wisdom*. Harvard Business School Press.

Manns, M. L., & Rising, L. (2005). *Fearless change: Patterns for introducing new ideas*. Addison Wesley.

Mintzberg, M., & Westley, F. (2001). Decision making: It's not what you think. *Sloan Management Review, 42*(3), 89–93.

Neuweg, H. G. (2004). Tacit knowing and Implicit Learning. In *European perspectives on learning at work: The acquisition of work process knowledge*.

Polanyi, M. (1966). *The tacit dimension*. University of Chicago Press.

Rosa, H. (2016). *Resonanz. Eine Soziologie der Weltbeziehung*. Suhrkamp.

Sailer, K., Stark, W., & Szogs, G. (2017). *Hochschule als Resonanzraum der Gesellschaft*. Manuscript (online) (January 2024). https://www.researchgate.net/publication/333149490_Hochschule_als_ResonanzRaum_der_Gesellschaft_Ausgangspunkte_gesellschaftlicher_Entwicklungen

Sandel, M. (2021). In S. Farrar, & Giroud (Eds.), *The tyranny of merit*.

Scharmer, O. (2009). *Theory U—Learning from the future as it emerges*. Berett-Koehler.

Scharmer, O. (2019). *Vertical literacy: Reimagining the 21st-century university*. Field of the Future Blog, Medium.

Schonbrun, Y., & Schwartz, B. (2020). *How practical wisdom helps us to cope with uncertainty*. Behavioral Scientist, online. https://behavioralscientist.org/how-practical-wisdom-helps-us-cope-with-radical-uncertainty/. Downloaded January 2024.

Schön, D. (1983). *The reflective practitioner: How professionals think in action*. Routledge.

Schuler, D. (2008). *Liberating voices: A pattern language for communication revolution*. MIT Press.

Schümmer, T., Haake, J., & Stark, W. (2014). Beyond rational design patterns. In *Proceedings of the 19th European Conference on Pattern Languages of Programs* (EuroPLoP '14). ACM.

Schwartz, B., & Sharpe, K. (2010). *Practical wisdom. The right way to do the right things*. Penguin.

Sliwka, A., Klopsch, B., & Beigel, J. (2022). *Deeper learning in der schule*.

Small, A., & Schmutte, K. (2022). *Navigating ambiguity: Creating opportunity in a world of unknowns*. Ten Speed Press.

Stark, W. (2014). Implizites Wissen der Improvisation für innovative Organisationskulturen verstehen und nutzen. *Praeview – Zeitschrift für innovative Arbeitsgestaltung und Prävention*, *1*, 12.

Stark, W. (2021). Innovation patterns and improvisation in organisations. *In The Power of Music Thinking—Podcast mit Christof Zürn*, Njimegen, NL (online) (January 2024). https://musicthinking.com/innovation-patterns-and-improvisation-in-organisations/

Stark, W., & Dell, C. (2013). Tuning into the improvisational field. In R. Grossmann, K. Mayer, M. Lenglacher, & K. Scala (Eds.), *Learning for the future in management and organizations*. Information Age Publishing.

Stark, W., Dell, C., Vossebrecher, D., & Schmidhuber, H. (Eds.). (2017). *Improvisation und Organisation. Muster zur Innovation sozialer Systeme*. Transcript.

Vossebrecher, D. (2017). *Organisationspartituren* [Organisational Scores]. In W. Stark, C. Dell, D. Vossebrecher, & H. Schmidhuber (Eds.), *Improvisation und Organisation. Muster zur Innovation sozialer Systeme*. Transcript.

Wahl, D. C. (2016). *Designing regenerative cultures*. Triarchy Press.

Weick, K. (1995). Organizational redesign as improvisation. In G. P. Huber & W. H. Glick (Eds.), *Organizational change and redesign*. Oxford University Press.

Weick, K. E., & Westley, F. (1996). Organizational learning: Affirming an oxymoron. In S. R. Clegg, C. Hardy, & W. R. Nord (Eds.), *Handbook of organization studies* (pp. 440–458). Sage Publishers.
Wilber, K. (2000). *Integral psychology.* Shambhala.
Zenk, L., Pausits, A., Brenner, B.,Campbell, D. A., Behrens, D., Stöckler, E. M., Oppl, S., & Steiner, G. (2023). *Meta-competences in complex environments: an interdisciplinary perspective* (in press).
Zürn, C. (2022). *The power of music thinking.* BIS Publishers.

CHAPTER 17

Transdisciplinary Competencies for Transformation

Hussein Zeidan, Sarju Sing Rai, and Marjolein B. M. Zweekhorst

17.1 Introduction

Transdisciplinarity has been discussed in academic circles since its introduction back in the 1970s (Bernstein, 2015), and scholars have subsequently explored and established diverse approaches to apply it in research and academic work. These diverse approaches are typically informed by various philosophies and the conceptualization of transdisciplinarity. Julia Thompson Klein is among those who have devoted their academic careers to conceptualizing and actively engaging with transdisciplinarity. She distinguishes three major discourses of transdisciplinarity based on the underlying and prevailing elements that shape its comprehension and

H. Zeidan (✉) · S. S. Rai · M. B. M. Zweekhorst
Athena Institute, Vrije Universiteit Amsterdam, Amsterdam, The Netherlands
e-mail: h.zeidan@vu.nl

S. S. Rai
e-mail: s.s.rai@vu.nl

M. B. M. Zweekhorst
e-mail: m.b.m.zweekhorst@vu.nl

B. J. Regeer et al. (eds.), *Transdisciplinarity for Transformation*,
https://doi.org/10.1007/978-3-031-60974-9_17

affect its practice: transcendence, transgression, and the ability to solve complex problems (Klein, 2015). First, the element of 'transcendence' challenges reductionist disciplinary perspectives and promotes the integration of diverse perspectives from various cultural, national and ethical backgrounds to support a holistic view. Second, 'transgression' encourages collaboration between academic institutions and society towards a more inclusive approach incorporating socially relevant knowledge. And third, the problem-solving element is viewed as vital in guiding the efforts of the various stakeholders to solve social problems.

More recently, however, the limitations of transdisciplinarity as being able to make a meaningful impact on social issues (Brandt et al., 2013; Lang et al., 2012) have been acknowledged, alongside the potential benefits of offering additional guidance that aims to catalyse transformation and change. This has led to a growing tendency to foster a discourse of a transdisciplinarity that is purposeful and transformative, achieves an impact, and is radical, with the power to bring about profound change (e.g. Holm et al., 2013; Mitchell et al., 2015; Schneider et al., 2019).

Although this chapter will not delve into the distinctions among these proposed attributes of transdisciplinarity and attempt to identify their shared characteristics, it is nevertheless important to note that these discussions and the range of proposals converge in aiming to define transdisciplinarity through desired outcomes rather than treating it only as a foundational meta-method. The discourse surrounding 'transformation' and related concepts (purposeful, radical, etc.) therefore suggests additional criteria to complement, refine and further develop the earlier definitions. Mitchell and colleagues (2015) identified three key elements that distinguish purposeful transdisciplinarity: (1) improving the current situation; (2) generating and making knowledge from diverse sources accessible; and (3) enabling mutual and transformational learning among all actors involved. In a similar vein, Schneider et al. (2019) explored ways to promote transdisciplinarity that will achieve an impact and made three key suggestions:

- Decision-making processes should be informed by fostering a descriptive and explanatory understanding of the problem situation (system knowledge), defining desired future development through norms and values (target knowledge), and understanding how to make the transformation from the current to the desired state (transformation knowledge).

- Social learning should be encouraged to promote collective action among all actors involved.
- Competencies for reflective leadership should be enhanced to ensure that all stakeholders can critically reflect on their actions and make informed decisions towards achieving their goals.

It is noteworthy that although the discourse surrounding transformation places significant emphasis on the term 'change', this concept remains unclear, which also underscores the lack of precise definitions of related concepts like 'impact', 'meaningfulness', 'purposefulness', and 'transformative'. Despite this, academic discourses converge in regarding transformative research as being more value-driven than transdisciplinary research. To put it differently, transformation is centred around effecting change, an aspect less emphasized in transdisciplinary work. These differences are apparent in the roles that researchers assume in the two processes. In transdisciplinarity, researchers typically adopt a more passive role, acting as facilitators, with considerable engagement in reflection. Conversely, advocates of transformation see researchers as more proactively involved, taking an active stance and exerting pressure on systems to bring about change (e.g. Doring, 2002; Massingham, 2014). This change can materialize, for instance, through activism, lobbying, and campaigning (Jessani et al., 2022).

The discussions on transformation and transdisciplinarity offer distinct perspectives on the conduct expected of researchers, which in turn determines the understanding of what higher education should aim to teach and informs the teaching strategies to equip students with what is considered to be necessary. The readiness of individuals, especially university students, for active participation in transdisciplinary collaboration has been the subject of discussion in the scholarly literature. We have synthesized these discussions elsewhere (Zeidan et al., forthcoming), with a view to understanding the design of transdisciplinary courses, their learning approaches, and anticipated outcomes. Notably, to realize their objectives, many transdisciplinary courses focus on competencies such as reflectivity, communication, and teamwork. Likewise, Wiek et al. (2011) synthesized a substantial body of literature on sustainability with a focus on creating a comprehensive competency framework, which highlighted five essential competencies that graduate students should possess: systems thinking, futures thinking (anticipatory), values thinking (normative), strategic thinking, and interpersonal collaboration (see Box 13.1; this volume). While the framework primarily centred around 'sustainability', it captured a significant level of interest from those who design transdisciplinary courses, and who structured them around these competencies as intended outcomes (Zeidan et al., forthcoming).

However, the evolution of the 'transformation' discourse began to approach preparing students from a different angle, aiming to empower them as 'change agents' and engaged researchers (Kay et al., 2010). This calls for a change agent mindset and also elicited a specific set of competencies perceived as essential for the next generation, emphasizing the role of higher education in preparing these competencies. Redman and Wiek (2021) revisited and upgraded the framework in the light of the transformation discourse 10 years after their initial framework. The upgraded framework suggested three new competencies complementing the initial sustainability competencies and advancing a more transformational aspect: intrapersonal, implementation, and integration competencies (see Box 13.1; this volume).

There are growing academic discussions on preparing individuals to engage effectively in transdisciplinarity and/or collaborative transformation. Researchers are delving into specific instances, highlighting the competencies that align with the unique problem contexts, themes, roles, methods employed, and more. Despite the advantages these approaches offer, the literature has become overwhelmed with lists of competencies. This presents challenges in terms of understanding their convergence, equilibrium, the feasibility of mastering such an extensive array of skills, and the dynamic trade-offs involved in possessing various competencies simultaneously.

These aspects and challenges have been acknowledged in earlier chapters, particularly concerning the positionality of researchers and the shifting of roles. As these chapters drew insights from practical experiences, we recognize the significance of re-examining the discourse on competency development in conjunction with these practical experiences and reflections. We aim to address the question of 'what' are we preparing individuals for, which involves revisiting, challenging, and questioning the practice of continually adding new competencies to the list of prerequisites of both transdisciplinarity and transformation. We go on to explore 'how' higher education is expected to nurture these competencies. Lastly, we contemplate an element that is missing from these discussions by suggesting that instilling a sense of purpose in students could be more significant in empowering them to proactively engage in their own trajectory and have ownership of their development in transformation processes. This, in turn, would equip them to better position themselves and adeptly navigate the complexities of real-world challenges.

17.2 Transformational Preparation and Transdisciplinarity

In recent decades, there has been significant attention devoted to exploring approaches for preparing (under-)graduate students in the realms of transdisciplinarity. The lack of a consensus on the meaning of 'transdisciplinarity education' has created a fertile ground for diverse perspectives and interpretations. Consequently, university faculties have developed various approaches and curricula, each offering a unique perspective on how to prepare students to be able to work in these complex and interconnected realms.

In our research, we found that the curriculum designed to cultivate transdisciplinarity can be classified according to three distinct aspects: cross-disciplinary collaboration, social engagement, and problem-solving (Zeidan et al., forthcoming). These aspects align with Klein's (2015) categorization of transdisciplinarity into three types: transcendence, transgression, and problem-solving. Some courses aim to develop students' ability to navigate and integrate diverse disciplinary and non-disciplinary forms of knowledge (Zeidan et al., forthcoming). Others prioritize to equip students with the skills necessary for participatory approaches to social engagement, or emphasize problem-solving, focusing on equipping students with the tools to address complex challenges. Similarly, 'transformation' has received its fair share of scholarly attention. However, a clear distinction between 'education for transdisciplinarity' and 'education for transformation' remains elusive, beyond the latter's emphasis on creating a meaningful impact. The question, then, is how the educational approach ensures that individuals will indeed contribute to a making meaningful impact. Some scholars have perceived transdisciplinary approaches as being effective in promoting learning for transformation among students in higher education, valuing and recognizing the capacity of transdisciplinary education to expose them to a wide range of knowledge and perspectives, thereby facilitating a profound comprehension of complex real-life issues (Baumber, 2022; Leal Filho et al., 2018). Furthermore, transformation, which transdisciplinarity also endorses, emphasizes the cultivation of reflexivity rather than mere knowledge transfer. This empowers students to critically assess their own biases and facilitate the connection with various forms of knowledge.

Nonetheless, the focus of 'transformation' on achieving desired change and generating significant social impact highlights the importance of individuals playing the role of 'change agent', who are able to participate effectively in the process of change and help bring about its realization (e.g. Doring, 2002; Massingham, 2014). In this regard, the crucial distinction between education for transformation and education for transdisciplinarity revolves around equipping individuals to serve as effective change agents, capable of assuming responsibility, reconciling tensions and dilemmas, and fostering the creation of new values (OECD, 2018).

In a similar vein, Popa and colleagues (2015) argue that transformative and meaningful transdisciplinarity involves deliberating on the normative and epistemic orientation of research, framing problems in socially relevant ways, generating reflexivity on values and norms in problem-solving, and the ability to reflect on normative commitments and ideological orientations in processes of social transformation. Similarly, scholars have recognized that the specificity of transformation necessitates emphasizing certain competencies to address the roles required in these processes. For instance, the notion of a self-reflexive scientist as an additional actor in the transformation process highlights the need for individuals to be competent in reflecting on their own positionality and normativity while navigating a dynamic environment in order to facilitate change (see Fazey et al., 2018; Wittmayer & Schäpke, 2014). Others recognized the need for a comprehensive or umbrella competency, such as 'co-productive agility', which involves the ability to bridge knowledge and action, embrace diverse perspectives, adapt to changing goals and information, and effectively navigate inherent tensions in collaborative settings (Maas et al., 2022). These lists of competencies differ in their framing and terminology, but essentially they are in alignment. To facilitate our discussion, we will draw on the work of Redman and Wiek from both 2011 and 2021.

In relation to training and preparing, umbrella competencies are tricky to cater for as they involve a complicated and intertwined range of skills, knowledge, and attitudes. These overarching competencies may benefit being broken down into trainable aspects. However, should education solely prioritize teachable competencies, or should it—particularly within the realm of transformation—emphasize instilling in students a set of guiding principles on which they can depend to supervise and take ownership of their competency development? This is an important question this chapter addresses.

17.3 Revisiting Competencies Through a Practical Lens

In this section, we aim to enrich the scholarly discourse by connecting these conceptual discussions with practical insights from researchers who have engaged in transformative collaborations. Rather than starting from scratch, we draw on the personal accounts and experiences of some of the contributions to Part III of this volume to redirect, refine, and channel their experiences to tackle and address the question of what prepares an individual for transformational processes in a meaningful manner. To give a solid foundation for our exploration, we project these personal accounts onto the influential competency framework proposed by Redman and Wiek (2021). Chapters 14 and 15 explored two significant aspects that researchers encounter in transformational collaborations: the researchers' positionality and the dynamic shift in roles they experience in these collaborations. We first examine the interplay and the tensions arising from positionality and the required competencies, before going into analysing how role shifting, with its accompanying synergies and tensions, interacts with the proposed competencies.

17.3.1 Competencies vis-à-vis Positionality

Despite the extensive discussions surrounding transformation processes, there is still no consensus regarding the definition of 'successful' transformation and the key factors that contribute to it. Furthermore, the lack of a clear definition contributes to the ambiguity of the criteria and guidelines required for achieving successful transformations. This creates a space that permits agendas to be moulded by the perspectives, values, positions, and expertise of the individuals or groups engaged in these transformation processes.

The chapter by Gunn, Hoffmann, Sager, Wittmayer, and Zuiderent-Jerak (this volume) presented a reflective account of how the complex interplays of different forms of knowledge could challenge researchers' perspectives and positionalities. Their chapter highlights the fact that transformative processes are not value-neutral but 'normatively charged'.

This normativity is rooted in the influences of ethical, moral, social, and cultural values and judgements that arise during the process (Copp, 1995). These elements inform decision-making throughout the transformative process alongside empirical evidence. Furthermore, political ideologies, cultural and religious beliefs, as well as disciplinary epistemology, all contribute to shaping the 'epistemic commitment' that is evident in individuals' actions in collaborative spaces (Frodeman, 2011; Granjou & Arpin, 2015). Thus, positionality is displayed as the embodiment of the intertwined layers of normative and epistemic characteristics.

By delving into the interplay between positionality and competencies, we perceive potential and valuable insights to effectively equip individuals to navigate transformative processes. From the literature, we observed a subtle distinction between preparing students for 'transdisciplinarity' and preparing them for 'transdisciplinarity for transformation'. Training for transdisciplinarity comes across as rather passive regarding the direction of the change and deliberating on what is valuable and meaningful. The literature on transdisciplinary education highlights equipping individuals with problem-solving, communication, collaboration, and teamwork skills to enable them to participate in transdisciplinary collaborations (e.g. Acevedo-Osorio et al., 2020; Balsiger, 2015; Barrett et al., 2019; Dlouha & Burandt, 2015). There is also a notable emphasis on improving higher education curricula and crafting innovative courses to cultivate students' reflective competencies, allowing them to critically assess their perspectives, disciplinary expertise, and inherent biases (e.g. Barret et al., 2019; Fortuin & Van Koppen, 2016). Furthermore, there is a growing recognition of the value of instilling epistemic humility as a fundamental competency (e.g. Lake et al., 2016).

While acknowledging the need for the skills mentioned above, there remains a gap in covering the specific ideals that advocates of transformation endorse. Transformation revolves around meaningful changes and perceives researchers as key contributors in shaping the course of this change, actively involving their values in the processes. In the previous chapter, Wolfgang Stark argues that while transdisciplinarity embodies the intellectual aspect (head), it must also be complemented by practical skills (hand) and the translation of passion and values into behaviour (heart) (see also Sipos et al., 2008). This classification places greater emphasis on the translation of individual values into attitudes and behaviour, alongside normative elements such as comprehending sustainability and fostering global citizenship to facilitate transformational processes.

Redman and Wiek (2021) emphasize 'values-thinking' competencies, which prioritize an individual's normative values over general norms. That highlights the significance of considering positionality and recognizing the interplay between normative and epistemological perspectives when shaping competencies. 'Values-thinking' competencies emphasize the ability to identify, map, assess, negotiate, reconcile, and reflect on the variety of norms and knowledge of the diverse participants' portfolios; supporting them in understanding the conflicts and trade-offs that they will be facing (Redman & Wiek, 2021). Although the 'values-thinking' competency may appear to be the most critical for positionality and addressing normatively and epistemically charged spaces, competencies are interconnected and they 'need to be integrated for advancing sustainability transformation' (Redman & Wiek, 2021, p. 5).

Based on the experiences described by Gunn and colleagues in Chapter 15, it is evident that in transformational processes, it is also crucial to empower researchers to identify and actively enact their positionality. It is essential to equip them with the necessary skills and knowledge to effectively navigate the inherent tensions that arise in (1) ambiguous settings, and (2) normatively and epistemologically charged spaces, while recognizing how their positionality (3) is not solely determined by the narrow confines of their areas of expertise but also by their values, beliefs, culture, ideologies, etc. Suggested competencies such as reflexivity and humility have the potential to encompass these three aspects by empowering individuals to contemplate their own and others' values and biases. However, they do not directly cater for preparing individuals to formulate a sense of positionality and effectively communicate their perspectives, values, and commitments in the context of transformation and change. These attributes are 'must-haves' for 'change agents' and need to be supported by competencies, such as decision-making, self-confidence, self-expression, and a sense of responsibility (Akin et al., 2017). Likewise, the Organisation for Economic Co-operation and Development (OECD) acknowledges the importance of individuals functioning as change agents, and that in order to navigate contemporary complexities and uncertainties, they must actively exercise agency. This involves cultivating transformative competencies, including the capacity to establish values, address tensions and dilemmas, and assume responsibility (Organisation for Economic Co-operation and Development, 2018).

Therefore, within transformational collaboration, the competencies that hold value and significance are those that enable individuals to contribute effectively to the change they aspire to achieve. This goes beyond merely facilitating transdisciplinary participation; it involves incorporating the person's values and envisaging change that is influenced by a significant reservoir of personal knowledge, beliefs, and traits.

17.3.2 Competencies and Roles

In response to the ongoing transformation processes, some academics are departing from their conventional roles to embrace the more active roles that are influenced by the discussions surrounding transformation. This shift challenges the conventional expectations of academics, as the transformational processes enable them to assume varied roles, based on the phase and nature of the transformation. Our colleague Alanya den Boer (this volume) explored these roles and shared insights from her experience of participating in a transformational Living Lab. Below, we seek to understand competencies in relation to the roles that scientists assume in the transformation process. We explore the interplay between competencies and shifting roles of scientists in transformation processes; and how the transformation framework adopted shapes roles and the required competencies.

Recently, there has been a surge in attention being paid to the roles academics assume in transformative processes, highlighting both the synergies and tensions involved (e.g. Bulten et al., 2021; Fazey et al., 2018; Gisler & Schicktanz, 2009; Sarkki et al., 2014; Schuijer et al., 2021; Wittmayer & Schäpke, 2014). Scholars have made significant efforts to explore the conflicts and trade-offs inherent in these roles, aiming to bring clarity to the expectations placed on researchers when contributing to transformation. This includes a quest to identify the specific activities that researchers should undertake and the corresponding responsibilities they should assume to effectively contribute to transformative endeavours.

It is beyond the scope of this chapter to explore the intricacies of why and how role shifts take place, but our subsequent discussion acknowledges that the transformation of roles in transformative processes is influenced by various factors. These factors include, among others, the project level, the project's nature and theme, the specific working packages in which academics are engaged (project activities, etc.), and the identities of other actors involved. Moreover, the role shifts are connected to academics' individual characteristics, such as their background and expertise. These differences have resulted in a lack of consensus on the specific typologies and roles that academics should assume. This is not the consequence of a shortage of people to fulfil the necessary tasks, but rather stems from academic integrity motivating researchers to willingly take on responsibilities and conscientious work (Levin, 2012). This is evident in den Boer's account in Chapter 14, in which she recognizes the need to shift from a primary focus on problem-solving and project work to a more critical examination and reflective research on practices, to ensure that she maintains a high level of integrity in her work.

To facilitate the discussion in this chapter, we use the typology adopted by den Boer (Table 14.1, this volume), which enables us to entertain the competencies required for specific roles. Den Boer's reflective account highlighted the interplay of synergies and tensions she encountered as she assumed different roles—the scientist, change agent, capacity builder, process and reflexive facilitator, knowledge broker and project worker—in the transformation process in a Living Lab. In her reflective account, it became apparent that roles cannot be neatly separated with clear boundaries or considered in isolation. Rather there are significant overlaps and interconnections between them (Chilvers, 2013). Thus, it might be difficult to assert that each role requires a distinct set of competencies. Yet, certain roles still dominate in specific competencies, each displaying varying levels of proficiency and expertise.

To illustrate this, let us consider the roles that den Boer adapted to portray the diverse requirements within a Living Lab. Within this context, there emerges a need for individuals to take on the responsibilities of capacity builders and knowledge brokers throughout the process in order to facilitate the acquisition of specific skills or knowledge among participants through training and the exchange of expertise. When projected onto the early framework of Wiek et al. (2011), it seems that to undertake these roles individuals will heavily rely upon normative competencies. In this scenario, they will be engaged in sharing knowledge and providing training to participants on concepts such as sustainability, justice, responsibility, harm, among others. However, these competencies appear somewhat obscured and less evident in the updated iteration of the framework (2021). Equally, the roles of process facilitator and project worker primarily focus on overseeing the logistical and practical aspects of the project, which encompass tasks like arranging work sessions and coordinating activities. For these roles, we highlight the significance of implementation and integration competencies from Redman and Wiek's framework (2021). Conversely, reflexive facilitators expect to foster reflexivity 'both from using new knowledge from research as it emerges and by asking critical and challenging questions to keep ambitions for transformative change high' (Fazey et al., 2018, p. 64). This role can benefit from individuals with 'value-thinking' and 'system-thinking' competencies as defined in Redman and Wiek's 2021 framework. The significance of interpersonal and intrapersonal competencies cannot be ignored as they play a supportive role in bolstering the other competencies.

Scholars consider certain roles to be crucial in driving transformation processes, particularly those of change agent and capacity builder (e.g. Massingham, 2014; Stephens et al., 2008). Wittmayer and Schäpke (2014) highlight that change agents have the agency to exercise control over their level of involvement and how they define their self-concept in relation to their roles and contributions. Their role involves cultivating a sense of importance and inspiring and empowering participants to address problems, while also becoming an active part of both defining the problem and its solution (Wittmayer & Schäpke, 2014). Empowering participants involves building their capacities. Yet, the sense of importance and purpose is intricately tied to the researcher's own beliefs and values, and thus may not be easily compartmentalized as a competency.

Examining Redman and Wiek's competency framework in conjunction with the role of change agent, it becomes hard to pinpoint a specific competency that defines a change agent. Nonetheless, all the competencies included are in one way or another relevant to supporting change agents in their roles. The competencies of system thinking, future thinking, values thinking, and strategic thinking are unquestionably essential for supporting research throughout transformation processes. Researchers can draw on these competencies to establish a significant connection with the subject of transformation, thereby fostering meaningful engagement. Similarly, implementation and integration competencies can support the change agent in navigating the logistics in the transdisciplinary and transformation set-ups. Furthermore, interpersonal and intrapersonal competencies can help change agents deal effectively with other people and with themselves. However, we are concerned that Redman and Wiek's competency framework is somewhat limiting in addressing the specificity of change agent competencies, as it depicts individuals as static facilitators who are not actively contributing their own values to the transformation processes. Conversely, we envisage change agents as drawing on their own values and worldviews to leverage effectively their diverse competencies, using them to drive the anticipated change and achieve transformative outcomes.

The previous points highlight that certain competencies may have a more dominant role in specific roles. However, to emphasize the synergies in shifting roles and mitigate the accompanying tension, Schuijer et al. (2021) proposed the need for 'navigation skills' that can assist researchers in moving between roles, while also complementing role-specific competencies. We are particularly intrigued by the term 'navigation', especially in view of the metaphorical significance of relying on a compass for effective navigation.

Considering the above, the current challenge lies in defining the competencies that are essential for distinct roles, when these roles are not static but rather dynamically influenced by factors such as the situation, context, project emphasis, the individual's characteristics, and other variables. This complexity makes it problematic to assert that a specific competency exclusively dictates one's capability to function in a specific role in a transformational context. Thus, we recognize the importance of adopting a lifelong learning approach, wherein individuals can cultivate the skills they need to effectively navigate the dynamic landscape of transformation.

17.3.3 Roles as a Reflection of a Framework

In the realm of transformation, scholars have presented substantial frameworks to guide or inspire change. The diverse frameworks discussed in the broad literature used varied terminology and assume distinct interpretations of 'meaningfulness', yet, they all converge on the central notion of effecting change *with* society. Each framework is constructed on a specific comprehension of what constitutes meaningful change, which governs the processes, steps, and dimensions to be considered in achieving this change. These processes, in turn, determine the specific roles that participants must assume in order to facilitate the change. While this chapter does not delve into an exhaustive assessment of the effectiveness of these frameworks, this section highlights the fact that different frameworks require unique blends of skills to effectively accommodate their various phases. Below we will illustrate that by using two frameworks as examples.

We start with Responsible Research and Innovation (RRI)[1] as one of the influential concepts on 'meaningful' transformations. We have adopted the RRI framework proposed by Stilgoe et al. (2013), which comprises four interconnected phases: anticipation, inclusiveness, reflexivity, and responsiveness. To effectively incorporate these phases into practical transformation processes, it is crucial to translate them into actionable tasks and define specific roles. In den Boer's case (Chapter 14), we observed how she deconstructed the various phases of an RRI-oriented Living Lab into tasks and assigned them to the different roles that she fulfilled. However, it is important to note that specific phases may require a variety of roles. For instance, inclusiveness stresses the involvement of diverse stakeholders in the different stages while addressing power imbalances. If we were to apply this to the roles that den Boer undertook, she would need to function as a project worker by planning the various milestones, facilitating the process through organizing the necessary work sessions and inviting participants, building the capacity of stakeholders to contribute to the co-production exercises, mediating between different perspectives as a knowledge broker, and serving as a reflexive facilitator who encourages reflexive practices. Similarly, the other phases of RRI would benefit from a different combination of roles. Thus, the roles required to enact RRI in practice are determined by its various phases, and den Boer's account highlights how assuming different roles involves both synergies and tensions.

For the effective implementation of the inclusivity aspect in RRI, the roles of project worker, process facilitator, knowledge broker, capacity builder, and reflexive facilitator are considered essential. When examining these roles through the lens of the competencies framework by Redman and Wiek (2021), the project worker and process facilitator roles require strong interpersonal and intrapersonal competencies, while the knowledge broker and reflexive facilitator roles are grounded in the integration competency and the ability to think in systems. Capacity building ensures the transfer of these competencies to other actors and players in the project, thereby empowering them. Den Boer's active participation in the Living Lab also entailed assuming the role of a change agent. This role is characterized by positionality, embodying normative and epistemic

[1] For a comprehensive exploration of the evolution and ongoing dialogue concerning the RRI framework see Burget et al. (2017).

aspects, assisting in tackling relevant problems and working towards transformation. Therefore, in this case, it is hard to view the change agent as a distinct and independent role. Rather, it is more of a mindset and set of values that inform and shape the way an individual approaches each role without compromising their specific boundaries and requirements. A change agent can gain substantial advantages through the synergistic combination of competencies such as future-oriented thinking, values, and strategic reasoning.

Despite our attempts to simplify and categorize the roles and competencies required for RRI, it remains difficult to disentangle them into distinct and compartmentalized categories. This is because these roles and competencies are interconnected and intertwined, with synergies and tensions emerging between them. Thus, it may be more effective to approach RRI as a holistic and integrated practice, recognizing the interdependence of its derived roles and competencies.

Examining an alternative framework for transformation, we encounter the 'four collaboration pathways' developed by Chambers et al. (2022). This framework outlines the sequential steps towards achieving transformation as follows: (1) elevating marginalized agendas in ways that maintain their integrity and broaden struggles for justice; (2) questioning dominant agendas by engaging with power in ways that challenge assumptions; (3) navigating conflicting agendas to actively transform interlinked paradigms, practices, and structures; and (4) exploring diverse agendas to foster learning and mutual respect for a plurality of perspectives. While both the 'four collaboration pathways' and RRI encompass similar aspects in a holistic sense and ultimately converge towards the same direction, they exhibit notable structural disparities that can influence their practical implementation. When translating the phases into tangible roles, the initial stage of the 'four collaboration pathways' involves elevating marginalized agendas and empowering them in their quest for justice. This step surpasses mere inclusivity and incorporates normative and epistemic deliberations. The vocabulary used is different for RRI and has an epistemic direction that is of value. Thus, it urges and requires the researchers to play mainly the role of change agent while relying on their normative and epistemic stance to guide the other roles (facilitator, project worker, etc.).

To translate this into Redman and Wiek's competencies framework (2021), we would need to emphasize competencies in values thinking to inform strategy, future thinking, as well as systems thinking. Moreover, the successful implementation of this step would depend on strong competencies in integration, interpersonal and intrapersonal communication. However, the scholars who developed the 'four collaboration pathways' framework identified 'co-production agility' as a main competency that supports their approach. Co-production agility involves being open to different viewpoints, responsive to changing objectives or new knowledge, and able to constructively navigate tensions that arise in co-productive processes. It also emphasizes that knowledge and action are intertwined, recognizing that outcomes of co-productive processes go beyond formal knowledge, thus, incorporating pluralism and humility (Maas et al., 2022).

These brief examples serve to illustrate that the competencies essential in transformative processes cannot be rigidly predefined as static elements. The dynamic interplay of tensions and synergies is something that individuals involved in these processes must continually navigate. From that perspective, higher education plays a crucial role in establishing the essential foundation of knowledge and skills. It is important for students to have agency in developing their competencies, aligning them with their values, aspirations, and their envisaged impact on society. At the same time, it is vital for higher education to provide a platform for students to engage with real-world problems, encouraging them to scrutinize and question their own biases and values through interactions with a wide range of diverse knowledge and experiences. Nonetheless, higher education institutions also have to recognize the limitations of their role in fostering these competencies. Transformative processes demand the ability to embrace uncertainty and navigate through it, and while higher education does have a part to play in nurturing some of these skills, it occupies a specific niche within students' broader character development. There are numerous other factors—both direct and indirect—that contribute to shaping students' character. These include their parents, peers, the environments they inhabit, the ideologies they embrace, their values and beliefs, and their ways of making sense of and interpreting the world around them.

17.4 Piecing It Together

In the preceding section, we extended the discourse on competencies to our colleagues' experiences, seeking insights into 'what' are the competencies needed for individuals engaged in such transformative endeavours. We now want to revisit 'how' higher education is training and preparing individuals for transdisciplinarity and transformation.

In the endeavour to prepare students for transdisciplinarity, transdisciplinary education is assuming diverse forms. New approaches have been suggested by scholars that embed in courses learning activities that aim to expose students to both mirror real-life complex problems and also create spaces, either virtual or physical, where diverse types of knowledge and experiences can interplay (Zeidan et al., forthcoming). Approaches such as problem-based learning, challenge-based learning, and community service learning (CSL) are being proposed as viable options. These approaches are shifting higher education courses away from the traditional model of transmitting knowledge from teacher to students, and emphasizing experiential learning or learning by doing. In this new model, students actively engage with real-life problems, seeking to interact and acquire the knowledge that supports them in addressing these challenges (McGregor, 2017). In these courses, learning not only draws from disciplinary knowledge but also can thrive on interactions among students from diverse ethnicities, religions, cultures, ideologies, socio-economic classes, and academic backgrounds.

Regardless of the specific learning approach adopted in transdisciplinary education, we consistently observe the creation of diverse knowledge spaces and inputs to which students are exposed. According to Sabina Hoffman (in Chapter 15), these spaces provide the opportunity to embrace vulnerability, ask questions, and learn to be comfortable with the tension and ambiguity of the transformation process. They serve as 'safe spaces' where individuals can accept not knowing and support learning from each other and nurture new forms of reflexive scholarship and knowledge production. These spaces can serve as valuable environments for individuals to learn how to comfortably navigate uncertain situations and effectively leverage their diverse competencies to confront the challenges posed by uncertainty. Moreover, these spaces encourage students to recognize that in a transdisciplinary environment, their expertise must expand beyond their specific disciplinary training. They must also embrace the uncertainty surrounding their identity, as a transdisciplinary identity is rooted in the capacity to harness both external (such as situations) and internal (such as competencies) resources to enhance the transdisciplinary process.

While these courses may intend to benefit from this mirroring real-life environment, they often overlook the fact that it is an uncontrollable space that can yield varying outcomes in shaping students' growth and competencies. Thus, students entering a classroom do not necessarily emerge with the predefined set of competencies outlined in the course objectives. While it can be difficult to measure the cultivation of competencies such as critical thinking in these uncontrolled and unpredictable environments, it is important to recognize that competencies evolve over an extended period and through diverse inputs (Bajis et al., 2020).

The narrative of 'safe spaces' in transdisciplinary education mostly describes an inclusive and contained 'conflict zone' in which the participants are keen on working towards shared values. Thus, the predominant set of competencies that are usually suggested revolves around soft skills like communication, teamwork, empathy, humility, flexibility, and adaptability. While these competencies have some significance, they often correspond more with the role of a facilitator rather than that of a change agent, and they alone may not adequately equip individuals to effectively navigate the complex and multifaceted nature of transformation processes. For these competencies, scholars have suggested various approaches that have proved beneficial (e.g. Acevedo-Osorio et al., 2020; Balsiger, 2015; Barrett et al., 2019; Dlouhá & Burandt, 2015; Fortuin & Van Koppen, 2016; Lake et al., 2016).

The accounts of practical experience by den Boer and Gunn et al. (this volume) highlighted that transformation processes can be characterized as spaces filled with tension. These tensions arise because the spaces are imbued with normative and epistemic significance, while also stemming from the trade-offs in the shifting roles that researchers perform. Nonetheless, these tensions represent an integral aspect of transformation and are closely linked to personal values and worldviews, rather than mere technical or soft skills. Consequently, the preparation of individuals to navigate these tensions should focus on empowering them to express, debate, convey, and communicate their perspectives effectively. Simultaneously, it necessitates fostering a mindset that maintains one's values while remaining receptive to understanding, interacting, and negotiating with other ideas and viewpoints. These competencies represent a distinct set that need to be harmonized with transdisciplinary ones.

The significance of moving to a better understanding of individuals' development is in tapping into their reservoir of past experiences and knowledge, enabling them to develop new competencies, swiftly generate ideas, seamlessly adjust to unfamiliar circumstances, and decisively respond with action. Over time, this evolution will manifest in the form of 'improvisation' as discussed by Wolfgang Stark (this volume). Improvisation arises from the accumulation of a broad range of experiences and exposure to various situations over time, allowing a person to act spontaneously and instinctively (Berk & Trieber, 2009). This 'intuition' typically does not entail a sudden realization that occurs after a brief period of knowledge incubation or reflection. Rather, it emerges as a manifestation of the implicit, tacit, multifaceted learning experiences the individual has undergone (Shirley & Langan-Fox, 1996). Moorman and Miner saw that improvisation is built on a repertoire of know-how more than on know-what (1998), which supports individual 'to deal with a circumstance for which no script appears to be immediately to hand' (Mangham & Pye, 1991, p. 41).

When considering a transformation, it is important to explore how improvisation can be harnessed to support transformational goals. Reflecting on our discussion about positionality and role shifting, we realize that the transformation realm requires substantial incorporation of normative and epistemic knowledge and learning. This type of improvisation can be referred to—in the management terminology—as 'deep smarts', which represent a form of 'gut knowledge' that experts have developed over time (Leonard & Swap, 2004). At this level of mastery, individuals can apply their skills, attitudes, mindsets, experiences, and knowledge effortlessly and without conscious thought. This would make it difficult to classify improvisation into specific categories within the Redman and Wiek framework. Rather, we view improvisation as a manifestation of an individual's ability to unconsciously leverage their various competencies to tackle specific situations.

This once again prompts the question of what is the ideal blend of competencies to accommodate transformation. It is important to acknowledge that there is no 'one-size-fits-all' blend of competencies that guarantees success in transformation processes. Numerous factors determine the required competencies, starting with the transformation framework that dictates the necessary roles and respective competencies to accommodate them. Moreover, factors like context, the theme of the transformation, the stakeholders involved, etc., determine the competencies. For instance, engaging with Indigenous knowledge demands a different skill set than collaborating with technical experts or government institutions. Therefore, education for transformation should embrace and capitalize on an adaptable blend of competencies, achievable only through a commitment to continuous development and lifelong learning. That requires individuals to have ownership over the development of their own competencies' development. To foster this ability, educators should move beyond designing standalone courses that claim to instil transdisciplinary and transformational competencies, but rather view their courses as integral components of a larger chain or scaffolding for various courses that tap into students' normative learning and competency development while promoting a lifelong learning attitude. This approach involves connecting courses and blocks of knowledge and competency development to work collectively towards achieving the overarching goal. Within the realm of transformation, we observe the significant value of constructing basic knowledge blocks, enabling individuals to cultivate confidence in an uncertain world, while stimulating a perspective that is comfortable in pursuing the skills that enable them to navigate these uncertainties. To cultivate this mindset, it is essential not to perceive it as a static skill but rather recognize that it is connected to and driven by the individual's own sense of importance and purpose.

Second, we find it difficult to view competency development as being concentrated only in the realm of higher education. Much of the scholarly discourse on competencies stems from frameworks designed to facilitate transdisciplinarity, emphasizing the necessary competencies for effective engagement in such processes. However, these discussions often overlook the intricate interplay between competency development and an individual's normative, epistemic, institutional, social, cultural, religious, and ideological identities, among others. These aspects are developed beyond the university and are crucial elements of how individuals are shaped. While transdisciplinarity aims to incorporate the knowledge and experiences of diverse stakeholders, paradoxically to date it has failed to accommodate the complexity and challenges inherent to individuals involved in these processes. Consequently, it seems to treat students as rather passive entities whose behaviour can be predicted once they acquire the recommended set of competencies.

17.5 Moving Forward

This chapter showed that most literature on competencies tends to look at transdisciplinary and transformational processes in an ideal-typical way and propose generic competencies, overlooking the significant influence of individual factors such as education, background, culture, ideology, religion, living circumstances, and socio-economic class. These factors profoundly shape individuals, including their perceptions and values, as well as their actions and positions. Therefore, we argued that there is a vital aspect missing that guides individuals' epistemological commitment, channelling their efforts towards transformation.

Advocates of transformation emphasize the action towards change and the preparation of change agents. The sole difference in preparing a transdisciplinary individual versus a change agent lies in instilling in the latter the capacity to effectively use their competencies to bring about change and transformation. While the discussion of what is 'desirable change' and 'positive impact' is beyond the scope of this chapter, the concept of positive impact aligns with the aspiration to make a meaningful difference in society and contribute to the greater good, which is how 'sense of purpose' is defined (Staples & Troutman, 2010). A sense of purpose also sparks individuals' motivation and persistence, enabling them to overcome barriers and challenges on their path of learning and development (Sharma & Yukhymenko-Lescroart, 2018). And, ultimately a sense of purpose is perceived as the guiding force behind an individual's goals, reflecting their dedication and aspiration for achievement (Bronk, 2011).

Similarly, taking a closer look at the metaphorical implications of the term 'navigation' commonly used in discussions on transformation, it is interesting to explore the 'direction' we are navigating to and the 'compass' we are using to direct us. The notion of direction revolves around determining what constitutes desirable change, while the compass represents the guiding tool and the skill of interpreting it. Therefore, it is crucial for educational approaches to recognize students' own values and experiences, facilitating their ability to channel these experiences, critical thinking, and interactions with other learners. This aligns with Merizow's notion of transactional education, wherein sense-making serves as a guiding spark for transformative learning that students can employ throughout various phases of growth (1985). That extends beyond the classroom to the transformational spaces and projects, encompassing learning through interactions among diverse actors and their implicit knowledge.

Rather than viewing the development of competencies for transformative transdisciplinary research in isolation, we embrace the idea that it is an integral part of the evolution of a person's identity. The aim is to equip individuals with the tools and skills necessary to 'progress from a novice to independent, confident and agential individuals' who are 'adaptive, articulate and able to build a well-attuned portfolio that sets them up for building reach, impact and influence' (Debowski, 2022, pp. 10–12). This calls for a deeper insight into the role of higher education in nurturing and developing individuals. Acknowledging the involvement of other actors and factors in the process, it is important to emphasize that tertiary education is not the only way to achieve such nurturing but rather a part of a broader network of contributory factors. Training for transformation should be viewed as a dynamic interaction between values and worldviews, guided by the 'compass' of a 'sense of purpose'. In this context, higher education should accommodate students who are taking control over the cultivation of the competencies they deem valuable, and that enable them to effectively navigate transformative processes.

References

Acevedo-Osorio, Á., Hofmann-Souki, S., & Cruz Morales, J. (2020). Holistic competence orientation in sustainability-related study programmes: Lessons from implementing transdisciplinary student team research in Colombia, China, Mexico and Nicaragua. *Sustainability Science, 15*(1), 233–246. https://doi.org/10.1007/s11625-019-00687-8

Akin, S., Calik, B., & Engin-Demir, C. (2017). Students as change agents in the community: Developing active citizenship at schools. *Educational Sciences-Theory & Practice, 17*(3), 809–834. https://doi.org/10.12738/ESTP.2017.3.0176

Bajis, D., Chaar, B., & Moles, R. (2020). Rethinking competence: A nexus of educational models in the context of lifelong learning. *Pharmacy, 8*(2), 81. https://doi.org/10.3390/pharmacy8020081

Balsiger, J. (2015). Transdisciplinarity in the classroom? Simulating the co-production of sustainability knowledge. *Futures, 65*, 185–194. https://doi.org/10.1016/j.futures.2014.08.005

Barrett, M. J., Alphonsus, K. B., Harmin, M., Epp, T., Hoessler, C., McIntyre, D., Reeder, B., & Singh, B. (2019). Learning for transdisciplinary leadership:

why skilled scholars coming together is not enough. *BioScience, 69*(9), 736–745. https://doi.org/10.1177/1932202X211061121

Baumber, A. (2022). Transforming sustainability education through transdisciplinary practice. *Environment, Development and Sustainability, 24*(6), 7622–7639. https://doi.org/10.1007/S10668-021-01731-3/

Berk, R. A., & Trieber, R. H. (2009). Whose classroom is it, anyway? Improvisation as a teaching tool. *Journal on Excellence in College Teaching, 20*(3), 29–60. http://celt.miamioh.edu/ject/issue.php?v=20&n=3

Bernstein, J. H. (2015). Transdisciplinarity: A review of its origins, development, and current issues. *Journal of Research Practice, 11*(1). https://academicworks.cuny.edu/kb_pubs/37/

Brandt, P., Ernst, A., Gralla, F., Luederitz, C., Lang, D. J., Newig, J., Reinert, F., Abson, D. J., & Von Wehrden, H. (2013). A review of transdisciplinary research in sustainability science. *Ecological Economics, 92*, 1–15. https://doi.org/10.1016/j.ecolecon.2013.04.008

Bronk, K. C. (2011). The role of purpose in life in healthy identity formation: A grounded model. *New Directions for Youth Development* (132). https://doi.org/10.1002/YD.426

Bulten, E., Hessels, L. K., Hordijk, M., & Segrave, A. J. (2021). Conflicting roles of researchers in sustainability transitions: Balancing action and reflection. *Sustainability Science, 16*, 1269–1283.

Burget, M., Bardone, E., & Pedaste, M. (2017). Definitions and conceptual dimensions of responsible research and innovation: A literature review. *Science and Engineering Ethics, 23*(1), 1–19. https://doi.org/10.1007/S11948-016-9782-1/

Chambers, J. M., Wyborn, C., Klenk, N. L., Ryan, M., Serban, A., Bennett, N. J., Brennan, R., Charli-Joseph, L., Fernández-Giménez, M. E., Galvin, K. A., Goldstein, B. E., Haller, T., Hill, R., Munera, C., Nel, J. L., Österblom, H., Reid, R. S., Riechers, M., Spierenburg, M., ... Rondeau, R. (2022). Co-productive agility and four collaborative pathways to sustainability transformations. *Global Environmental Change, 72*, 102422. https://doi.org/10.1016/J.GLOENVCHA.2021.102422

Chilvers, J. (2013). Reflexive engagement? Actors, learning, and reflexivity in public dialogue on science and technology. *Science Communication, 35*(3), 283–310. https://doi.org/10.1177/1075547012454598/

Copp, D. (1995). *Morality, normativity, and society*. Oxford University Press.

Debowski, S. (2022). Shifting sands: Navigating being academic in an evolving sector. *Higher Education Research & Development, 41*(1), 7–20. https://doi.org/10.1080/07294360.2021.2008327

Dlouhá, J., & Burandt, S. (2015). Design and evaluation of learning processes in an international sustainability oriented study programme. In search of a new

educational quality and assessment method. *Journal of Cleaner Production, 106*, 247–258. https://doi.org/10.1016/j.jclepro.2014.09.096
Doring, A. (2002). Challenges to the academic role of change agent. *Journal of Further and Higher Education, 26*(2), 139–148. https://doi.org/10.1080/03098770220129415
Fazey, I., Schäpke, N., Caniglia, G., Patterson, J., Hultman, J., Van Mierlo, B. C., Säwe, F., Wiek, A., Wittmayer, J., Aldunce, P., Al Waer, H., Battacharya, N., Bradbury, H., Carmen, E., Colvin, J., Cvitanovic, C., D´Souza, M., Göpel, M., Goldstein, B. E., … Wyborn, C. (2018). Ten essentials for action-oriented and second order energy transitions, transformations and climate change research. *Energy Research & Social Science, 40*, 54–70. https://doi.org/10.1016/J.ERSS.2017.11.026
Fortuin, K. P. J., & Van Koppen, C. S. A. (2016). Teaching and learning reflexive skills in inter-and transdisciplinary research: A framework and its application in environmental science education. *Environmental Education Research, 22*(5), 697–716. https://doi.org/10.1080/13504622.2015.1054264
Frodeman, R. (2011). Interdisciplinary research and academic sustainability: Managing knowledge in an age of accountability. *Environmental Conservation, 38*(2), 105–112. https://doi.org/10.1017/S0376892911000038
Gisler, P., & Schicktanz, S. (2009). Introduction: Ironists, reformers, or rebels? *Technology & Innovation Studies, 5*. https://doi.org/10.17877/DE290R-8460
Granjou, C., & Arpin, I. (2015). Epistemic commitments: Making relevant science in biodiversity studies. *Science, Technology, & Human Values, 40*(6), 1022–1046. https://doi.org/10.1177/0162243915587361
Holm, P., Goodsite, M. E., Cloetingh, S., Agnoletti, M., Moldan, B., Lang, D. J., Leemans, R., Moeller, J. O., Buendía, M. P., Pohl, W., Scholz, R. W., Sors, A., Vanheusden, B., Yusoff, K., & Zondervan, R. (2013). Collaboration between the natural, social and human sciences in global change research. *Environmental Science & Policy, 28*, 25–35. https://doi.org/10.1016/J.ENVSCI.2012.11.010
Jessani, N. S., Ling, B., Babcock, C., Valmeekanathan, A., & Holtgrave, D. R. (2022). Advocacy, activism, and lobbying: How variations in interpretation affects ability for academia to engage with public policy. *PLOS Global Public Health, 2*(3), e0000034. https://doi.org/10.1371/journal.pgph.0000034
Kay, J., Dunne, E., & Hutchinson, J. (2010). Rethinking the values of higher education–students as change agents. *Quality Assurance Agency for Higher Education (QAA)*. https://dera.ioe.ac.uk/id/eprint/1193
Klein, J. T. (2015). Reprint of "Discourses of transdisciplinarity: Looking back to the future." *Futures, 65*, 10–16. https://doi.org/10.1016/J.FUTURES.2015.01.003

Lake, D., Fernando, H., & Eardley, D. (2016). The social lab classroom: Wrestling with—And learning from—Sustainability challenges. *Sustainability: Science, Practice and Policy, 12*(1), 76–87. https://doi.org/10.1080/15487733.2016.11908155

Lang, D. J., Wiek, A., Bergmann, M., Stauffacher, M., Martens, P., Moll, P., Swilling, M., & Thomas, C. J. (2012). Transdisciplinary research in sustainability science: Practice, principles, and challenges. *Sustainability Science, 7*, 25–43. https://doi.org/10.1007/s11625-011-0149-x

Leal Filho, W., Raath, S., Lazzarini, B., Vargas, V. R., de Souza, L., Anholon, R., Quelhas, O. L. G., Haddad, R., Klavins, M., & Orlovic, V. L. (2018). The role of transformation in learning and education for sustainability. *Journal of Cleaner Production, 199*, 286–295. https://doi.org/10.1016/J.JCLEPRO.2018.07.017

Leonard, D., & Swap, W. (2004). Deep smarts. *Harvard Business Review*. https://hbr.org/2004/09/deep-smarts

Levin, M. (2012). Academic integrity in action research. *Action Research, 10*(2), 133–149. https://doi.org/10.1177/1476750312445034

Maas, T. Y., Pauwelussen, A., & Turnhout, E. (2022). Co-producing the science–policy interface: Towards common but differentiated responsibilities. *Humanities and Social Sciences Communications, 9*(1), 1–11. https://doi.org/10.1057/S41599-022-01108-5

Mangham, I. L., & Pye, A. (1991). *The doing of managing*. B. Blackwell.

Massingham, P. (2014). The researcher as change agent. *Systemic Practice and Action Research, 27*, 417–448. https://doi.org/10.1007/s11213-013-9293-9

McGregor, S. L. (2017). Transdisciplinary pedagogy in higher education: Transdisciplinary learning, learning cycles and habits of minds. In P. Gibbs (Ed.), *Transdisciplinary higher education: A theoretical basis revealed in practice* (pp. 3–16). Springer.

Mezirow, J. (1985). A critical theory of self-directed learning. *New Directions for Continuing Education, 25*, 17–30. https://doi.org/10.1002/ace.36719852504

Mitchell, C., Cordell, D., & Fam, D. (2015). Beginning at the end: The outcome spaces framework to guide purposive transdisciplinary research. *Futures, 65*, 86–96. https://doi.org/10.1016/J.FUTURES.2014.10.007

Moorman, C., & Miner, A. S. (1998). Organizational improvisation and organizational memory. *Academy of Management Review, 23*(4), 698–723. https://doi.org/10.2307/259058

Organisation for Economic Co-operation and Development (OECD). (2018). *The future of education and skills: Education 2030*. OECD.

Popa, F., Guillermin, M., & Dedeurwaerdere, T. (2015). A pragmatist approach to transdisciplinarity in sustainability research: From complex systems theory to reflexive science. *Futures, 65*, 45–56. https://doi.org/10.1016/J.FUTURES.2014.02.002

Redman, A., & Wiek, A. (2021). Competencies for advancing transformations towards sustainability. *Frontiers in Education, 6*, 484. https://doi.org/10.3389/FEDUC.2021.785163/

Sarkki, S., Niemelä, J., Tinch, R., Van Den Hove, S., Watt, A., & Young, J. (2014). Balancing credibility, relevance and legitimacy: A critical assessment of trade-offs in science–policy interfaces. *Science and Public Policy, 41*(2), 194–206.

Schneider, F., Giger, M., Harari, N., Moser, S., Oberlack, C., Providoli, I., Schmid, L., Tribaldos, T., & Zimmermann, A. (2019). Transdisciplinary co-production of knowledge and sustainability transformations: Three generic mechanisms of impact generation. *Environmental Science & Policy, 102*, 26–35. https://doi.org/10.1016/J.ENVSCI.2019.08.017

Schuijer, J. W., Broerse, J., & Kupper, F. (2021). Juggling roles, experiencing dilemmas: The challenges of ssh scholars in public engagement. *NanoEthics, 15*(2), 169–189. https://doi.org/10.1007/S11569-021-00394-8/TABLES/4

Sharma, G., & Yukhymenko-Lescroart, M. (2018). The relationship between college students' sense of purpose and degree commitment. *Journal of College Student Development, 59*(4), 486–491. https://doi.org/10.1353/CSD.2018.0045

Shirley, D. A., & Langan-Fox, J. (1996). Intuition: A review of the literature. *Psychological Reports, 79*(2), 563–584. https://doi.org/10.2466/pr0.1996.79.2.563

Sipos, Y., Battisti, B., & Grimm, K. (2008). Achieving transformative sustainability learning: Engaging head, hands and heart. *International Journal of Sustainability in Higher Education, 9*(1), 68–86. https://doi.org/10.1108/14676370810842193/FULL/PDF

Staples, J. M., & Troutman, S. (2010). What's the purpose?: How urban adolescents of color interpret and respond to noble and ignoble purposes constructed in media texts. *Journal of Urban Learning, Teaching, and Research, 6*, 31–43.

Stephens, J. C., Hernandez, M. E., Román, M., Graham, A. C., & Scholz, R. W. (2008). Higher education as a change agent for sustainability in different cultures and contexts. *International Journal of Sustainability in Higher Education, 9*(3), 317–338. https://doi.org/10.1108/14676370810885916

Stilgoe, J., Owen, R., & Macnaghten, P. (2013). Developing a framework for responsible innovation. *Research Policy, 42*(9), 1568–1580. https://doi.org/10.1016/J.RESPOL.2013.05.008

Wiek, A., Withycombe, L., & Redman, C. L. (2011). Key competencies in sustainability: A reference framework for academic program development. *Sustainability Science, 6*(2), 203–218. https://doi.org/10.1007/S11625-011-0132-6/

Wittmayer, J. M., & Schäpke, N. (2014). Action, research and participation: Roles of researchers in sustainability transitions. *Sustainability Science, 9*(4), 483–496. https://doi.org/10.1007/S11625-014-0258-4/

Zeidan, H., Raj, S., & Zweekhorst, M. B. M. (Forthcoming). *Making sense of transdisciplinary education: Re-visiting courses conceptualization, configuration, and competencies.*

CHAPTER 18

Frame Reflection Lab: A Playful Tool to Reflect on Views of Science

Annemarie Horn and Marjoleine G. van der Meij

Me Ellen, You Jane?

Student 1: *"I don't know any Janes; do they really exist? Nobody thinks that way, do they?!"*

Student 2: *"I used to be a Jane, it's how science was taught in my Bachelor's education. In my Master's education, I was forced to become an Anthony. But now I want to be an Ellen; I want my research to be relevant straight away! So, where should I position myself? I think I am still an Antony, and deep down even a Jane".*

A. Horn (✉)
Liberal Arts and Sciences, Utrecht University, Utrecht, The Netherlands
e-mail: a.horn@uu.nl

M. G. van der Meij
Athena Institute, Vrije Universiteit Amsterdam, Amsterdam, The Netherlands
e-mail: m.g.vander.meij@vu.nl

B. J. Regeer et al. (eds.), *Transdisciplinarity for Transformation*,
https://doi.org/10.1007/978-3-031-60974-9_18

This is a typical conversation when students or early career researchers reflect on research characters representing different views of science in the Frame Reflection Lab (FRL) videos.

In this chapter, we present FRL and show how it can be used in practice. We share insights from our own experiences to inform its use in various contexts, such as inter- and transdisciplinary research teams and education for inter- and transdisciplinary research. We hope to inspire and offer some guidance for those engaging in or facilitating inter- and transdisciplinary teamwork to implement FRL and similar approaches in their own context. We invite readers to adapt and use FRL in their own research and teaching. All FRL materials are openly available online.[1]

18.1 Interdisciplinary Consciousness for Transdisciplinary Teamwork

We developed the FRL tool in the context of two inter- and transdisciplinary courses for Master's students (Tijsma et al., 2023). In these courses, namely, students from the full breadth of the VU University Amsterdam collaborated in teams to work on complex societal issues such as circular economy and digitalisation. As is characteristic of transdisciplinarity, this diversity was imperative in order to address those complex issues that are not confined to disciplinary boundaries.

At the same time, however, this very diversity also presented some of the main challenges to their teamwork. The students' differences were rooted in different values and assumptions of how to conduct research, what makes good scientific knowledge, and how science should contribute to addressing societal issues; in different 'views of science'. And often they were not aware of their own nor of their team mates' diverse views of science. This could cause miscommunications in their teamwork, resulting in either conflict or in the overgeneralisation of project work.

We aimed to better prepare collaborators for inter- and transdisciplinary teamwork, preventing miscommunications that are rooted in differences in how they view science and developed the FRL tool to scaffold this. The FRL tool supports the development of awareness of one's own and others' views of science—and hence of *interdisciplinary consciousness* (Horn et al., 2022).

[1] https://vu.nl/en/employee/toolbox-for-stakeholder-engagement/frame-reflection-tool-for-inter-and-transdisciplinarity.

18.2 What Is 'Frame Reflection Lab'?

We developed FRL in order to support the development of interdisciplinary consciousness. It is a playful tool (van der Meij et al., 2018), in which we aimed to make conversations about different views of science less cognitive and more personal, and to expose the diversity of views that is present in inter- and transdisciplinary teamwork in a structured and safe manner. In earlier attempts to spark reflection on views of science, we saw, for instance, that students often did so in an abstract, impersonal manner, took moderate stances in conversations that tended to conceal the diversity of views represented among students of widely varying backgrounds, and lacked the vocabulary to speak about views of science and underlying value systems.

The playful approach of FRL centres around three videos of four semi-realistic, fictitious researchers, who represent different views of science. Through those characters, different views of science are given names and faces. In interactive workshops, participants in FRL reflect on how they relate to the different characters to open up conversations about the views of science that they represent. In the following sections, we explain in more detail what the videos and workshop format look like.

18.2.1 The Frame Reflection Lab Videos

FRL centres around three videos that portray four fictitious researchers who all work on research projects related to climate change: Anthony, Ellen, Jane, and Marc. Those characters were based both on experiences with students leading to the development of FRL and on science philosophical theories (for more details, see: Horn et al., 2022). Figure 18.1 gives a simplified insight into the characters and their main characteristics.

In three videos, the fictitious researchers gradually reveal more about themselves, their research work, and their views of science by answering three questions:

Video 1: *What is your climate change research about?*
Video 2: *What do you consider 'good scientific knowledge'?*
Video 3: *How do you think knowledge should be generated to make a social impact?*

Fig. 18.1 Introducing the four FRL characters—Anthony, Ellen, Jane, and Marc—and their core messages

The videos serve as the starting point to reflect on views of science in an interactive workshop and in written reflection exercises before and after the workshop.

18.2.2 The Frame Reflection Lab Workshop

Groups of four to six participants watch the three FRL videos and do the interactive group exercises in supervised workshops of 90 to 120 minutes.

During the workshops, participants position themselves relative to the ideal-typical characters (Anthony, Ellen, Jane, and Marc) on a flipchart or online canvas, and make sense of the corresponding views of science in their own words and by using cards with predefined values and assumptions and knowledge production strategies (see Fig. 18.2 for the cards). The instructions on how to conduct the workshop are shown in Box 18.1,

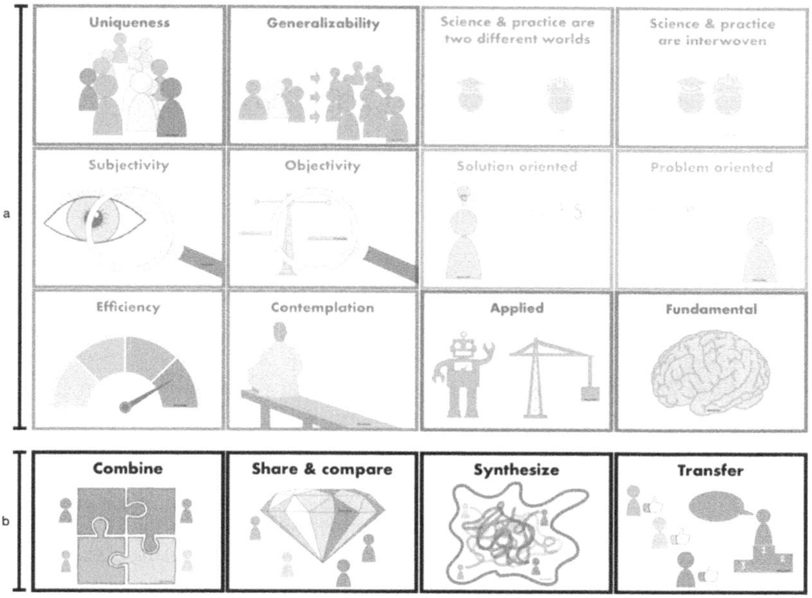

Fig. 18.2 Cards for FRL workshop; (**a**) value and assumptions cards for step (4) of the workshop; (**b**) knowledge production strategy cards, corresponding to step (6) of the workshop

and the FRL website provides a more elaborate manual for how to run the workshop, as well as printable versions of the cards, and a digital template for running the workshop online.

Box 18.1: Step-by-step instructions for conducting the FRL workshop

Watch video 1

Step 1: Exploration: as a group, briefly describe and define what keywords could be assigned to each character.
Step 2: Position yourself: individually, position yourself on the canvas relative to the four characters and explain to each other the position that you have chosen.

Watch video 2

Step 3: Differences and similarities: as a group, make sense of the similarities and differences between the characters that stood out to you. Indicate them on the canvas by using, for instance, text, arrows, circles, or lines.
Step 4: Value and assumption cards: as a group, place the 12 assumption cards on the canvas relative to the characters (see Fig. 18.2).
Step 5: Reposition yourself: return to your initial positioning of step (2). After discussing the characters in more detail, is that still where you would place yourself? Explain to each other.

Watch video 3

Step 6: Knowledge production strategy cards: as a group, place the four knowledge production strategy cards, each with one of the characters.
Step 7: Reposition yourself: return to your positioning of step (5). After discussing the characters in more detail, is that still where you would place yourself? Explain to each other.

We designed and implemented FRL in online, offline, and hybrid formats. In classrooms, we use flipcharts as a canvas, and printed value and assumption cards. In online settings, the participants collaborate remotely in the online workspace Mural. And in hybrid settings, in-class and online participants work in Mural on their own devices while viewing the videos in plenary or via their computer. Figure 18.3 shows some impressions of the online materials and the on-site workshops.

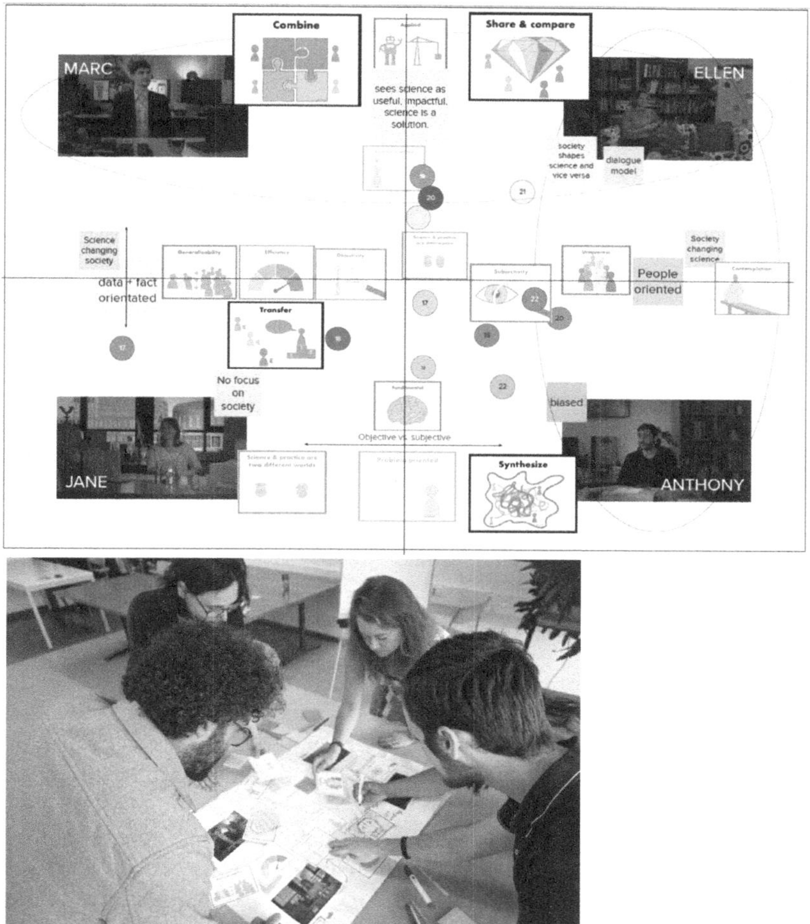

Fig. 18.3 Mural canvas from an online FRL workshop (top) and flipcharts and cards of an in-person workshop (bottom)

18.3 Using Frame Reflection Lab in Practice and Lessons Learned

Between 2020 and 2024, we have developed and run FRL workshops in a university context with over 400 participants. This included Bachelor's, Master's and PhD students, and on some occasions more experienced researchers, from social to natural science-related (inter-) disciplinary backgrounds. We derive the following four main lessons from the experiences of teachers and students.

Lesson 1: Integrate FRL into Interdisciplinary Teamwork Practice

When we embedded FRL as just one intervention in the context of a larger inter- and transdisciplinary learning process, rather than as a one-off event, it helped bring home the lessons from the workshop(s) and further support inter- and transdisciplinary learning. We gave participants written reflection assignments before and after the workshop, which helped to both broaden and deepen their engagement with the topic beyond the workshop(s).

In one course, we integrated reflection on views of science in the full five months of the course duration. We explicitly connected the FRL approach to teamwork in a practical project and returned to the FRL lessons and terminology repeatedly. In this case, we offered FRL as two interactive workshops covering steps 1–5 and 6–7, respectively, and complemented the workshops with repeated individual written reflection assignments. Figure 18.4 gives a schematic timeline of a longer-term FRL process, in which FRL activities are implemented at different time points.

Students reported that the integration of FRL into the course, and explicitly linking the lessons from FRL to their teamwork, shaped how they made sense of the interactions in their team. As one student put it:

> I think that it [engaging in the FRL workshops] made me very aware of the different ways of doing research and I also think that that subconsciously contributed to the collaboration. I could place others relative to the FRL characters and understand their perspectives better.

The terms from FRL knowledge production strategies cards (see Fig. 18.2) gave the teams a shared understanding and the vocabulary to discuss desired and actual levels and forms of knowledge integration.

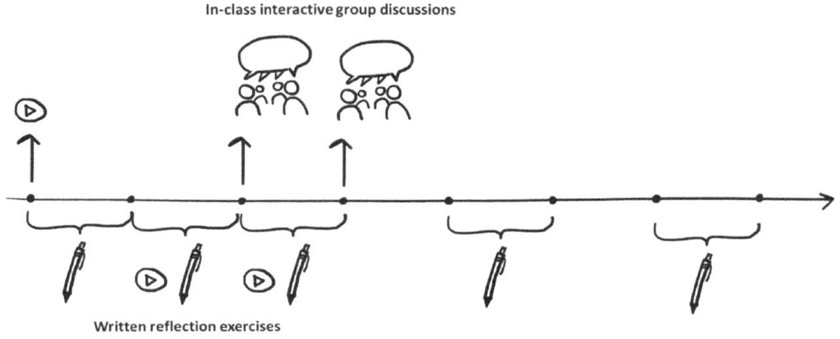

Fig. 18.4 Schematic timeline of implementing FRL as a continuous process over a longer period of time: videos, written reflection assignments, and interactive group discussions are combined and incorporated at different points throughout the process as a means to leverage one another

We also used reference to the FRL terminology in discussing the forms and levels of integration in the products on which the students worked, for instance in instruction, making sense of the assignment rubric, and feedback. Beyond the conversation about values and beliefs in the FRL workshop, the knowledge production strategy cards and corresponding terms in this case also served as a shared understanding to link explicitly to their joint work.

Lesson 2: Tailor to Different Learning Goals and Needs

FRL can achieve different learning outcomes (see also Horn et al., 2022) and can thus be used for different purposes. We found that at Bachelor's level, FRL contributed mostly to developing basic awareness about the existence of diverse views of science and to providing a vocabulary to reflect on such views. This growing awareness is well illustrated in this Bachelor's student's words:

> I learned that there are different approaches to research. I mostly thought that all research is close to Jane's.

At the Master's and PhD level, FRL was more likely to stimulate deeper reflection on academic identities. When we used FRL among PhD candidates in inter- and transdisciplinary projects, we adapted the workshop and asked them to position their past, current, and aspired future selves, rather than choosing a single position. The following quote from a PhD student illustrates how this sparked reflection on identities, as well as changes and tensions in identity:

> Ellen is a side that I have been running away from, because I think it is very hard, so I need to have the skills, and I need courage.

So, we found that FRL could be applied in a range of settings to achieve various learning goals provided it is adapted appropriately.

Lesson 3: Facilitate the Discussions

In all contexts, we found that facilitation of the group dialogues supported more substantial reflection. Follow-up questions about the meaning of statements and the reasoning behind certain decisions (e.g. in relation to positioning on the canvas) helped participants to be more explicit about and further question their reasoning. To give an example, we saw that participants quite commonly identified with Ellen because they aspired to make a contribution to addressing major issues in society, while at the same time they argued from quality criteria for scientific rigour more in line with Jane's view, which clashed with Ellen's. In such instances, we posed follow-up questions aimed at stimulating deeper engagement with the views of science that the characters represent. Other topics that commonly required such further disentanglement included the difference between societally relevant and applied research, and between personal beliefs and one's understanding of the role of science and scientists.

Lesson 4: Create and Nurture a Safe Space

FRL invites participants to disagree with one another, even when they hardly know each other, or, conversely, when they know each other very well and assume that they all think alike. At the same time, we saw that not all participants were eager to express disagreement or share potentially contested views. As the participants' willingness to express their views even when those may be different from those of others, and the openness to question and potentially change their views are imperative to the FRL learning experiences, facilitators must create a safe space in which participants can disagree and explore different perspectives. In running the workshops, we emphasised that there are no right or wrong answers and that different perspectives and approaches can exist alongside and need each other.

In addition, in rethinking their own positioning on the FRL canvas, psychological safety seemed to play a role. Some participants indicated that revising their position felt like admitting a mistake in their initial position. Once we observed this dynamic, we framed the repositioning stages explicitly as an opportunity to view one's position differently based on the newly acquired information rather than to 'correct' an earlier 'mistake'. Students said that this framing made them more inclined to rethink and adapt their positioning.

We should note that the experience of a safe space could also depend significantly on online or offline participation. We saw that online workshops best exposed differences when the participants knew each other well. Interestingly, a student who indicated to often struggle with social situations mentioned that an offline session in which each student used their own device (instead of a canvas and cards) felt safer, because it did not require eye contact with other students.

18.4 Conclusion

Based on our experiences, we believe that FRL is a useful instrument to spark conversation about views of science and train (teams of) researchers in interdisciplinary consciousness, as a key competency for transdisciplinary collaboration. Returning to the example with which we opened this chapter, we see that FRL provided a framework to talk about views of science, and the identity-first language—"*I used to be a Jane*"—which indicates that participants also experienced an opening to reflect on academic identities, and go beyond a merely cognitive engagement with their views of science. We are pleased to make the materials for the workshops openly accessible for others to use in their own contexts. In presenting what we have learnt, we have attempted to provide insight into how the FRL format can be adapted to other contexts—and invite readers to do so.

References

Horn, A., van der Meij, M. G., Willems, W., Kupper, F., & Zweekhorst, M. B. M. (2022). Developing interdisciplinary consciousness for sustainability: using playful frame reflection to challenge disciplinary bias. *Sustainability: Science Practice & Policy, 18*(1), 515–530. https://doi.org/10.1080/15487733.2022.2095780

Tijsma, G., Horn, A., Urias, E., & Zweekhorst, M. B. M. (2023). Training students in inter-and transdisciplinary sustainability education: Nurturing cross-faculty staff commitment and continuous community collaboration. *International Journal of Sustainability in Higher Education, 24*(4), 765–787. https://doi.org/10.1108/IJSHE-02-2022-0049

Van der Meij, M. G., Heltzel, A. A. L. M., Broerse, J. E. W., & Kupper, F. (2018). Frame reflection lab: A playful method for frame reflection on synthetic biology. *NanoEthics, 12*(2), 155–172. https://doi.org/10.1007/s11569-018-0318-9

PART IV

Concluding Remarks

CHAPTER 19

Transdisciplinarity for Transformation: What's Next?

Barbara J. Regeer, Pim Klaassen, and Jacqueline E. W. Broerse

As many of the chapters in this volume have illustrated, there are no single or easy answers to the question of 'what is transdisciplinarity?' and 'how can I, through research or everyday inquiry, contribute to transformation?'. Everyone who was involved in this book project went on their own quest, their own journey, sometimes together with others, sometimes alone. And it is all of us, as individuals and collectives, who together create momentum for transformation. Let's therefore start this concluding chapter by means of an example of an individual; it concerns the experience of a NALAM worker in Southern India. NALAM means

B. J. Regeer (✉) · J. E. W. Broerse
Athena Institute, Vrije Universiteit Amsterdam, Amsterdam, The Netherlands
e-mail: b.j.regeer@vu.nl

J. E. W. Broerse
e-mail: j.e.w.broerse@vu.nl

P. Klaassen
Stichting Stadslab, Amsterdam, The Netherlands
e-mail: pim@hetstadslab.nl

B. J. Regeer et al. (eds.), *Transdisciplinarity for Transformation*,
https://doi.org/10.1007/978-3-031-60974-9_19

'well-being' in Tamil, and NALAM workers are lay mental healthcare workers, trained by The Banyan, a non-government organization (NGO) to offer community-based services. The NALAM worker narrated the following story (Dijkxhoorn, 2020, p. 187):

> A 21-year-old woman's husband committed suicide, leaving her a widow with three young children. She developed depression and suicidal tendencies, which she shared in a community support group meeting. Instead of referring her to the clinic, I spent time with her and discovered that she had grave financial issues due to a loan her husband had taken on a motorbike. The lenders harassed her frequently and her sister-in-law had taken possession of the motorbike. She did not have money to feed herself and her children. I decided to speak to the panchayat leader, who was very supportive. He spoke to the lenders, who waived the loan amount, and to the sister-in-law, who agreed to return the motorbike, so it could be sold. We enrolled the children in school. After solving these issues, she was doing much better and did not need any medication.[1]

The actions the community mental health worker took seem very logical—spending time with the woman in order to gain a deeper understanding of the immediate and underlying issues and taking steps to address these. However, in highly specialized, and often fragmented, public-sector systems that are commonplace in many western countries, this course of action would be highly exceptional. First, because she works for a mental health NGO, but also because she does not approach the woman with a set of mental health interventions. Rather, she takes a holistic, problem-oriented approach. Second, she spent time with the woman to work out what was going on. In highly specialized work settings, especially when the question of financing treatments is a concern, spending this kind of time upfront is not possible. Third, her intervention is highly tailored to the specific situation. No protocol, blueprint or specific standards are applicable. She needs to be creative and brave rather than working from specific (mental health) knowledge.

We could see the approach taken by the NALAM worker as privileging her attachment (Puig de la Bellacasa, 2017) to a real-world problem over her attachments to her professional or epistemic community, with a commitment to making a positive change for the woman involved, and

[1] This example by no means serves to say that mental health problems should be solved without medication.

those around her. One could argue that this is what transdisciplinarity is all about; it is about gearing knowledge production to problems of the life-world, rather than disciplinary boundaries (Mittelstraβ, 1992, cited in Pohl & Hadorn, 2008). This implies putting our normative commitments—to our epistemic community, to our professional context—at risk (Moats & Seaver, 2019; Zuiderent-Jerak, 2015). Transdisciplinarity is about a commitment to making a change, to contributing to transformation to 'develop knowledge and practices to promote what is perceived to be the common good' (Pohl & Hadorn, 2008, p. 112), while recognizing that both 'the problem' and 'the common good' are negotiated time and again.

The example of the NALAM worker might not be considered representative of a transdisciplinary *research* practice, but it does bring to the fore what we might call a spirit of 'purpose-ledness' (again, notwithstanding the diverse perceptions of 'purpose' that are inherently part of each transdisciplinary endeavour); an orientation, not so much towards 'what is', but towards 'what ought to be', that pervades the thinking and doing (while, again, putting that normative commitment at risk). The past decades of increasing engagement with transdisciplinary research have led to multiple epistemological, methodological and ethical advances. There are tested methodologies to organize meaningful knowledge integration; principles and heuristics to guide multi-actor innovation processes; strategies to address power dynamics and to overcome systemic barriers; and approaches to sustain and upscale processes and outcomes. However, the increasingly urgent need to create impact in view of ever more devastating health and sustainability challenges points towards a gap in the current focus of the literature on knowledge co-creation. The apparent need for a comprehensive understanding of the intricacies associated with multi-actor innovation processes introduces the potential danger of perceiving collaborative efforts, co-creation, social learning and reflexivity, along with their political and power dimensions, as an end in themselves, rather than a means to an end. Or, otherwise stated, the (undeniably important) focus on these process criteria tends to blur the 'purpose-ledness' of the endeavour. Bringing purpose back to the centre is what we call Transdisciplinarity for Transformation.

19.1 Agenda for Action, Learning and Research

In this final chapter, we will draw up an agenda for action, learning and research for transdisciplinarity for transformation. Upon reflection of the broad spectrum of experiences and profound insights shared by our colleagues in their diverse engagements with transdisciplinarity for transformation, we have identified what we consider to be key items on this agenda. We elaborate on each of these and formulate concrete questions. A thread throughout these reflections is how to keep the focus on the envisaged transformative change—the purpose of transdisciplinarity—while juggling the many complexities and challenges in the transdisciplinary research process.

19.1.1 Making Frameworks Actionable

The first item on the agenda builds on Chapter 3 (Regeer et al.; this volume), in which we argue that accompanying transdisciplinary practices with deliberate and repeated cycles of action and reflection can help to keep purpose centre stage. As Gjefsen et al. (this volume, Chapter 4, p. 125) put it: transdisciplinary projects with transformative ambitions might not be *'a matter of "planning then doing", but rather a matter of "planning by doing"'*. Sophisticated generic frameworks for designing and evaluating transdisciplinary research (see Chapter 3) have been developed in recent decades: they can be seen as disembodied and decontextualized sets of knowledges and mindsets, i.e. years of highly situated experiences went into these frameworks, in different empirical domains and in different countries across the globe. The decontextualization and disembodiment that took place in constructing these frameworks allow for them to travel into different (academic or transdisciplinary) spaces. A question on the agenda for transdisciplinarity for transformation concerns the 'landing' of these frameworks in transdisciplinary spaces:

> How can these knowledge-rich frameworks become recontextualized and re-embodied in situated transdisciplinary spaces, where they can guide practitioners/participants in keeping purpose centre stage and addressing challenges as they come along?

We already gave some pointers in Chapter 3, where we argued that these frameworks become an asset for 'planning by doing' by becoming themselves one of the actors in the messy entanglements that characterize transdisciplinary practices—they become part of the conversation. The Dynamic Learning Agenda (in itself yet another materiality) can be used to foster the embodiment ('How can I ...') and contextualization ('while...?') of the more generic guiding questions that accompany frameworks for transdisciplinary research,[2] where and when issues, to which the questions pertain, arise (see also Regeer et al., Chapter 1, this volume). Frameworks thus become actionable in the guiding of transdisciplinary practice. The second pointer we gave in Chapter 3 is the use, and further development, of approaches, such as Reflexive Monitoring in Action (Van Mierlo et al., 2010) and accompanying research (Defila & Di Giulio, 2018; Schäpke, this volume, Chapter 6), but also, from the field of STS, situated interventions (Zuiderent-Jerak, 2015), to support reflection-in and reflection-on action (Schön, 1983), to keep purpose centre change (or address the challenge that often *'ambitions are diluted because [...] people are distracted by everyday details'* (Van Mierlo et al., 2010, p. 17)), as well as to help understand the diverse associations (Grijseels et al., under review, Latour, 1986) and commitments that come with bringing a diversity of people into spaces for transformation—spaces in which there is still a need for work on putting in place conditions conducive to transformation (Holle, Ponzoni & Ghorashi, Chapter 11, this volume). Given the often-significant power differentials between people from highly diverse backgrounds, and the fact that transdisciplinary research can be seen as an attempt to breach existing power structures more widely, continuous reflection is crucially important to prevent perpetuating prevailing *'hierarchical, academic, postcolonial knowledge orders'* (Schmidt & Neuburger, 2017, p. 65) in the practice of transdisciplinary research (see also Strumińska-Kurta & Scholl, 2022).

19.1.2 Dealing with Institutional Settings

A second item on the agenda for transdisciplinarity for transformation concerns dealing with (un)conducive institutional settings. Professionals (practitioners, academics) are not risk-free if they privilege occupying

[2] See, for instance, Lang et al., 2012, Table 1 for a set of 12 guiding questions to put design principles for transdisciplinary research into practice. And see, for example, Roura, 2021, Table 1 for a set of 34 monitoring questions to guide the assessment of power dynamics using a framework for power-sensitive participatory health research.

a problem and action space that seems remote, in terms of routines, values and discourse from their institutional home. As Holle, Ponzoni and Ghorashi (Chapter 11, this volume, p. 319) describe in their reflection on a co-creative research collaboration with queer refugees, artists and academics: *'acts of decentring make scholars vulnerable in the process of knowledge production'*. In some cases, professionals' transdisciplinary mode of doing and thinking might align very well with the institutional culture and structure, but this is rare. The organization of the NALAM worker at The Banyan (a mental health care NGO in Southern India) does provide an example; its evolution over 25 years demonstrates the careful emergence of a responsive mental health system with both a user-centred and a service-integration focus (Narasimhan et al., 2019), which optimally supports the purpose-led approach of NALAM workers. The case of child mental health and the law described in Ramaswamy, Seshadri and Bunders (Chapter 10, this volume) arguably shows something similar—if only in relation to the success of realizing transformation as a result of consistent and concerted efforts at multiple levels and spread out over several years. In most other cases, however, professionals experience a greater tension between institutionalized ways of thinking and doing and what is required for transformation. For instance, Gunn, Hoffmann, Sager, Wittmayer and Zuiderent-Jerak (Chapter 15, this volume, p. 427) explore tensions between academic home-bases and doing transdisciplinary work, and ask *'What sort of institutional spaces or frameworks enable all actors involved to adapt, learn and transform?'* In the scholarly literature on transdisciplinary research, this is reflected in a growing ambition for contributing to systemic change, beyond individual project outputs or outcomes (e.g. Clark et al., 2016, Marshall et al., 2018, Van Breda & Swilling, 2019, see also literature on system innovation, transition studies and sustainability transformations research: Geels, 2005, Grin, 2020, Lam et al., 2020, Pereira et al., 2015). Institutional settings that are more conducive to a transdisciplinary mode of thinking and doing—embracing epistemological plurality, facilitating multi-stakeholder collaboration across boundaries (whether they are disciplinary, sectoral or multi-level), and purpose-oriented governance—can be argued to be better equipped to respond to (super)wicked problems. So, the question for transdisciplinarity for transformation could be:

> How to re-imagine organizational, cultural and systemic conditions to shape conducive ecosystems for a transdisciplinary mode of doing and thinking?

One direction that both action and research can take is that of institutionalization, or 'normalization' (May & Finch, 2009) of a transdisciplinary mode of thinking and doing (see also Vienni-Baptista & Klein, 2022). This can include building new institutional spaces, or infrastuctures, such as the Making and Doing programme in the STS field (Gunn et al., Chapter 15, this volume, referring to Downey & Zuiderent-Jerak 2021). But it can also build on case studies that have been undertaken in the context of research organizations (e.g. Verwoerd et al., 2023) and Higher Education Institutions (HEIs) (Tijsma et al., 2023). Naturally, this is not limited to systemic change in the institutional ecosystems of academics (including research funders, see Defila & Di Giulio, Chapter 5, this volume)—the same applies to professionals in other sectors. Here we can build upon case studies on institutionalizing reflexivity (e.g. in perinatal care, medicine development, Schuitmaker-Warnaar et al., 2021), normalizing system- and client-oriented approaches in youth protection (Van Veelen et al. 2017, Bunders et al., 2023) and combining community listening with organizational listening (Zachariah et al., 2023). These and other initiatives take a purpose-led orientation to heart.

Beyond these case studies, we can take pointers from other scholars that call for explicit directionality, or purpose orientation in governance, research and innovation. Emphasizing the difference between collaborative governance and problem-oriented governance, Mayne and colleagues (2020, p. 33), for instance, state: '*Even if multiple organizations coordinate their individual efforts, there is no guarantee that the whole is more than the sum of the parts. Problem-oriented governance takes the problem rather than institutions as point of departure*'. Moreover, this type of governance is '*fundamentally outward-looking in its efforts to shape both long-term strategy and day-to-day working arrangements around problems as they manifest themselves*' (ibid., p. 34). Similarly, Van der Steen and colleagues define purpose-led working in a policy context as follows: '*It means that the societal issue is taken as a starting point [...]. In such an approach, one's own organization and capacity is folded around the issue*' (Van der Steen et al., 2020, p. 5, authors' translation). This tallies with the idea of mission-oriented innovation policy (Mazzucato, 2017) and the associated concept of Mission-oriented Innovation Systems (MIS,

Hekkert et al., 2020). While it is recognized that studying the impact of directionality provided by a clearly defined mission is important (Hekkert et al., 2020), and that it requires from public institutions that they are willing to experiment, to change day-to-day routines and to implement processes to build dynamic organizational capabilities (Mazzucato, 2018), the focus on the research agenda (see Hekkert et al., 2020) is more on the processes involved in formulating missions and selecting and integrating possible solutions. The need to study the interplay between agents and institutions and their adaptive capacities, taking missions (purpose) as focal point, is much less articulated in this body of literature. While the research on MIS is still young, the explicit mission-*oriented* approach appears relevant.

In these approaches, as we have seen, purpose is taken as point of departure. Taking this seriously may imply fundamental re-imagining of 'structuring' in organizations; the interactional spaces available to professionals; the types of standards devised to coordinate and professionalize working routines; and the accountability mechanisms put in place to recognize quality and progress, and to monitor compliance with laws and regulations. This leads to an additional direction for the agenda for transdisciplinarity for transformation:

> How can standards and accountability practices be re-imagined and (re-) materialized to serve as boundary objects with transformative powers?

Scholars of transdisciplinary research have long recognized that standard academic metrics[3] do not provide a conducive ecosystem for academics to engage in transdisciplinary research (Fischer et al., 2012). The same goes for professionals in other spaces. Prevailing accountability practices are often counter-productive and ask to reduce 'meaningful' practices to 'measurable' standards, ridding these practices of engagement, lived experience and situatedness. As de Weger and colleagues (Chapter 12, this volume) show, where engagement practices are concerned there is also plenty of room for improving metrics. A key question is how such accountability practices can be upended, and then re-imagined and altered

[3] Including measuring the number and academic impact of publications—or, where it comes to 'practising inclusion', conventional demographics focused on socio-economic status, gender or age as discussed by de Weger et al. (Chapter 12, this volume).

and serve as boundary objects (Star & Giesemer, 1989) between purpose-oriented micro-practices and existing regimes. And more than boundary objects, they may carry great transformative potential in relation to our systems of governing and in relation to professional practices, supporting reflective practice (Schön, 1983) in relation to purpose.

19.1.3 Looking Simultaneously 'Inward' and 'Outward'

From within the conceptual and practical realm where we find challenges for inclusion, diversities and positionality, we can easily latch onto the case made above—i.e. the case for focusing research efforts towards developing ever-better understandings of, and ever-better tailored practices to, what is needed in order to institutionalize the orientation of transdisciplinarity towards whichever purpose (or purposes) it is (or are) most worthwhile to contribute. To be clear, this requires both of transdisciplinarians as individuals and of transdisciplinary collectives that they dedicate the necessary time and efforts to (re-)negotiation, opening up, closing down, decentring and recentring, zooming out and zooming in—all in order to *not* lose sight of the purpose, and to connect multiple levels of reality that are variably accessible from within different aspects of anyone's multiple identities (be it as a trained economist, a father, a civil servant or whatever). This brings us to the following question, which we think is central to the next steps of dealing with the challenges of inclusion, diversities and positionality in transdisciplinarity for transformation:

> How can we carve out the necessary space for looking inward, while looking outward —namely, space for self, space for being human, space for the beautiful and the ugly?

Let us elaborate what we mean by this. It starts by recognizing an often-overlooked boundary that demands to be crossed if transdisciplinarity is to live up to its promise; the boundary not only between Science and Society, but also the Self (including self-care and reflexivity in terms of a researcher's own role and position); a point made by four PhD students reflecting on their experiences with using a transdisciplinary approach (Sellberg et al., 2021). This issue was exemplified by den Boer (Chapter 14, this volume) in her autoethnography. She adeptly recounted her challenges in manoeuvring through the dynamics of the science system, project logic, the imperative for societal impact and the intricacies of her own positionality, all while recognizing the importance of self-care. Chapter 15 (this volume, p. 435) gives a beautiful account of the different ways in which experienced transdisciplinary scholars reflect on '*what [...] it mean[s] for us to be alive to the world in the research process*', exemplified by the following quote '*Sometimes it feels like I am surrounded by heroes and I then ask myself: where are the human beings, with all their failures and successes?*'. Consistent with this, we argue that the focal point of future research in transdisciplinarity for transformation in dealing with challenges associated with inclusion, diversities and positionality is what we would like to call *looking outward and looking inward*, echoing Lloro-Bidart and Finewood (2018).

Looking inward means that transdisciplinarians working on social transformation ought to confront explicitly all sorts of challenges that come with the question of how to organize transdisciplinary work, thinking through and materially enacting what everyone's (intersectional), and their own, positionality means for their role(s) in the transient transdisciplinary collective they are part of, and for how to collaborate within such a collective. The Frame Reflection Lab as a tool for self-reflection on views of science (See Horn and Van der Meij, Chapter 18, this volume) is one example of how looking inward can be facilitated. Pressing questions around participatory research collectives that are raised by scholars of critical Participatory Action Research are of relevance here, e.g.: '*When does it make sense for collectives to be "participatory contact zones" (Torre, 2005) in other words, teams that bring together very differently positioned people to research together? And when might it be important to have separate spaces for marginalized and oppressed groups to research together?*' (Torre, 2014, p. 1325). This is where the Self takes centre stage, as the node connecting transdisciplinary roles and everyone's, and one's own, multiple identities and which is needed for tapping into purpose and navigating, or breaking away from, the rules and institutions and one's own epistemic culture, for instance. This requires looking inward. As Fals-Borda (2013, pp. 165–166) aptly points out: '*Let us recall that the paradigms that so far have moulded our professional training are sociocultural constructs originating in Europe. Today we try to take inspiration from our own surroundings and to construct more flexible paradigms of a holistic and participatory nature. Academic arrogance is an obstacle to achieving these goals; it should be removed from circulation*'.

Looking outward means that throughout the transdisciplinary projects, transdisciplinarians can never take for granted how research questions are asked, which research questions are asked, which groups or individuals are made part of the conversation on this, who are or are not engaged in the relevant implementation of research, experimentation or exploration, and how this is done.

If these two themes of looking inward and looking outward are not already hard enough in and of themselves, arguably the biggest challenge arises when attempting to combine the two. If anything, then, the above question and its supporting conceptual work constitute a call to action for transdisciplinarians to do the hard work of looking inward while looking outward.

Several chapters suggest interesting avenues for further strengthening our understanding and practices of co-creative knowledge and solution building that follow this conceptual signpost. The chapter by Brouwers, Egberts and de Hoop (this volume, Chapter 9), for instance, underscores the contextual and embodied nature of knowledge and knowledge exchange, the acknowledgement of and reflection on which constitutes one tangible way of looking inward for finding better ways of looking outward. Consistent with this, yet underscoring slightly different aspects of our embodied identities, Stark's contribution (this volume, Chapter 16) invites us to engage with head, hand and heart to break with academia's consistent and exclusive focus on disembodied cognition in knowledge practices (this volume, Chapter 16). As has been argued elsewhere, there is a marked affinity between such an approach and decolonialism and traditional and indigenous knowledge practices (Penna & English, 2022). For instance, a theme such as deep and generative listening (Scharmer, 2009), also prominent in the contribution to this volume by Bruhn and colleagues (this volume, Chapter 7) and Stark, runs through all such literatures and arguably constitutes a call to work on appreciating everyone's intersecting positionalities, consistent with how we just now explained looking inward and looking outward. When considering how to develop the competencies for both looking inward and outward, Zeidan et al. (this volume, Chapter 17) argue that it is crucial to acknowledge students' inherent values and experiences, and facilitate their ability to effectively channel these experiences, engage in critical thinking and foster meaningful interactions with fellow learners. Ultimately, looking inward and outward is very much about finding ways of making available '*the experiences of marginalized peoples, nonhuman entities, and the environment more broadly'* (Lloro-Bidart & Finewood, 2018, p. 149).

However, we have to acknowledge that this is much easier said than done. Indeed, whoever tries to do it, immediately runs the risk of reproducing forms of inequity and (epistemic) injustice. Making the experiences of marginalized peoples, nonhuman entities and the environment available to others might well amount to forms of extractivism. For instance, if we realize that, while *'comprising less than 5% of the world's population, indigenous people protect 80% of global biodiversity'* (Raygorodetsky, 2018, p. 1), it is clear the challenge is immense for those other 95% of the world's population. This relates as much to the types of solutions to today's complex problems to which transdisciplinarians might contribute—think of the energy transition that constitutes a direct threat to so many Indigenous peoples, because of the environmental destruction that comes with mining all sorts of minerals and metals necessary for electrifying our energy system—as to the appreciation of and practices in different forms of knowledge production. Interestingly, *'Indigenous people and traditional knowledge keepers seem to be the ones currently leading by example in the struggle of protecting the Earth for future generations, [and] they should not only be counted as stakeholders but possibly even as guides/leaders in the development of sustainable initiatives and discourse'* (Breidlid & Krøvel, 2020, cited in Silvestru, 2023, p. 2).

That many already recognize this becomes clear in the learning questions so many of our graduate students formulate in response to their own learnings on the road to becoming transdisciplinarians working in fields like environmental sustainability or global health. This is reflected in their learning questions that express an aspiration to foster equitable and mutually beneficial partnership and an acknowledgement of their own positionality, values and normativity. Again, learning questions such as these are not so much expecting the one unique correct answer, but rather questions that remind us of the immensity of challenge confronting us, and that we need to hold up for ourselves as we muddle through, trying to make a meaningful impact while trying to co-create knowledge.

19.2 Concluding Remarks

If, as an early-career researcher embarking on your journey into mastering transdisciplinarity for transformation, you have made it this far, we assume you are utterly overwhelmed and confused, both conceptually and practically as well as personally. Unfortunately, there are no short-cuts, other than to start somewhere and slowly work your way through, carefully crafting your own learning questions and personal development path as Zeidan and colleagues (this volume, Chapter 17) so strongly emphasize. One specific mindset that is key in this respect is to 'be at ease with unease'. Transdisciplinarity for transformation demands that you let go of your perfectionist tendencies; the messy reality out there will certainly make you 'fail' many times. Do not forget: learning by doing and learning from mistakes are the best teachers. Be a 'true academic' in the sense of being the innovative, creative 'out of the box' thinker that a scientist is supposed to be (but in reality rarely is), rather than a passive rule-follower ('cultural dope') (Geels & Schot, 2007, p. 403). Challenge the known and dive into the unknown with curiosity and humility, without losing your ideals to make the world a more just, sustainable and safe place for all.

References

Breidlid, A., & Krøvel, R. (Eds.). (2020). *Indigenous knowledges and the sustainable development agenda*. Routledge.

Bunders, A. E., de Wit, E. E., Dinkgreve, M. A. H. M., Broerse, J. E. W., & Regeer, B. J. (2023). Toward a new 'normal': An analysis of emerging leadership practices in a changing child and youth protection organization. *Child & Youth Services*, 1–32.

Clark, W. C., van Kerkhoff, L., Lebel, L., & Gallopin, G. C. (2016). Crafting usable knowledge for sustainable development. *Proceedings of the National Academy of Sciences, 113*(17), 4570–4578. https://doi.org/10.1073/pnas.1601266113

Defila, R., & Di Giulio, A. (2018). What is it good for? Reflecting and systematizing accompanying research to research programs. *GAIA-Ecological Perspectives for Science and Society, 27*(1), 97–104.

De La Bellacasa, M. P. (2017). *Matters of care: Speculative ethics in more than human worlds* (Vol. 41). University of Minnesota Press.

Dijkxhoorn, M. A. (2020). *Of hardship, growth and support. Family caregivers of people with mental ill health in Chennai, India* (PhD thesis). Vrije Universiteit Amsterdam.

Fals-Borda, O. (2013). Action research in the convergence of disciplines. *International Journal of Action Research, 9*(2), 155–167.

Fischer, J., Ritchie, E. G., & Hanspach, J. (2012). Academia's obsession with quantity. *Trends in Ecology and Evolution, 27*, 9. https://doi.org/10.1016/j.tree.2012.05.010

Geels, F. W. (2005). Processes and patterns in transitions and system innovations: Refining the co-evolutionary multi-level perspective. *Technological Forecasting and Social Change, 72*(6), 681–696. https://doi.org/10.1016/j.techfore.2004.08.014

Geels, F. W., & Schot, J. (2007). Typology of sociotechnical transition pathways. *Research Policy, 36*(3), 399–417.

Grijseels, M., Regeer, B., & Zuiderent-Jerak, T. (under review) Empiricizing Transitions: From Empirical Detail to Transformation. Available at SSRN 4693130.

Grin, J. (2020). 'Doing' system innovations from within the heart of the regime. *Journal of Environmental Policy & Planning, 22*(5), 682–694.

Hekkert, M. P., Janssen, M. J., Wesseling, J. H., & Negro, S. O. (2020). Mission-oriented innovation systems. *Environmental Innovation and Societal Transitions, 34*, 76–79.

Lam, D. P., Martín-López, B., Wiek, A., Bennett, E. M., Frantzeskaki, N., Horcea-Milcu, A. I., & Lang, D. J. (2020). Scaling the impact of sustainability initiatives: A typology of amplification processes. *Urban Transformations, 2*(1), 1–24.

Lang, D. J., Wiek, A., Bergmann, M., Stauffacher, M., Martens, P., Moll, P., Swilling, M., & Thomas, C. J. (2012). Transdisciplinary research in sustainability science: practice, principles, and challenges. *Sustainability Science, 7*, 25–43.

Latour, B. (1986). The powers of association. In J. Law (Ed.), *Power, action and belief: A new sociology of knowledge?* (Vol. 32, pp. 264–280). Routledge & Kegan Paul.

Lloro-Bidart, T., & Finewood, M. H. (2018). Intersectional feminism for the environmental studies and sciences: Looking inward and outward. *Journal of Environmental Studies and Sciences, 8*, 142–151.

Marshall, F., Dolley, J., & Priya, R. (2018). Transdisciplinary research as transformative space making for sustainability. *Ecology and Society, 23*(3).

May, C., & Finch, T. (2009). Implementing, embedding, and integrating practices: An outline of Normalization Process Theory. *Sociology, 43*(3), 535–554. https://doi.org/10.1177/0038038509103208

Mayne, Q., De Jong, J., & Fernandez-Monge, F. (2020). State capabilities for problem-oriented governance. *Perspectives on Public Management and Governance, 3*(1), 33–44.

Mazzucato, M. (2017). *Mission-oriented innovation policy* (UCL Institute for Innovation and Public Purpose Working Paper, 1).

Mazzucato, M. (2018). Mission-oriented innovation policies: Challenges and opportunities. *Industrial and Corporate Change, 27*(5), 803–815. https://doi.org/10.1093/icc/dty034

Moats, D., & Seaver, N. (2019). 'You social scientists love mind games': Experimenting in the 'divide' between data science and critical algorithm studies. *Big Data & Society, 6*(1), 2053951719833404.

Narasimhan, L., Gopikumar, V., Jayakumar, V., Bunders, J., & Regeer, B. (2019). Responsive mental health systems to address the poverty, homelessness and mental illness nexus: The Banyan experience from India. *International Journal of Mental Health Systems, 13*(1), 1–10.

Penna, K. N., & English, J. P. (2022). The power of indigenous peoples in decision-making processes of mining projects: The Pilbara region. *International Journal of Humanities and Social Sciences, 16*(12), 782–789.

Pereira, L., Karpouzoglou, T., Doshi, S., & Frantzeskaki, N. (2015). Organising a safe space for navigating social-ecological transformations to sustainability. *International Journal of Environmental Research and Public Health, 12*, 6027–6044. https://doi.org/10.3390/ijerph120606027

Pohl, C., & Hadorn, G. H. (2008). Methodological challenges of transdisciplinary research. *Natures Sciences Sociétés, 16*(2), 111–121.

Raygorodetsky, G. (2018, November 16). Indigenous peoples defend Earth's biodiversity—But they're in danger. *National Geographic*. https://www.nationalgeographic.com/environment/article/can-indigenous-land-stewardship-protect-biodiversity

Roura, M. (2021). The social ecology of power in participatory health research. *Qualitative Health Research, 31*(4), 778–788.

Scharmer, O. (2009). *Theory U—Leading from the future as it emerges*. Berrett-Koehler Publishers.

Schmidt, L., & Neuburger, M. (2017). Trapped between privileges and precariousness: Tracing transdisciplinary research in a postcolonial setting. *Futures, 93*, 54–67.

Schön, D. A. (1983). *The reflective practitioner: How professionals think in action*. Basic Books.

Schuitmaker-Warnaar, T. J., Gunn, C. J., Regeer, B. J., & Broerse, J. E. (2021). Institutionalizing reflexivity for sustainability: Two cases in health care. *Sustainability, 13*(21), 11712.

Sellberg, M. M., Cockburn, J., Holden, P. B., & Lam, D. P. (2021). Towards a caring transdisciplinary research practice: Navigating science, society and self. *Ecosystems and People, 17*(1), 292–305.

Silvestru, A. (2023). *Weaving relations: Exploring the epistemological interaction between indigenous & traditional ecological knowledge and Euro-western paradigms in education for sustainable development-an umbrella review* (Master's thesis). University of Gothenburg.

Star, S. L., & Giesemer, L. (1989). Institutional ecology, 'translations', and boundary objects: Amateurs and professionals on Berkeley's Museum of Vertebrate Zoology. *Social Studies of Science, 19*, 387–420.

Strumińska-Kutra, M., & Scholl, C. (2022). Taking power seriously: Towards a power-sensitive approach for transdisciplinary action research. *Futures, 135*, 102881.

Tijsma, G., Urias, E., & Zweekhorst, M. (2023). Embedding engaged education through community service learning in HEI: A review. *Educational Research, 65*(2), 143–169.

Torre, M. E. (2014). Participatory action research. In T. Teo (Ed.), *Encyclopedia of critical psychology*. Springer. https://doi.org/10.1007/978-1-4614-5583-7_211

Van Breda, J., & Swilling, M. (2019). The guiding logics and principles for designing emergent transdisciplinary research processes: Learning experiences and reflections from a transdisciplinary urban case study in Enkanini informal settlement, South Africa. *Sustainability Science, 14*, 823–841.

Van der Steen, M., Van Delden, M., & Van Schaik, E. (2020). De opgave aan tafel. *De praktijk van werken vanuit maatschappelijke opgaven*. Den Haag: Nederlandse School voor Openbaar Bestuur (NSOB).

Van Mierlo, B. C., Regeer, B., van Amstel, M., Arkesteijn, M. C. M., Beekman, V., Bunders, J. F. G., De Cock Buning, T., Elzen, B., Hoes, A.-C., & Leeuwis, C. (2010). *Reflexive monitoring in action. A guide for monitoring system innovation projects*. Communication and Innovation Studies, WUR, Athena Institute, VU.

Van Veelen, J. S., Regeer, B. J., Broerse, J. E. W., Van de Poel, S. F. P., Dinkgreve, M. A. H. M. (2017). Embedding the notion of child-and family-centered care into organizational practice: Learning from organizational visioning. *Journal of Public Child Welfare, 11*(2), 231–259.

Verwoerd, L., Brouwers, H., Kunseler, E., Regeer, B., & de Hoop, E. (2023). Negotiating space for knowledge co-production. *Science and Public Policy, 50*(1), 59–71.

Vienni-Baptista, B., & Klein, J. T. (Eds.). (2022). *Institutionalizing interdisciplinarity and transdisciplinarity: Collaboration across cultures and communities*. Routledge.

Zachariah, B., Bunders-Aelen, J., & Regeer, B. (2023). Listening as a tool for transformative change in families and neighbourhoods: The case of SALT. In G. D. Bodie, D. L. Worthington, & Z. Beyene (Eds.), *Listening, community engagement, and peacebuilding: International perspectives* (pp. 55–78). Taylor & Francis.

Zuiderent-Jerak, T. (2015). *Situated intervention: Sociological experiments in health care*. MIT Press.

Index

B. J. Regeer et al. (eds.), *Transdisciplinarity for Transformation*,
https://doi.org/10.1007/978-3-031-60974-9

U

V

W

SPRINGER NATURE

GPSR Compliance

The European Union's (EU) General Product Safety Regulation (GPSR) is a set of rules that requires consumer products to be safe and our obligations to ensure this.

If you have any concerns about our products, you can contact us on ProductSafety@springernature.com

In case Publisher is established outside the EU, the EU authorized representative is:

Springer Nature Customer Service Center GmbH
Europaplatz 3
69115 Heidelberg, Germany

The manufacturer's authorised representative in the EU is Springer Nature Customer Service Centre GmbH, Europaplatz 3, 69115 Heidelberg, Germany. If you have any concerns regarding our products, please contact ProductSafety@springernature.com

Printed and bound by CPI Group (UK) Ltd, Croydon, CR0 4YY
01/10/2026
02225903-0018